CHANGES IN A CALIFORNIA ESTUARY

A Profile of Elkhorn Slough

Edited by
Jane Caffrey, Martha Brown, W. Breck Tyler, and Mark Silberstein

Elkhorn Slough Foundation

The Elkhorn Slough Foundation is a community-supported non-profit organization dedicated to the conservation and restoration of Elkhorn Slough and its watershed. Since 1982, the Elkhorn Slough Foundation has developed programs in research, education, interpretation and stewardship focused on understanding and protecting the natural resources of this remarkable estuary. The Foundation and the Department of Fish and Game staff of the National Estuarine Research Reserve have formed a strong partnership to conserve the slough. The Foundation took responsibility as a land trust in 1997 and today manages over 2,500 acres of slough lands—the largest conservation holdings in the watershed. Currently the Foundation is actively acquiring and restoring key wetlands and habitats in the slough and developing long-term stewardship for these lands.

The Elkhorn Slough National Estuarine Research Reserve is owned and managed by the California Department of Fish and Game. The Reserve is one of 26 sites around the nation's coastline that operates in a state-federal partnership with the National Oceanic and Atmospheric Administration. The Reserve manages an active research and monitoring program implemented by staff, volunteers, and university faculty and students. Education programs focus on teacher training, school field trips, docent-led tours, and workshops for coastal decision makers. The public can explore the Reserve via award-winning visitor center exhibits, hiking trails, boardwalks, and overlooks. The stewardship program addresses issues such as hazardous spill prevention, erosion control, and habitat restoration. Over 100 volunteers assist with all functions of the Reserve. Reserve programs seek to address key Elkhorn Slough watershed issues including habitat degradation and loss, tidal scour, invasive species, and water quality.

Elkhorn Slough Foundation
Moss Landing, California
www.elkhornslough.org

Major funding for this publication was provided by:
National Oceanic and Atmospheric Administration,
Grant # NA770R0469
Patricia Price Peterson Foundation
Acacia Foundation
Lysbeth Anderson
Monterey Bay Aquarium

Cover art: Bill Fenwick
Design: Kirsten Carlson

Printed in Hong Kong on recycled paper.

ELKHORN
SLOUGH
FOUNDATION

Changes in a California Estuary
A Profile of Elkhorn Slough

TABLE OF CONTENTS

Chapter 4. Hydrography (continued)

Chapter 5. Soils

Marc Los Huertos and Carol Shennan

Chapter 6. Archaeology and Prehistory

Terry Jones

Chapter 6. Archaeology and Prehistory (continued)

Chapter 7. History of Land Use

Mark Silberstein, Chela Zabin, Louise Newberry, Daniel Mountjoy, Les Strnad, and Jane Caffrey

Chapter 7. History of Land Use (continued)

Chapter 8. Primary Producers

Richard C. Zimmerman, Jane M. Caffrey

Chapter 9. Invertebrates

Kerstin Wasson, Rikk Kvitek, James Nybakken, Caren Braby, Mark Silberstein

Chapter 10. Fishes

Mary Yoklavich, Gregor M. Cailliet, Dion Oxman, James Barry, David Lindquist

Chapter 11. Birds and Mammals

James Harvey, Sarah Connors

Chapter 12. Biogeochemical Cycling

Jane Caffrey

Chapter 14. Management Issues (continued)

Chapter 15. Summary

Jane Caffrey and Chela Zabin

CHANGES IN A CALIFORNIA ESTUARY

Preface

Such an immense ecosystem reveals itself only by hard-won bits and pieces, even as it changes all the while . . . to do the place justice, it ought to be fairly overrun from now on with enchanted and enchanting researchers building on what has been done and addressing yawning gaps and fascinating opportunities. There cannot be many such places on earth anymore where this prospect is even imaginable.

- Todd Newberry, Emeritus Professor of Biology
University of California, Santa Cruz

This book describes the "hard-won bits and pieces" that have emerged from decades of scientific research at Elkhorn Slough. Scientists have been drawn to the slough and its watershed since the 1920s. In the past thirty years, the burgeoning of scientific and educational institutions ringing Monterey Bay, combined with the slough's importance as a critical wetland, has focused research efforts on Elkhorn Slough's rich natural resources and the effects of human activities on its ecosystems. We have envisioned this project as a summary of that research, a chronicle of the natural and human impacts that have shaped the slough ecosystem, and a snapshot of conditions at the beginning of the twenty-first century.

This edited volume is part of a series profiling the estuaries of the National Estuarine Research Reserve System, which is administered by the National Oceanic and Atmospheric Administration (NOAA). Each chapter was developed by a different author or set of authors who were invited to take part in the project based on their expertise and experience—in some cases spanning several decades—working in and around Elkhorn Slough. The opinions and recommendations are those of the authors and reflect their insights into the opportunities and challenges facing researchers and managers at Elkhorn Slough and throughout the watershed.

Depending on the topic, the frame of reference shifts throughout the book. Some chapters focus on Elkhorn Slough proper (hydrography, benthic invertebrates, fishes), some cover the entire watershed (soils, primary producers, birds and mammals, biogeochemistry), and some encompass the larger Monterey Bay region (geology, climate, archaeology, history). Although the authors have sought to include the most recent studies, the dynamic nature of the slough and ongoing research efforts mean that conditions change rapidly and our knowledge base is constantly evolving.

Despite the extensive scientific work at Elkhorn Slough, there are still "yawning gaps and fascinating opportunities" to address and pursue. For example, there is scant information on terrestrial invertebrates, reptiles and mammals, upland vegetation communities, and the relationship between slough ecosystems and those of Monterey Bay. To that end, each chapter (except for the introduction, management, and synthesis) closes with a description of research efforts needed to increase our basic understanding and in particular to generate information vital to the management community. In pointing out these gaps, it is our hope that future researchers will take up where others have left off to develop a comprehensive picture of Elkhorn Slough. With its range of habitats, plant and animal communities, environmental challenges, and human impacts, the slough can serve as a model for future studies of critical, dynamic ecosystems.

Jane M. Caffrey, Martha T. Brown, W. Breck Tyler, Mark Silberstein
Editors

*A note on English and metric units: A table of English and metric equivalents is provided on page 278. In chapters geared to the more general readership (chapter 1, "Introduction," chapter 7, "History of Land Use," and chapter 14, "Management Issues") or where we report on studies where English units were used, English equivalents are provided for some metric units.

Changes in a California Estuary

Acknowledgments

This book was published in part with funding from the National Oceanic and Atmospheric Administration Estuarine Reserves Division Grant # NA77OR0469.

The Elkhorn Slough Foundation also thanks The Acacia Foundation, Lysbeth Anderson and the Patricia Price Peterson Foundation and Monterey Bay Aquarium for generous support of the publication of this work.

This work would have been impossible without the generous and patient participation of all the authors who donated their time to the effort and to Kirsten Carlson for the design and layout of this book. Thank you all very much.

The editors would like to acknowledge the assistance of the many reviewers whose generous efforts helped improve the text—

Larry Allen, California State University, Northridge
Julie Amft, Skidaway Institute of Oceanography
Alan Baldridge, Hopkins Marine Station
Jane Borg, Pajaro Valley Historical Society
Don Roberson, Monterey Peninsula Audubon Society
Walter Boynton, University of Maryland
Matt Brennan, Stanford University
Gary Breschini, Archaeological Consulting
Ken Bruland, University of California, Santa Cruz
Randy Chambers, Fairfield University
Nora Deans, Birchtree Cove Studio
Linda Deegan, Marine Biological Laboratory, Woods Hole
Mike Foster, Moss Landing Marine Laboratories
Scott Hennessey, California State University, Monterey Bay
Bill Hildebrant, Far Western Anthropological Research Group
Karen Holl, University of California, Santa Cruz
John Hunt, University of California, Davis
Rich Iverson, Florida State University
Christine Jong, University of California, Santa Cruz
Bob Lea, California Department of Fish and Game
Steve Maki, Monterey County Department of Planning
Eric Mielebrecht, University of California, Santa Cruz
Michelle McKenzie, Monterey Bay Aquarium
Mike Murrell, Environmental Protection Agency
Todd Newberry, University of California, Santa Cruz
John Oliver, Moss Landing Marine Laboratories
John Pearse, University of California, Santa Cruz
Dave Peterson, US Geological Survey
Steve Ross University of North Carolina, Wilmington
Don Smith, University of California, Santa Cruz
Bess Ward, Princeton University
Gerry Weber, University of California, Santa Cruz

We make special acknowledgment to some of the pioneers who blazed a trail for diverse research in the slough and to those who put their shoulder to the wheel of slough conservation—George MacGinitie, Roy Gordon, John Oliver, Tom McCarthy, Bernice Porter, John Warriner.

CHANGES IN A CALIFORNIA ESTUARY

Contributors

James P. Barry (Fishes) is an associate scientist with the Monterey Bay Aquarium Research Institute. His research on Elkhorn Slough centered on the distribution and feeding ecology of fishes in the tidal creeks and main channel of the slough. His Master's thesis, completed in 1983, investigated the role of tidal creeks in the ecology of Elkhorn Slough fishes, including their importance as nursery habitats. He received his Ph.D. from University of California, San Diego in 1988.

William W. Broenkow (Hydrography) is professor of oceanography at Moss Landing Marine Laboratories, where he has taught since 1969. He and his students made the first hydrographical studies of Elkhorn Slough in the early 1970s and have continued these studies to the present using modern acoustic Doppler current meters. He teaches physical oceanography, satellite oceanography, applications of computers in oceanography and marine instrumentation at Moss Landing Marine Laboratories. During the 1970s he worked on local oceanographic problems in Monterey Bay and Elkhorn Slough; during the 1980s he participated in John Martin's VERTEX program that led to Martin's now- famous iron hypothesis; and throughout the 1990s he has worked on the Marine Optical Buoy (MOBY) program that provides ground truth to the SeaWiFS and MODIS ocean color satellites. He received his Ph.D. in 1969 from the University of Washington.

Martha Brown (Editor) writes and edits articles, books, and interpretive material on natural history and sustainable agriculture topics. She has worked as a naturalist in Baja California and on Midway Atoll, and has assisted with seabird research projects in Alaska, California, and Hawaii. She received her graduate degree in Science Communication from the University of California, Santa Cruz in 1982.

Jane M. Caffrey (Editor, Introduction, Climate, Hydrography, History of Land Use, Biogeochemical Processes, Management Issues, Summary) is research assistant professor for the Institute of Marine Sciences, University of California, Santa Cruz, and the Center for Environmental Diagnostics and Bioremediation, University of West Florida. She began studying biogeochemical processes in Elkhorn Slough in 1993, served as research coordinator at the Elkhorn Slough National Estuarine Research Reserve from 1995–1998, and was the primary editor for this book. She continues to examine how human influences, as well as biological and physical processes, control nutrient and oxygen dynamics in Elkhorn Slough. She earned her Ph.D. from the University of Maryland.

Gregor M. Cailliet (Fishes) is professor of biology specializing in marine icthyology at Moss Landing Marine Laboratories and California State University, Fresno, where he has taught since 1972. His main involvement in Elkhorn Slough started with the studies he and his graduate students did in the mid-1970s on the slough's fish assemblages and aspects of their ecology (feeding, reproduction, growth). These have continued to the present. He was an initial member of the Elkhorn Slough

Estuarine Research Reserve (formerly Sanctuary) Advisory Committee and chaired the Research Subcommittee for many years. He also was a founding member of the Elkhorn Slough Foundation. He earned his doctorate in 1972 from the University of California, Santa Barbara.

Sarah Connors (Birds and Mammals) is a Master's student at Moss Landing Marine Laboratories, where she is studying seasonal distribution patterns of shorebirds in the Elkhorn Slough watershed. She was a recipient of the National Estuarine Research Reserve System's Graduate Research Fellowship, which funds her thesis research.

Bill Fenwick (Cover Art) "Before I begin a painting I do several small outdoor sketches of a scene that inspires me. In this phase of constructing a painting, I work out abstract shapes and initial color that come to me from my first impression. After completion of the initial sketch I put the image away for a week or more. When I start to work on a larger painting, I bring out the small outdoor sketch to see if I get the same feeling that originally transpired when I painted it. At this point I am not trying to copy the smaller painting but rather capture the feelings that I felt when I was standing in the landscape. If the light, color, and shapes are right in the painting then the emotional content or mood is there. I enjoy listening to comments of art aficionados as they view my work. Their response to visual elements is important information that confirms for me the success of the painting." www.fenwickatecier.com

Jim Harvey (Birds and Mammals) is associate professor of marine science at Moss Landing Marine Laboratories (MLML) and San Jose State University. His course topics include biology of marine birds and mammals, statistics, experimental and sampling design, and scientific writing. He has studied marine bird and mammal ecology for the past 27 years, with an emphasis on the ecology of harbor seals along the west coast of North America. He was the major advisor for more than 50 graduate students who have earned their master's degree at MLML, many of whom studied fishes, birds, and mammals in Elkhorn Slough. He currently serves on the Reserve Advisory Committee for the Elkhorn Slough National Estuarine Research Reserve and on the Research Advisory Panel for the Monterey Bay National Marine Sanctuary. He earned his doctorate from Oregon State University in 1987.

Gary Ichikawa (Land Use and Contaminants) is an Environmental Scientist working for the California Department of Fish and Game (CDFG) at Moss Landing Marine Laboratories (MLML). He has worked on the California State Mussel Watch Program since 1978. This program initiated the first long-term pollution monitoring project in Elkhorn Slough. In 1998 he started the Coastal Fish Contamination Study, a project to examine pollutants in fish from the entire California coast including Elkhorn Slough.

Terry Jones (Archaeology) is assistant professor of anthropology at California Polytechnic State University, San Luis Obispo, where he teaches archaeology and introductory anthropology. A professional archaeologist for over twenty years, most of his research has focused on the central coast of California in Santa Cruz, Monterey, and San Luis Obispo Counties, including excavations at a number of sites at Elkhorn Slough. His dissertation research focused on the prehistory of Big Sur, where he conducted field work between 1983 and 1995. He received his B.A. in Anthropology and Environmental Studies from the University of California, Santa Cruz in 1978, completing his senior project on the archaeology of Elkhorn Slough. He holds an M.A. in cultural resources management from Sonoma State University, and an M.A. and Ph.D. in anthropology from the University of California, Davis.

Rikk G. Kvitek (Invertebrates) is associate professor of earth systems science and policy at California State University Monterey Bay (CSUMB). He first got his foot caught in the mud of Elkhorn Slough in 1982 as a master's student at Moss Landing Marine Laboratories, and has not managed to get free of it yet. His research in the slough has included the feeding ecology of resident sea otters, distribution and change of invertebrate populations, salt marsh erosion, tidal scour and habitat alteration, and the impacts of human invertebrate harvest. He is frequently found plying the slough's waters with his students aboard CSUMB's hydrographic survey vessel, the R/V *MacGinitie*. He earned a doctorate in 1990 from the University of Washington.

David Lindquist (Fishes) is a research associate at Louisiana State University's Coastal Fisheries Institute, and did his graduate work at Moss Landing Marine Laboratories, where he received his master's degree. He enjoys being out in Elkhorn Slough, and is most impressed by the diversity of organisms that is packed into a relatively small area.

Marc Los Huertos (Soils) is a post-doctoral researcher at the University of California, Santa Cruz, where he is working on water quality monitoring projects in central coast watersheds. He did his doctoral work at Elkhorn Slough, where he tested the potential for vegetative buffers to improve the slough's water quality. While getting his master's degree at San Francisco State University in plant ecology, he became interested in the role that soils have in plant communities and biogeochemistry. Currently, understanding the role of soil biogeochemistry in soil fertility and water quality is a central part of his research. He received his Ph.D. in Environmental Studies from UC Santa Cruz in 1999.

Daniel Mountjoy (History of Land Use, Management Issues) is an area resource conservationist with the U.S. Department of Agriculture–Natural Resources Conservation Service (NRCS). A cultural ecologist, he has worked in the Elkhorn Slough watershed for the past ten years. He conducted his dissertation research on the cultural and economic issues affecting agricultural land management and was subsequently hired by the NRCS to develop the Elkhorn Slough Watershed Project. The project provides culturally relevant technical outreach and assistance to local farmers to help them prevent soil erosion and improve water quality. He is now promoting similar efforts in watersheds throughout California's central coast counties. He received his doctorate from the University of California, Davis.

Louise M. Newberry (History of Land Use) is a docent at the Elkhorn Slough National Estuarine Research Reserve. She organized and curated the 1998 Elkhorn Observed: Paintings, Photographs, Prints, Drawings exhibition at Watsonville's Pajaro Valley Gallery. In 2001 she co-curated with Mary Warshaw the Gallery's exhibition, A Watershed Experience: Discovering the Watsonville Wetlands. She is former Curator of the Smith Gallery at the University of California, Santa Cruz, and former Curator of Exhibitions at the Museum of Art and History, Santa Cruz. She holds bachelor's degrees in art from Stanford University and art history from UC Santa Cruz.

James W. Nybakken (Invertebrates) is emeritus professor of biological sciences at Moss Landing Marine Laboratories. His principal research interests have been in the ecology and systematics of predatory neogastropod mollusks and nudibranch molluscs, and the ecology of benthic and deep sea invertebrates. A prolific researcher and author, he has published thirty-seven scientific papers, authored or co-authored six textbooks and eleven major reports, and is a Fellow of the California Academy of Sciences as well as a member of a dozen scientific societies, having served as president, vice president, and secretary in four of them. He has been a peer reviewer for National Science Foundation proposals in invertebrate zoology, oceanography, systematics, and ecology and has reviewed manuscripts for 11 different scientific journals. He received his doctorate in 1965 from the University of Wisconsin, Madison.

Bryn M. Phillips (Land Use and Contaminants) is a research specialist with the Department of Environmental Toxicology at the University of California, Davis, working at the Marine Pollution Studies Laboratory at Granite Canyon in Monterey County. His Elkhorn Slough research experience includes work on several pollution monitoring projects examining the biological effects of potentially toxic sediment and water. He received his master's degree from Moss Landing Marine Laboratories.

David Schwartz (Geology) is professor of geology and oceanography at Cabrillo College in Santa Cruz County, where he has worked since 1986. Prior to that he taught at San Jose State University and Hartnell College in Salinas, California. He earned his master's degree in 1983 from Moss Landing Marine Laboratories, where he did his thesis work on the geologic history of Elkhorn Slough. His research on the slough appeared in 1986 in Estuarine, Coastal and Shelf Science.

Carol Shennan (Soils) is director of the Center for Agroecology and Sustainable Food Systems at the University of California, Santa Cruz, where she also is professor of Environmental Studies. She has a long-standing interest in agriculture and wetlands interactions, and conducted research for a number of years at Tulelake in northern California. She now leads an extensive research and education program to improve the ecological health of agricultural landscapes on the central coast of California, with a focus on Monterey Bay watersheds. She received her doctorate in botany from Cambridge University.

Mark Silberstein (Editor, Introduction, History of Land Use, Invertebrates, Management Issues) is a graduate of the Moss Landing Marine Laboratories, where he completed a degree in marine zoology and ecology. He was part of the team of students and faculty that undertook the first quantitative ecological studies of Elkhorn Slough in the early 1970s. He has worked widely in marine systems from the edge of the Arctic Ocean to Antarctica, but settled in the mid latitudes. After establishing research, education and interpretive programs at the Elkhorn Slough National Estuarine Research Reserve in the 1980s, he became the director of the Elkhorn Slough Foundation where he remains today.

Mark Stephenson (Land Use and Contaminants) is an environmental scientist working for California Department of Fish and Game at Moss Landing Marine Laboratories (MLML). He started the California State Mussel Watch program in 1977 and through that program initiated the first long-term pollution monitoring project in Elkhorn Slough. Over the past twenty years he has participated in almost every pollution research project in the slough. He earned a master's degree in 1975 from MLML.

Les Strnad (Introduction, History of Land Use). For over 24 years, Les Strnad devoted his personal and professional life to coastal protection, environmental education, enhanced public recreation opportunity, and wetland and marine wildlife protection within California's central coast region. He served with the California Coastal Commission's Central District from 1973 to 1996, taking early medical retirement from his job as Deputy District Director on advice from his doctors. He has received numerous commendations for his work, including the John Martin *Iron Man Award for Excellence in Marine Science* and the Monterey Bay National Marine Sanctuary's *Sanctuary Reflections Award for Conservation*. While no longer working, Strnad is much sought after as an advisor on coastal zone management issues and is helping with the development of one of his dreams—S.E.A. Lab Monterey Bay, a marine science education camp. He earned a bachelor's degree in Geopolitical Planning in 1974 from the University of California, Santa Cruz.

W. Breck Tyler (Editor) teaches ornithology, marine biology, and natural history for the University of California, Santa Cruz (UCSC), and Northeastern University's East-West program. A behavioral ecologist, he has coordinated seabird monitoring projects at Midway Atoll and Año Nuevo Island, conducted numerous at-sea surveys of seabirds and marine mammals, and currently directs the UCSC aerial survey program for oil spill response. He received his master's degree in Marine Sciences from UCSC.

Kerstin Wasson (Invertebrates) is research coordinator at the Elkhorn Slough National Estuarine Research Reserve, where she carries out broad long-term monitoring programs as well as focused studies on threats to estuarine ecosystems. Trained as an evolutionary ecologist, with interests in strategies of sex and growth in colonial invertebrates, she now directs her attention to estuarine conservation. In recent years she has become interested in estuarine invasions by exotic species, examining invertebrate invaders in the slough, testing novel ballast water treatments for reducing marine introductions, tracking invasions of exotic plants at the marsh-upland ecotone, and developing a vision for a national monitoring program for estuarine invasions. She earned her doctorate in biology from the University of California, Santa Cruz.

Mary Yoklavich (Fishes) works as a research fishery biologist for the National Oceanic and Atmospheric Administration—National Marine Fisheries Service's Santa Cruz Laboratory. She also is a research associate at Moss Landing Marine Laboratories (MLML) and the University of California, Santa Cruz, often including students in her studies. She received an M.S. in marine sciences from San Francisco State University and MLML, and published her graduate thesis on the energetic requirements and habitat use of English sole in Elkhorn Slough. Since then she has published a series of papers on distribution, abundance, and feeding habits of fish assemblages in the slough. She conducts research from California to Alaska on a variety of biological problems in marine fisheries and oceanography, and is well known for her research on reproduction, age, growth, and habitat assessments of West Coast rockfishes. She serves as a scientific advisor to several national, state, regional, and local committees on fishery conservation issues, and is a member of the Elkhorn Slough National Estuarine Research Reserve's Research Committee.

Chela J. Zabin (Introduction, History of Land Use, Management Issues, Summary) is a graduate student at Kewalo Marine Lab and the Department of Zoology, University of Hawaii, Manoa. She became interested in Elkhorn Slough while working as a reporter for the Watsonville newspaper, Register-Pajaronian, and ended up writing a number of stories about the slough and about other local wetlands. Writing about science was so much fun that she decided to go back to school and become a biologist herself.

Richard Zimmerman (Primary Producers) is adjunct professor at Moss Landing Marine Laboratories. He has been interested in the ecology of Elkhorn Slough primary producers since 1988, and has studied the dynamics of light availability, characterization of potential habitat for submerged aquatic vegetation, and the genetic structure of eelgrass populations in the slough. Current research interests involve the impact of climate change on benthic primary productivity and remote sensing of submerged aquatic vegetation. He received his doctorate from the University of Southern California.

1

Chapter One

Introduction

Jane Caffrey, Chela Zabin, Mark Silberstein, and Les Strnad

This chapter introduces Elkhorn Slough and the contents of the book. It briefly describes the slough's environmental and geographic setting, habitats, and plant and animal communities, and discusses the book's major themes.

Elkhorn Slough has been highly modified by human activities and it is only in the last thirty years that substantive efforts have been made to conserve its fragile habitats. The various governmental agencies that have jurisdiction over Elkhorn Slough and its watershed, particularly those playing critical roles in protecting and restoring slough resources, are introduced in Appendix 1.1.

Environmental Setting

Drive along the west coast of North America and you'll often find yourself winding along a mountain face looking nearly straight down at surf crashing against sheer cliffs. This meeting of land and sea takes place most dramatically along the Big Sur coast south of Monterey, where the Santa Lucia Mountains rise almost straight up from the ocean. Tectonic activity along the margins of the Pacific and North American plates has built mountain ranges that lie close to the coast, limiting the size and number of the region's estuaries. As a result, estuaries make up only 10–20% of the Pacific coastline compared to 80–90% of the Atlantic and Gulf coasts (Kjerfve 1989), which lie along a passive continental margin with a broad coastal plain and continental shelf. Estuaries along both continental margins formed when sea level rose following the last glaciation (about 10,000 years ago). Because the Pacific continental shelf is relatively narrow compared to that of the Atlantic, these Pacific estuaries were very small during the ice ages' periods of low sea level and are therefore young in terms of fauna compared to Atlantic and Gulf coast estuaries (Nichols and Pamatmat 1988).

Elkhorn Slough is relatively small when compared with San Francisco and Tomales Bays, the state's two largest estuaries (table 1.1). San Francisco Bay, with a watershed area of 153,000 square kilometers (59,073 sq mi), drains the entire Central Valley. In contrast, Elkhorn Slough's watershed area is 182 square kilometers (70 sq mi), about one-third that of the

Table 1.1. Geostatistics of Elkhorn Slough and San Francisco and Tomales Bays

Statistic	Elkhorn Slough	San Francisco Bay	Tomales Bay
Area (m^2)	3.25×10^6	1.04×10^9	28.2×10^6
Volume (m^3)	2.24×10^6	6.66×10^9	88.0×10^6
Average Depth (m)	1.4	6.1	3.1
Watershed Area (km^2)	132	153,000	570
Inflow (m^3/s)	0–3.8	600	1.25

Sources: Information on San Francisco Bay is from Conomos, Smith, and Gartner 1985; information on Tomales Bay is from Smith et al. 1987.

watershed surrounding Tomales Bay. The surface area of the slough is also about one-third the area of Tomales Bay and one-hundredth that of San Francisco Bay.

Elkhorn Slough, with an average depth of 1.4 meters (4.6 ft), is also relatively shallow compared to Tomales Bay, which averages about 3 meters (10 ft), and San Francisco Bay, which averages about 6 meters (20 ft). Surface water inflows are comparable between Elkhorn Slough and Tomales Bay, which receive most of their runoff from creeks, but these are a fraction of the flow into San Francisco Bay from the Sacramento and San Joaquin Rivers.

Geographic Setting

Elkhorn Slough opens into Monterey Bay at Moss Landing, a small fishing, tourist, and marine research community 145 kilometers (90 mi) south of San Francisco and 32 kilometers (20 mi) north of Monterey (see fig. 1.1). The slough is a shallow estuary that extends inland east from Moss Landing Harbor for approximately 6.4 kilometers (4 mi) before turning and curving north for another 5 kilometers (3.1 mi). Two hundred meters (656 ft) wide at its widest point and 7.5 meters (25 ft) deep at the Highway 1 bridge at *mean lower low water*, the main slough channel grows narrower and shallower as it travels inland. The winding branches of the slough encompass more than 1,420 hectares (3,506 ac) of marsh and tidal flats (fig. 1.1).

Surrounding Elkhorn Slough are the hilly uplands and marine terraces that lie between the Pajaro and Salinas Valleys in Monterey and San Benito Counties (fig.1.2). Planted in strawberries and other row crops and used for cattle grazing and housing, these areas drain into the slough through Carneros Creek at the head of the estuary and numerous small, ephemeral creeks.

The slough flows under state Highway 1 into Moss Landing Harbor, a man-made small-craft harbor that supports a commercial fishing fleet, recreational craft, and research vessels. The smaller and largely seasonal Moro Cojo and Tembladero Sloughs also empty into the harbor from the south.

Just off the harbor mouth lies the Monterey Bay Submarine Canyon. Extending 110 kilometers (68 mi) offshore and 3 kilometers (1.8 mi) below the surface, this underwater canyon forms the largest gorge along the west coast of North America—broad and deep enough to hold the Grand Canyon of the Colorado River. Monterey Bay and its surrounding waters make up the 13,700-square-kilometer (5,288 sq mi) Monterey Bay National Marine Sanctuary, administered by the National Oceanic and Atmospheric Administration.

Located near the harbor are marine-related businesses and two marine research organizations, the Monterey Bay Aquarium Research Institute and Moss Landing Marine Laboratories. A power plant owned by Duke Energy (purchased from Pacific Gas & Electric in 1998) lies directly inland from the harbor and pumps in 50 cubic meters (6[illegible] cu yds) of slough water per second for cooling while discharging heated water into Monterey Bay. Until 2000, National Refractories, a manufacturing plant located adjacent to the energy plant, extracted magnesium from seawater to make heat-resistant bricks.

Several areas surrounding Elkhorn Slough are protected by public agencies and private conservation groups. Elkhorn Slough National Estuarine Research Reserve (ESNERR) includes 583 hectares (1,439 ac) of slough and surrounding lands located on the slough's eastern shore. The reserve is part of NOAA's Estuarine Reserve Division and is managed by the California Department of Fish and Game (DFG). The primary mission of the reserve, which hosts 50,000 visitors each year, is to promote research, education, and stewardship of Elkhorn Slough. Public access to the slough is also available through Kirby Park, which is maintained through the Moss Landing Harbor District. Additional wildlife areas include 283 hectares (698 ac) in the upper slough owned by The Nature Conservancy, and the 259-hectare (639 ac) Moss Landing Wildlife Area on the slough's north side, managed by DFG. The Elkhorn Slough Foundation is a local land trust actively working at the slough. The foundation is acquiring lands for conservation and currently owns more than 600 hectares (1,500 ac) in the watershed.

Elkhorn Slough is engaged in a dynamic interaction with the upland areas, the harbor, and the ocean beyond, including both human and natural activities that take place in these systems. Although this document focuses on the estuarine environment, information from research on the surrounding watershed and bay is also included.

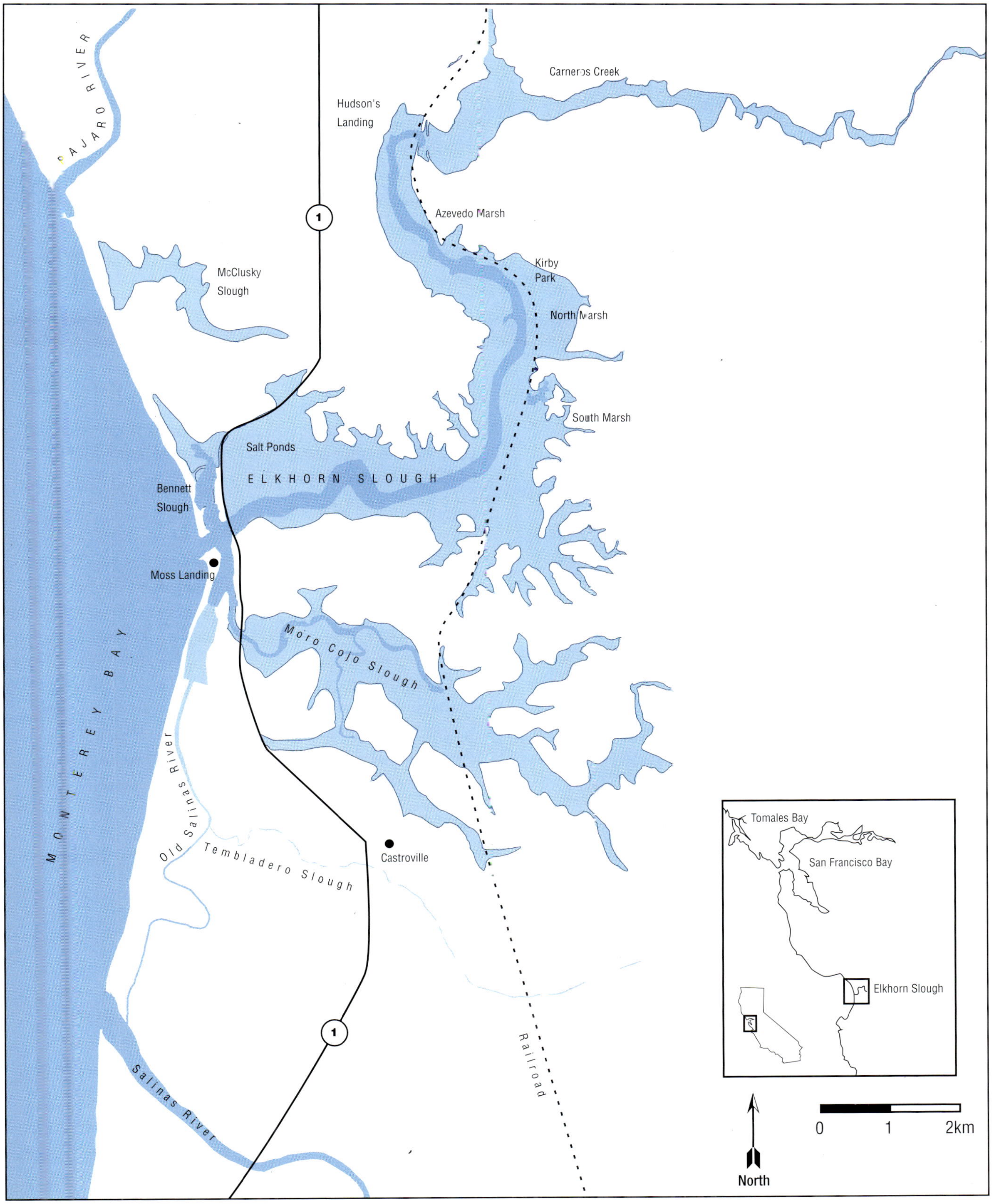

Figure 1.1 Map of slough and California locator map

Habitats and Communities

The Elkhorn Slough watershed encompasses a mosaic of habitats. They include the slough's channels, mudflats, eelgrass beds, salt marsh, and hard substrate; the adjacent harbor, coastal dunes, and open beaches; and the grasslands, oak woodlands, chaparral, and other upland areas. Each habitat supports diverse plant and animal communities. This volume describes our current understanding of these communities, how they are connected to one another, and how they have been affected by human impacts through time.

With its daily tidal flushing and soft mud bottom, Elkhorn Slough provides food and shelter for a rich invertebrate fauna (see chapter 9, "Invertebrates"). Few sites along the Pacific coast can match the levels of diversity and abundance of benthic (bottom- and soft-substrate-dwelling) invertebrates found at the slough. Several areas within the slough that feature hard substrate, such as riprap, stone bridges, docks, and drainage pipes, are populated with invertebrate assemblages that are a mix of rocky intertidal and harbor fouling communities. Researchers have reported more than 500 species of marine invertebrates from Elkhorn Slough; populations of several of these species

Table 1.2. Marine science research organizations in the Monterey Bay area.

California Department of Fish & Game, Oiled Wildlife Rescue & Rehabilitation Facility

rescue and rehabilitation of oiled wildlife and aquatic toxicology research
1451 Shaffer Rd., Santa Cruz, CA 95060; phone: (831) 469-1719; fax: (831) 469-1723

Department of Oceanography, Naval Postgraduate School

acoustical oceanography; coastal and nearshore oceanography; ocean modeling; air/sea interaction
1 University Circle, Monterey, CA 93943; phone: (831) 656-2441; www.nps.navy.mil/

Elkhorn Slough National Estuarine Research Reserve & Elkhorn Slough Foundation

ecology; biology; estuarine management and restoration. open to the public: hiking trails, bird-watching, school tours, visitors center, docent-led walks
Reserve: *1700 Elkhorn Road, Watsonville, CA 95076; phone: (831) 728-2822*
Elkhorn Slough Foundation: *P.O. Box 267, Moss Landing, CA 95039; phone: (831) 728-5939; fax: (831) 728-1056; www.elkhornslough.org*

Fleet Numerical Meteorology & Oceanography Center

provision of oceanographic and atmospheric services to the Department of Defense
7 Grace Hopper Ave., Monterey, CA 93943-5501; phone/fax: (831) 656-4875; fnmoc.navy.mil/

Hopkins Marine Station, Stanford University

marine invertebrate biology and ecology; molecular marine biology; biological oceanography
Ocean View Blvd., Pacific Grove, CA 93950; phone: (831) 655-6200; fax: (831) 375-0793; www.marine.stanford.edu

Institute of Marine Sciences, University of California, Santa Cruz; Long Marine Laboratory and The Seymour Center

marine mammal biology; environmental toxicology; nearshore ecology; marine invertebrate biology; molecular marine biology; marine geology and geophysics; ocean processes and paleoceanography; coastal processes and hazards. Seymour Center open to the public: aquarium, school programs, and docent-led tours
100 Shaffer Road, Santa Cruz, CA 95060; phone/fax: (831) 459-2883 (831) 459-3800 Seymour Center: fax (831) 459-3383; natsci.ucsc.edu/ims (Institute of Marine Sciences); www2.ucsc.edu/seymourcenter/ (The Seymour Center)

Marine Pollution Studies Laboratory, California Department of Fish and Game (CDFG)

research on the effects of marine pollution on living marine resources, including California Mussel Watch Program; development of methods for testing marine toxicity Located at three sites: *UC Davis Department of Environmental Toxicology, Moss Landing Marine Laboratories, California Department of Fish & Game's Granite Canyon Marine Laboratory*

Monterey Bay Aquarium

public education and marine research relevant to exhibit program, open to the public: aquarium, school tours, docent-led tours
886 Cannery Row, Monterey, CA 93940; phone: (831) 648-4800; fax: (831) 648-4810; www.montereybay-aquarium.org

have declined dramatically over the past several decades. Given their key role in the slough's food chain, protecting invertebrate populations is of critical concern to slough managers.

More than 100 species of fish forage, breed, and find shelter in the slough (see chapter 10, "Fishes"). Bat rays, leopard and smoothhound sharks, Pacific herring, starry flounders, staghorn sculpin, shiner perch, jacksmelt, topsmelt, and pile perch mate and bear their young there. Species important to sport and commercial fisheries include surf perch, halibut, Pacific herring, English sole, and northern anchovy. Many of these use the slough as a nursery. As tidal currents and erosion have increased, researchers have noted significant changes in the slough's fish assemblages.

While it supports a rich diversity of invertebrates and fishes, Elkhorn Slough is perhaps best known for its bird populations. Long recognized as a critical habitat for both resident and migratory bird species, Elkhorn Slough is one of the few remaining significant saltwater wetlands on the Pacific flyway. In the spring of 2000, the slough was designated a Globally Important Bird Area by the American Bird Conservancy, and a

Monterey Bay Aquarium Research Institute

ocean engineering, including undersea robotics, buoy technology, and development of micro-sensor technology for natural and pollutant chemical detection; ecology; oceanography of Monterey Bay; advanced scientific information management programs
7700 Sandholdt Rd., Moss Landing, CA 95039; phone: (831) 775-1700; fax: (831) 775-1620; www.mbari.org

Monterey Bay National Marine Sanctuary Office, National Oceanic & Atmospheric Administration

protection and management of the sanctuary
299 Foam St.., Suite D, Monterey, CA 93940; phone: (831) 656-1725; fax: (831) 647-4225; www.mbnms.nos.noaa.gov/

Moss Landing Marine Laboratories, California State University

physical, chemical, biological, and geological oceanography; marine ecology; marine ichthyology; marine birds; marine mammals; marine invertebrates
P.O. Box 450, 8272 Moss Landing Rd., Moss Landing, CA 95039; phone: (831) 632-4400; fax: (831) 632-4403; www.mlml.calstate.edu

National Weather Service Forecast Office, National Oceanic & Atmospheric Administration

provision of comprehensive weather, flood warning, and forecast services to the general public, pilots, and boaters
21 Grace Hopper Ave., Stop 5, Monterey, CA 93943; phone: (831) 656-1725; www.nws.mbay.net/home.html

Naval Research Laboratory, Marine Meteorology Division

atmospheric research and forecasting; advancement of marine-related technology
7 Grace Hopper Ave, Stop 2, Monterey, CA 93943; phone/fax: (831) 656-4758; www.nrlmry.navy.mil/

Ocean Applications Branch, National Oceanic & Atmospheric Administration

development, exchange, integration, and dissemination of oceanographic data, services, and products in support of NOAA marine programs
7 Grace Hopper Ave., Stop 1, Monterey, CA 93943-5501; phone: (831) 647-4206; fax: (831) 647-4225

Pacific Fisheries Environmental Laboratory, National Oceanic & Atmospheric Administration

assessment of effects of natural environmental variability on living marine resources
1352 Lighthouse Ave., Pacific Grove, CA 93950-2097; phone: (831) 648-8515; fax: (831) 648-8440; www.pfeg.noaa.gov

Santa Cruz Laboratory, Southwest Fisheries Science Center, National Oceanic and Atmospheric Administration

salmon and groundfish populations; ecology of fish and invertebrate communities; impacts on listed species
110 Shaffer Road, Santa Cruz, CA 95060; phone: (831) 420-3900; fax: (831) 420-3977; www.pfeg.noaa.gov/tib/

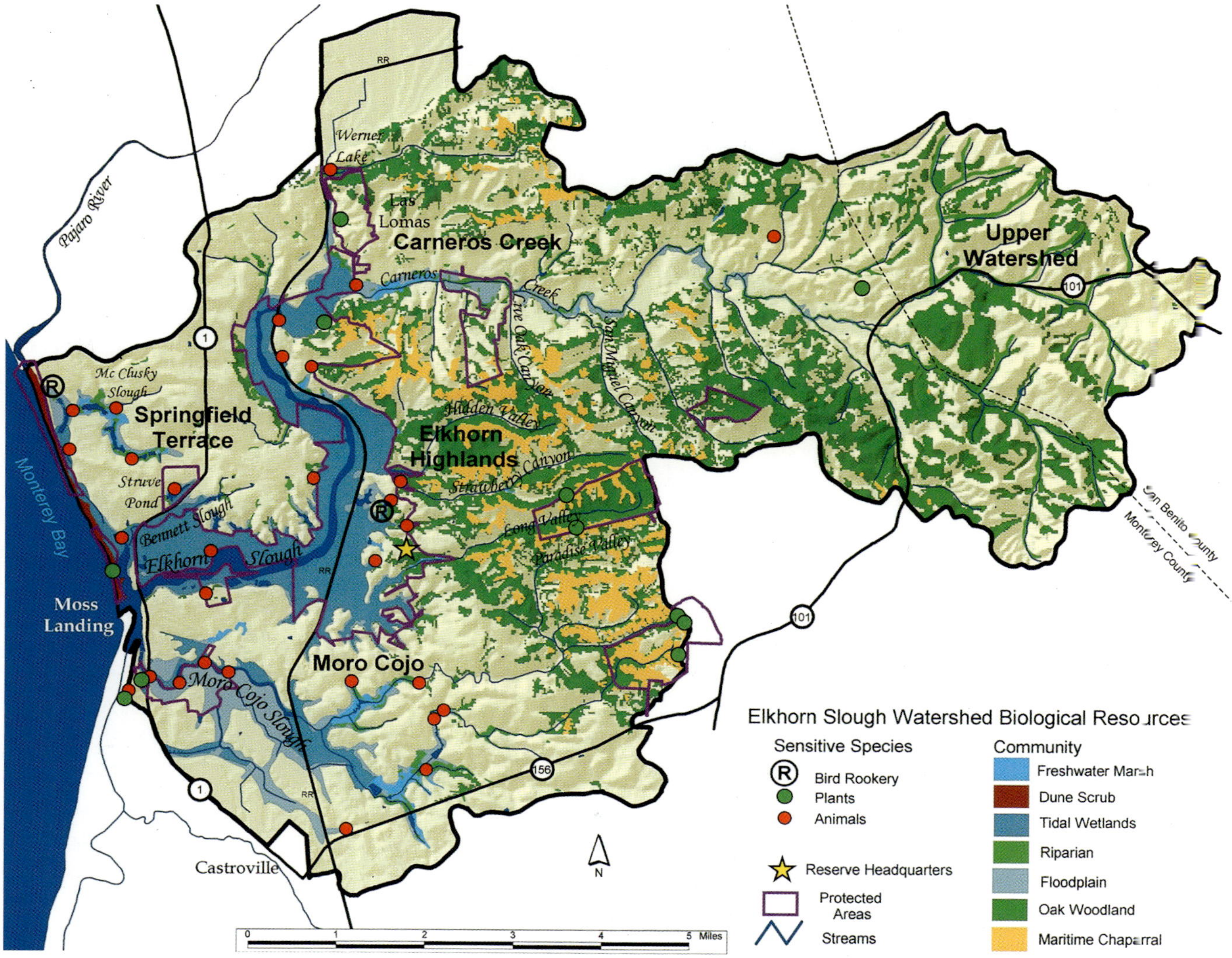

Figure 1.2 The Elkhorn Slough watershed drains 182 square kilometers (70 mi²) and encompasses a great diversity of habitats.

Western Hemisphere Shorebird Reserve by the Manomet Bird Observatory, in recognition of its importance to both migrating and resident birds. More than 265 species of birds have been sighted in the Elkhorn Slough area, including five species with rare or endangered status: the Peregrine Falcon, Snowy Plover, Clapper Rail (now considered extirpated), Brown Pelican, and Least Tern.

Many of the bird species recorded at the slough are seasonal visitors: tens of thousands of birds use the slough as a brief resting and feeding stop on their annual migration. Others overwinter each year. Many of these migratory birds are not obligate estuary users, but suitable habitat—wetland and otherwise—has been lost to development, draining, and other activities throughout the flyway. Although biologists once believed that how well migratory bird populations fared depended solely on conditions in their breeding grounds, it is now thought that the availability of overwintering areas is key to survival (USFWS 1978, in Onuf 1978). There is also evidence that migrating species that prefer freshwater habitats will use estuaries during drought years (USFWS 1978, in Onuf 1978).

At least 40 bird species make the slough their year-round home. These include Great Blue Herons and Great Egrets that nest in a grove of Monterey pines within the ESNERR. The slough's bird life is discussed in chapter 11, "Birds and Mammals."

Anyone visiting Elkhorn Slough by boat, canoe, or kayak in the past few years has likely encountered the California sea otters that regularly visit the slough and prey on its abundant invertebrate fauna. As the tidal influence on the slough increases, a colony of harbor seals has also established itself along the slough's seaward edges. Sea lions, harbor porpoise, and the occasional juvenile gray whale have been sighted along the lower reaches of Elkhorn Slough, and more than 50 species of terrestrial mammals are potentially found in the slough's upland habitats (see chapter 11).

Diverse upland plant communities characterize the Elkhorn Slough watershed (see chapter 8, "Primary Producers"). Coastal maritime chaparral in the easterly hills of the watershed is home to a variety of native woody shrubs and relatively free of introduced species. Oak woodlands support diverse wildlife, although many acres have been lost due to cutting for firewood or development. Both tan oak and coast live oak are now also threatened by the plague of sudden oak death syndrome; although no cases have been reported in the Elkhorn Slough watershed, there have been reports of the fast-spreading disease from nearby Santa Cruz and Monterey County locations. Cattle grazing and the introduction of nonnative grass species have significantly altered grassland or savannah habitats since settlement by the Spanish in the mid 1700s. In recent years, much of this habitat has been converted to agriculture or developed for commercial uses and housing.

Regional Research and Education Institutions

Monterey Bay lies between subtropical and temperate climatic zones and is characterized by seasonal periods of deep-ocean upwelling interspersed with warming currents. These factors, along with diverse habitats such as the mudflats of Elkhorn Slough and the rocky outer coasts at each end of the bay, have resulted in a complex ecosystem and a rich natural flora and fauna (Griggs 1995). This diversity has drawn marine researchers to the area since the 1920s, beginning with George MacGinitie, who catalogued invertebrate species in Elkhorn Slough, and Edward "Doc" Ricketts of John Steinbeck's *Cannery Row* fame.

Easy access to the submarine canyon from Moss Landing and advances in deep-sea exploration technology have since bolstered the area's reputation for marine research. There are now seventeen marine-research institutions located in or planning to relocate to the Monterey Bay area, employing some 1,500 scientists, graduate students, and support staff. Many of these organizations such as the Monterey Bay Aquarium and the Seymour Center at Long Marine Laboratory, have a public-education component and receive thousands of visitors daily. The impact of these research and educational organizations on the slough is discussed below. Major areas of research, addresses, phone numbers, and Web sites are listed in table 1.2.

Our understanding of Elkhorn Slough has been enlarged by the research conducted by members of these institutions over the years. One goal of this book is to summarize and synthesize the work of the many graduate students and researchers who have worked in the slough and surrounding watershed.
The first scientific studies began in the 1920s with George MacGinitie, a master's student from Stanford University, who described the benthic invertebrate communities in Elkhorn Slough (MacGinitie 1927, 1935). Shortly thereafter, Ida Hayward, another Stanford University master's student, conducted the area's first survey of freshwater plant communities (Hayward 1931). Occasional studies followed over the years until the 1970s, when the pace of research on Elkhorn Slough and its watershed greatly increased, thanks in part to the work of researchers at Moss Landing Marine Laboratories (fig. 1.3). William Broenkow, professor of physical oceanography, and his students Rich Smith and Lee Clark, studied slough hydrography from 1970 through 1976 (Clark 1972; Smith 1973; Broenkow 1977). Between 1974 and 1976, faculty and students of Moss Landing Marine Laboratories

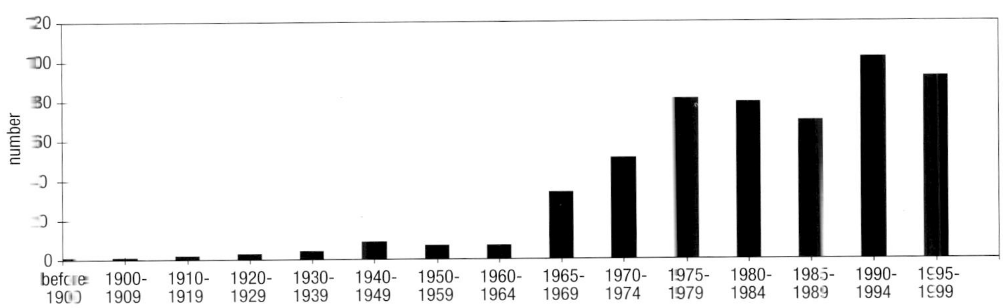

Figure 1.3. Number of published research articles and gray literature reports on Elkhorn Slough since 1900 based on bibliographic references (Caffrey, et al. 1998; Caffrey, pers. comm.).

An aerial view of Elkhorn Slough shows dramatic juxtapositions of critical habitats with a plethora of human uses. Photo credit: Mark Silberstein.

carried out the first major assessment of slough biota, in a study funded by Pacific Gas & Electric Company, and made comprehensive collections of slough plants and animals (Nybakken, Cailliet, and Broenkow 1977).

In subsequent years, research increasingly focused on nonpoint source pollution and restoration. The first monitoring programs were established in the 1980s to examine pesticide uptake by mussels (Mussel Watch Program, California State Department of Fish and Game) and water quality (John Oliver, Deb Frost, and the Elkhorn Slough Foundation). John Oliver of Moss Landing Marine Laboratories and his students initiated several projects to examine plant, invertebrate, and bird communities following the restoration of salt marshes in the National Estuarine Research Reserve in 1983 (King et al. 1986; Mayer 1986). Declines in eelgrass habitat in the slough led UCLA researchers Randy Alberte and Dick Zimmerman to restore eelgrass communities on the reserve, though these attempts were ultimately unsuccessful (Silberstein 1989; Zimmerman, Smith, and Alberte 1990).

In the 1980s, increasing awareness of the connection between land use and water quality prompted a number of studies at the slough. An aerial photo archive of the Elkhorn Slough watershed was first compiled in the mid-1980s by John Oliver and used to examine changes in wetlands associated with the opening of Moss Landing Harbor (Oliver and Schwartz 1988). Subsequent acquisitions of aerial photographs have made this one of the best resources available to track changes in habitat and land use from 1931 to the present. Other key studies during this period include the first major quantitative archaeological excavation at Elkhorn Slough as a result of the Caltrans Highway 1 bridge alignment (Hildebrant, Jones, and Hall 1983; Dondero 1984; Dietz, Hildebrant, and Jones 1988; see chapter 6, "Archaeology and Prehistory") and the Soil Conservation Service Strawberry Hills study that focused on erosion from sandy cultivated slopes (USDA 1984; see chapter 5, "Soil").

Basic research continued unabated in the late 1980s and 1990s; in addition, management plans were developed fo different habitats and regions. The Elkhorn Slough wetlands management plan was completed in 1989 (ABA Consultants 1989). In 1993, Daniel Mountjoy completed his study on farming practices and made recommendations to the Association of Monterey Bay Area Governments about ways to reduce soil erosion from farm fields in the Elkhorn Slough watershed (Mountjoy 1993). This was also the year that an uplands water quality plan was developed (Blankinship and Evans 1993). In 1996 the Moro Cojo Slough management and enhancement plan was completed by Monterey County. The Nature Conservancy and the Elkhorn Slough Foundation finalized the Elkhorn Slough watershed conservation plan in 1999 (Scharffenberger 1999).

The mid-1990s saw an expansion of marine and environmental studies programs at local universities. The University of California, Santa Cruz, began the Monterey Bay Regional Studies Program, which included a demonstration project focused on Elkhorn Slough with support for graduate and undergraduate research projects. Moss Landing Marine Laboratories established the Watershed Institute, which greatly expanded Geographic Information Systems (GIS) capabilities and restoration efforts in the area. California State University (CSU) Monterey Bay, which opened in 1995 on the former Fort Ord military base, has developed a research and education program focused on integrating basic research, policy, and technological innovations with environmental issues. More than ten years after the Loma Prieta earthquake devastated Moss Landing Marine Laboratories and forced them into temporary trailers in Salinas, the labs returned to new facilities at Moss Landing in 2000. Moss Landing Marine Laboratories' long-term commitment to research in Elkhorn Slough can only be strengthened by being on site.

Overview of Themes

Two themes will be developed throughout this volume. The first is that of change: estuaries are dynamic environments shaped by such factors as changing tides, freshwater inflow, climate change, anthropogenic inputs (the result of human impacts), and introduced species. Elkhorn Slough is no exception—it responds to a spectrum of natural changes that occur over different time frames, from uplift on geologic timescales of thousands of years to hourly changes in solar insolation. The changing physical environment sets boundaries for the survival of the flora and fauna of the region. Some of these phenomena are predictable, such as the twice-daily rise and fall of the tides; others, such as earthquakes, are not, while many, such as storms, fall in between these extremes. All of these changes affect the biotic communities and how the ecosystem functions.

We begin this book with a description of the physical environment, concentrating on how geologic, climatic, hydrologic, and edaphic (soil-related) processes have shaped and continue to shape Elkhorn Slough. Subsequent chapters describe the flora and fauna of the slough and surrounding watershed, emphasizing the ecological role of organisms, from bacteria in sediments to harbor seals. Whenever possible we describe patterns of distribution and abundance of major species and the factors controlling these patterns. For some, such as benthic invertebrates or sharks, collection records go back decades and we attempt to explain the long-term changes observed in these records

The other major theme of this book is that human activities have a significant impact on how estuarine and terrestrial ecosystems function. Elkhorn Slough's hydrology, chemistry, and biology have been highly modified by human activities over the last century. An important part of the story of Elkhorn Slough is how our perceptions, uses, and values of these habitats have changed during this period. Chapters on archaeology (chapter 6) and land use (chapter 7) describe the history of human impacts in this area from the beginnings of human settlement to the present day. The ever-increasing human population within the watershed has continued to make demands on local resources, leading to a degradation of the natural environment and increasing levels of pollutants in the slough. Chapter 13, "Land Use and Contaminants," describes the patterns in runoff of pollutants such as nutrients and pesticides and their effect on slough biota. This is an area in critical need of more research, particularly on ways to reduce inputs to the slough.

Changes in management strategies over the course of human use of Elkhorn Slough are described in chapter 14, "Management Issues." During the Spanish and Mexican periods, individual landowners managed their large ranches as they saw fit. Similarly, for over 100 years following American settlement, individuals managed their own land without governmental regulations or much thought to the consequences of their actions on the natural environment. Population growth in the watershed and region has brought greater competition for scarce resources and greater demands for fresh water, land for farming and housing, fish, and so on. This is not unique to Elkhorn Slough and represents another example of the "tragedy of the commons" (Hardin 1968), a situation in which there is no incentive for individuals to protect common resources, such as fisheries, topsoil, and water quality.

The rise of the environmental protection movement led to the passage of laws to protect common resources and placed most of the enforcement burden on government agencies. Although the governmental role continues and is essential for conserving fragile and valuable areas, private and community-based groups, as well as individuals, have also come to play a large part in these efforts. Nongovernmental organizations such as The Nature Conservancy, Elkhorn Slough Foundation, Monterey County Agricultural and Historic Land Conservancy, Monterey Bay Aquarium, and Carneros Creek Association work to protect natural resources in the Elkhorn Slough watershed. Their activities include land purchases, restoration, education, and research. Volunteers provide critical support for these programs at the Elkhorn Slough National Estuarine Research Reserve and the Watershed Institute at Moss Landing and CSU Monterey Bay. Individuals have also helped promote conservation activities. Some with large landholdings, such the late David Packard and the Porter/Cooley family, have established conservation easements and developed conservation plans for their land. Many farmers are working with the NRCS on soil management and erosion control practices on their slough-side farms. Thanks to these conservation, restoration, and protection efforts, Elkhorn Slough continues to offer habitat for wildlife, access and recreational activities for the public, and endless research opportunities for the scientific community.

Continued research will be critical to the future of Elkhorn Slough. In the face of increasing human pressure, applied research designed to answer specific management questions, combined with basic research and ongoing conservation efforts, is needed to protect and restore the slough and watershed's native habitats and communities. Our hope is that the chapters to follow will enhance understanding of Elkhorn Slough and the watershed while pointing out specific areas of research and management that should be addressed in the coming years.

Appendix 1.1
Regulatory Jurisdictions

Elkhorn Slough National Estuarine Research Reserve, Monterey Bay National Marine Sanctuary, National Oceanic and Atmospheric Administration

Elkhorn Slough National Estuarine Research Reserve is part of the National Estuarine Research Reserve System, which is administered by the U.S. Department of Commerce, National Oceanic and Atmospheric Administration (NOAA), National Ocean Service, Office of Ocean and Coastal Resource Management, Estuarine Reserve Division. NOAA provides programmatic support and financial assistance and ensures that the reserve is managed in accordance with the regulations and goals of the national program.

The National Estuarine Research Reserve System (NERRS) was created in 1972 in recognition of the fact that estuarine habitats were undergoing serious declines in quality and rapid, widespread destruction. Preserving the slough as a "natural field laboratory" was one of the primary goals of its designation within the NERRS. One function of the NERR system is to provide benchmark information to researchers, coastal communities, and ecosystem managers. Degrees of change in regional estuarine ecosystems can be measured by comparison to known, regularly monitored conditions in the reserve. Education is another fundamental function of the NERR system and at the reserve spans the breadth of our society, from school groups to the general public to the research and management communities.

Resource management and protection of Elkhorn Slough were intended from the beginning to involve the host of agencies that have jurisdiction in the slough and its watershed. It was also understood from the planning stages that protection and management would have to extend beyond the slough itself. Among the resources to be protected were air and water quality, historical and archaeological sites and artifacts, and an array of vegetation, wildlife, and both natural and modified habitats. The reserve's most recent management plan (in prep.) outlines four basic protection goals and two restoration goals: to enforce laws and regulations that protect the natural resources of the reserve; to protect the reserve from off-site activities that may have harmful impacts; to prevent detrimental effects of public use on natural resources; to minimize the degradation of reserve wildlife and habitats by nonnative and feral wildlife; to increase native biodiversity through the restoration of native habitats; and to complete the development of a habitat enhancement/ restoration plan. Additionally, to improve protection of the slough from off-site activities in the watershed, the Elkhorn Slough Foundation has developed a land-acquisition plan for purchasing parcels surrounding the slough.

The Monterey Bay National Marine Sanctuary, established in 1995, is part of the Marine Sanctuary Division within NOAA's Office of Ocean and Coastal Resource Management. The sanctuary has jurisdiction within Elkhorn Slough to the high water mark.

California Department of Fish and Game

The California Department of Fish and Game was established to administer and enforce the Fish and Game Code. It is empowered to review projects and recommend conditions for any activity proposed within inland waterways, and must be notified prior to any such action. Department authority is not restricted to a specified tide level in estuarine waters but may overlap the jurisdiction of other agencies. The department is in charge of the management of Elkhorn Slough National Estuarine Research Reserve in partnership with NOAA. Fish and Game is responsible for developing and carrying out a management plan for the reserve, including the establishment of a budget for operations and management, providing adequate staffing, establishing and enforcing regulations, and coordinating all reserve activities.

California Coastal Commission

The State Coastal Commission was established under 1976 legislation, succeeding the temporary California Coastal Zone Conservation Commission and regional commission created by 1972 statewide initiative. The 1976 law required all local governments within the coastal zone to prepare a coastal plan for their jurisdiction that is consistent with state policies (see chapter 7 for details). Along with the Monterey County Planning Department, the Coastal Commission is responsible for issuing construction permits in the area.

Monterey County Planning Department

The Monterey County Planning Department is responsible for ministerial decisions involving planning, zoning, and development in the county. Recommendations involving more significant planning actions are forwarded to the county Planning Commission and the Board of Supervisors for further action. The department is also responsible for preparation of the local coastal plan for the Salinas subregion 5 of the Central Coast Basin, which includes Elkhorn Slough. Preliminary special area studies in the subregion have stated that successful agricultural use of coastal lands depends heavily on the maintenance of water quality. Seawater intrusion and agricultural and domestic wastewater discharge have led to degradation of water quality in many areas within the subregion. Ultimate implementation of planning department recommendations for the local coastal plan will be the sole responsibility of the Board of Supervisors. Any planning or zoning decisions made by the board, however, are subject to appeal to the state Coastal Commission.

Moss Landing Harbor District

The Moss Landing Harbor District was established in 1947 as a political subdivision of the state of California. The passage of Senate Bill 1116 granted the trust title of Elkhorn Slough and Moss Landing Harbor "tidal" lands to the district. The district, which claims coincidental jurisdictional boundaries with the State Lands Commission, is authorized to regulate and monitor commerce, fisheries, and navigational uses of these tidal lands.

Regional Water Quality Control Board—Central Coast, Region 3, State Water Resources Control Board

The State Water Resources Control Board (SWRCB), formed in 1967, is responsible for carrying out the Porter-Cologne Water Quality Control Act of 1969 and the federal mandates of the Clean Water Act of 1972. The SWRCB has authority over both water allocation and water quality protection and supervises the nine Regional Water Quality Control Boards (RWQCB) in the state. Boundaries of the RWQCB were established based on the major watersheds in California. The regional boards are responsible for writing waste discharge permits; cleaning up pollution and contamination that threatens public health, safety, and welfare; monitoring the health, quality, conditions, and beneficial uses of California's surface and ground waters; inspecting dischargers; and enforcing state and federal laws.

Other State and County Agencies

A variety of other state and local agencies have responsibilities to protect human health or provide water within the Elkhorn Slough. The Monterey County Environmental Health Department oversees refuse disposal and septic systems. It is also responsible for closing shellfish harvesting. California Department of Health Services establishes public drinking water standards and oversees drinking water of the reserve to ensure that it meets those standards. North Salinas Valley Mosquito Abatement District carries out mosquito abatement in the area in accordance with the California Health and Safety Codes. Pajaro Valley Water Management Agency is responsible for assessing water sources, planning for future needs, and regulating water use in the area.

References

ABA Consultants. 1989. *Elkhorn Slough Wetlands Management Plan.* Prepared for Monterey County and the California Coastal Conservancy. Capitola, CA: ABA Consultants.

Blankinship, M. S., and A. D. Evans. 1993. Elkhorn Slough 205(j) uplands water quality management plan. Final report. Marina, CA: Association of Monterey Bay Area Governments.

Broenkow, W. 1977. Water chemistry of Elkhorn Slough and Moss Landing. In J. Nybakken, G. Cailliet, and W. Broenkow, *Ecologic and Hydrographic Studies of Elkhorn Slough, Moss Landing Harbor, and Nearshore Coastal Waters, July 1974–1976*, pp. 392-465. Moss Landing, CA: Moss Landing Marine Laboratories Technical Publication.

Caffrey, J. M., A. P. DeVogelaere, J. E. Holte, and ABA Consultants. 1998. *Elkhorn Slough Bibliography.* 4th ed. Watsonville, CA: Elkhorn Slough Foundation.

Clark, L. R. 1972. Long period wave characteristics in Moss Landing Harbor and Elkhorn Slough. M.A. thesis, Moss Landing Marine Laboratories. San Jose State University.

Conomos, T. J., R. E. Smith, and J. W. Gartner. 1985. Environmental setting of San Francisco Bay. In J. E. Cloern and F. H. Nichols, eds., *Temporal Dynamics of an Estuary: San Francisco Bay,* pp. 1–12. Dordrecht, The Netherlands: Dr. W. Junk Publishers.

Dietz, S. A., W. Hildebrandt, and T. Jones. 1988. Archaeological investigations at Elkhorn Slough: CA-MNT-229. A middle period site on the central California coast. Three papers in Northern California Anthropology. Berkeley, CA.: Northern California Anthropological Group.

Dondero, S. 1984. Preliminary report on archaeological testing, CA-MNT-229, Elkhorn Slough, California. California Department of Transportation, Office of Environmental Analysis.

Griggs, G. 1995. Monterey Bay, national center for marine science. *Sea Technology*, May, 43–53.

Habitat Restoration Group. 1996. Moro Cojo Slough wetlands management and enhancement plan. Monterey County, CA: Habitat Restoration Group.

Hardin, G. 1968. The tragedy of the commons. *Science* 162: 1243–1248.

Hayward, I. R. 1931. The marsh and aquatic plants of the Pajaro Valley. Master's thesis, Stanford University.

Hildebrant, W., T. Jones, and J. Hall. 1983. Archaeological research of MNT-229, Elkhorn Slough, California. Rohnert Park, CA: Cultural Resources Center, Sonoma State University.

King, J. A., M. Mayer, J. S. Oliver, B. Ramer, D. Small, and M. Silberstein. 1986. Monitoring ecological changes in the new south marsh on Elkhorn Slough National Estuarine Sanctuary. Technical report. Silver Springs, MD: Ocean and Coastal Resource Management/Sanctuaries Program Division, NOAA.

Kjerfve, B. 1989. Estuarine geomorphology and physical oceanography. In J. W. Day Jr., C. A. S. Hall, W. M. Kemp, and A. Yañez-Arancibia, eds., *Estuarine Ecology*, pp. 47–78. New York: John Wiley.

MacGinitie, G. E. 1927. Ecological aspects of Elkhorn Slough. M.A. thesis, Stanford University.

———. 1935. Ecological aspects of a California marine estuary. *American Midland Naturalist* 16: 629–765.

Mayer, M. A. 1986. Recruitment of plants into a newly established salt marsh in Elkhorn Slough, California. Final report. Silver Springs, MD: Ocean and Coastal Resource Management/Sanctuaries Program Division, NOAA.

Mountjoy, D. C. 1993. Farming practices survey and outreach recommendations for the Elkhorn Slough water quality management plan: prepared for Association of Monterey Bay Area Governments. Final report. Marina, CA: Association of Monterey Bay Area Governments.

Nichols, F. H., and M. M. Pamatmat. 1988. *The Ecology of the Soft-Bottom Benthos of San Francisco Bay: A Community Profile.* U.S. Fish Wildlife Service Biology Report, 85, no. 7.19.

Nybakken, J., G. Cailliet, and W. Broenkow. 1977. *Ecologic and Hydrographic Studies of Elkhorn Slough Moss Landing Harbor and Nearshore Coastal Waters, July 1974–June 1976,* pp. 216–386. Moss Landing, CA: Moss Landing Marine Laboratories Technical Publication.

Oliver, J. S., and D. L. Schwartz. 1988. Photograph and map library for Elkhorn Slough Reserve: Erosion and sedimentation in Elkhorn Slough. Final report. Watsonville, CA: Elkhorn Slough National Estuarine Research Reserve.

Onuf, C. P. M. L. Quammen, G. P. Shaffer, C. H. Peterson, J. W. Chapman, J. Cermak, and R. W. Holmes. 1978. An analysis of the values of central and southern California coastal wetlands. In P. E. Greeson, J. R. Clark, and J. E. Clark, eds., *Wetland Functions and Values: The State of Our Understanding*, pp. 186–199. Minneapolis: American Water Resources Association.

Scharffenberger, T. 1999. Elkhorn Slough conservation plan. Elkhorn Slough Foundation and The Nature Conservancy. Unpublished report, on file at the Elkhorn Slough National Estuarine Research Reserve.

Silberstein, M. 1989. Seagrass research in west coast National Estuarine Research Reserves. Sixth Symposium on Coastal and Ocean Management, American Society of Civil Engineers, New York.

Smith, R. E. 1973. *The Hydrography of Elkhorn Slough, a Shallow California Coastal Embayment.* Technical Publication no. 73-02. Contribution from the Moss Landing Marine Laboratories No. 42. Moss Landing, CA: Moss Landing Marine Laboratories.

Smith, S. V., W. J. Wiebe, J. T. Hollibaugh, S. J. Dollar, S. W. Hager, B. E. Cole, G. W. Tribble, and P. A. Wheeler. 1987. Stoichiometry of C, N, P, and Si fluxes in a temperate-climate embayment. *Journal of Marine Research* 45: 427–460.

USDA, Soil Conservation Service. 1984. Strawberry hills erosion control program. Salinas and Aptos, CA: Soil Conservation Service.

Zimmerman, R. C., R. C. Smith, and R. S. Alberte. 1990. Seagrass revegetation: Developing a predictive model of light requirements for *Zostera marina*. In K. Merkel, R. Hoffman, and J. Stuckrath, eds., *California Eelgrass Symposium*, pp. 6–12. National City, CA: Sweetwater River Press.

2

Chapter Two

Geology

Dave Schwartz

Lagoons and estuaries such as Elkhorn Slough are relatively rare on the west coast of the United States, due to the influence of plate tectonics on coastal landscape features. The proximity of Elkhorn Slough to the San Andreas Fault—the plate tectonic boundary that defines California geology—is the fundamental control on the landscape, which is then further modified by erosion, deposition, and sea level history as well as by physical processes of wave and current activity (Inman and Norstrom 1971).

Modern-day Elkhorn Slough is an elongated, tidally influenced, coastal estuarine embayment that meanders approximately 11.5 kilometers through the western portion of the Elkhorn Valley. Numerous small tidal channels dissect the slough, which is flanked by extensive intertidal mudbanks and salt marshes.

Elkhorn Valley itself appears to be the westernmost remnant of what was once a much larger river valley. Its headwaters are thought to have originated in either the Santa Clara or Central Valley and flowed into Monterey Bay during the early Pleistocene (1.5 million–500,000 years before present [B.P.]). The headwaters of this ancestral river were apparently cut off by slow, steady movement along the San Andreas Fault zone, resulting in the present anomalous configuration of the Elkhorn Valley: a small stream, without significant headwaters, in a large, broad, deep valley.

Elkhorn Valley lies within the Salinian Block microplate, a wedge of granitic crust that originated south of the Sierra Nevada and was transported to its present location by movement along the San Andreas Fault zone. The Salinian Block is bordered on the east side by the San Andreas Fault zone and on the west by the San Gregorio and Nacimiento Fault zones. Right lateral, strike slip motion along these active fault zones continues to transport the Salinian Block northwestward at a current rate of 27–33 millimeters per year (Vedder, Howell, and McLean 1983).

This chapter examines the impact of tectonics on the Elkhorn Slough region's landscape (comparing east- and west-coast settings) and reviews the events that shaped the present-day Elkhorn Slough watershed. Relying on research that has been carried out in the region, I provide a general picture of the central Monterey Bay area's pre-Holocene (up to 11,000 years B.P.) geologic history and a relatively detailed picture of its more recent geologic past (since 8,000 B.P.). I also examine the human impacts that have reversed the natural geologic processes taking place at Elkhorn Slough. The chapter closes with suggestions for research projects and management strategies designed to preserve important habitat and better understand the area's early geologic history.

Passive vs. Active Tectonic Settings

The influence of tectonic setting is best illustrated by the difference in coastal geomorphic features between the eastern United States—a tectonically nonactive margin—and the tectonically active coast of central California, where the Pacific and North American plates meet and slide past each other along the San Andreas Fault zone.

From New York to Florida, the coast is located thousands of miles from the nearest active plate boundary. Mountain building is absent here, as both the oceanic and continental crust are old, flat, and dense. Large rivers slowly erode the Appalachians, carrying sediment toward the Atlantic and Gulf coasts. Upon entering the ocean the sediment is reworked by waves and currents, creating constructional landforms, such as long, broad beaches and barrier islands. The recent rise of sea level (15,000–5,000 B.P.) has drowned the old stream valleys, forming large estuaries and salt marshes that are often fronted by extensive barrier islands, spits, bay mouth bars, and tidal inlet deltas. There are few if any sea cliffs and no rocky intertidal shoreline.

In comparison, there is a striking lack of constructional landforms along the central California coast, which is mostly dominated by rapid uplift of the coastline due to movement along fault zones. Wave action and erosion combined with the late Pleistocene rise in sea level have created the region's steep, rocky shoreline, fronted by sea cliffs and numerous offshore rocks, stacks, and small islands. The typical "stepped" coastline of benches (marine terraces, fig. 2.1) and risers (modified sea cliffs) indicates that this process of a rising sea level notching the continent has occurred repeatedly during the past 800,000 to 1,000,000 years.

Beach width and length vary widely along the central coast. Beaches are typically found at the mouths of rivers and streams,

Plate Tectonics

According to the theory of plate tectonics, the earth's outer layers, or lithosphere, are broken into a number of rigid individual plates that move in response to convection taking place in the semiliquid outer mantle. The movement of these plates accounts for the seismic activity, volcanism, and mountain building that occur in relatively narrow zones along the boundaries of large plates, in this instance the Pacific and North American plates. Figure identifies major plates and approximate plate boundaries.

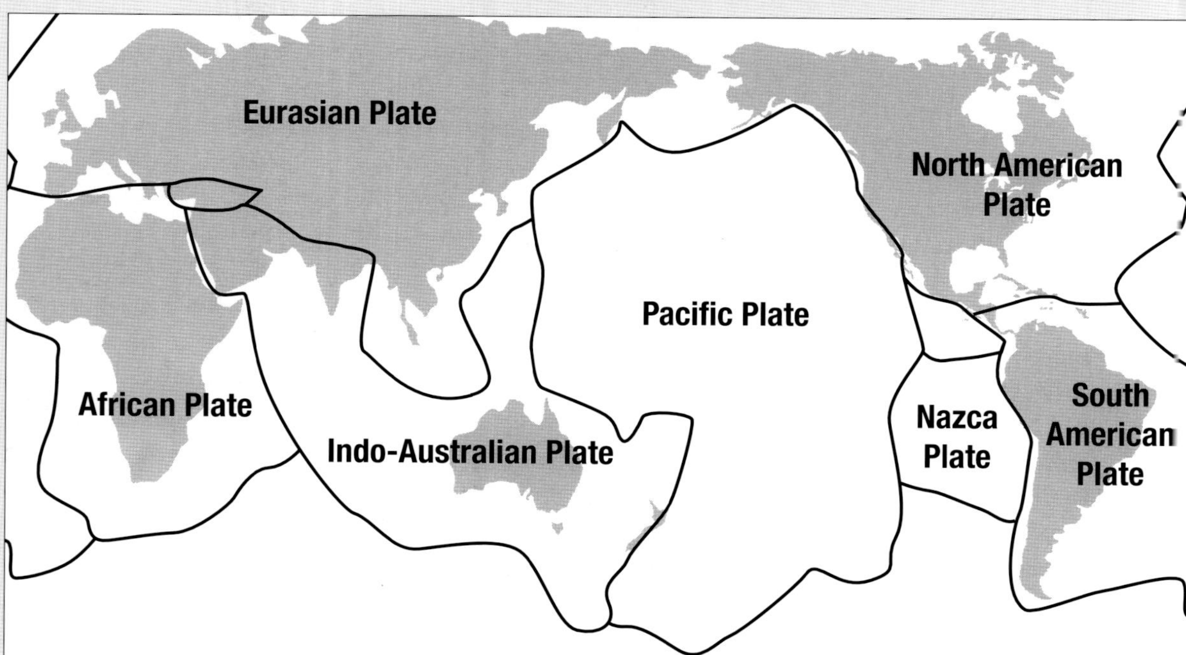

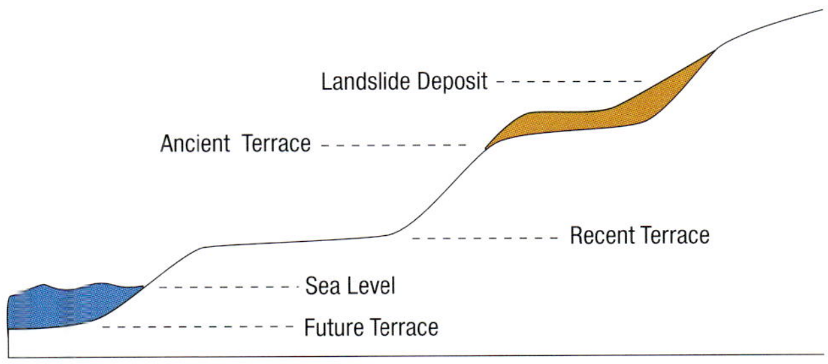

Figure 2.1. Marine terraces are ancient wave cut platforms, or sea floors, that have been uplifted above sea level and separated from each other by modified sea cliffs. Wave erosion that occurred during periods of sea level rise between glacial advances has combined with continuous uplifting to create the central coast's "stepped" profile. Adapted from *The Natural History of Big Sur*, Paul Henson and Donald J. Usner.

and as small, isolated pockets of sediment between rocky headlands (pocket beaches). Only a few relatively small estuaries and lagoons occur in the region. After San Francisco Bay and Tomales Bay, Elkhorn Slough is the third largest estuary in the state. All three owe their existence to a complex history of lateral and vertical movements along the faults of the San Andreas Fault system and the episodic sea level fluctuations (on about a 100,000-year cycle) that are directly related to the waxing and waning of the large continental glaciers in polar and subpolar regions during the Pleistocene epoch (Atwater, Hedel, and Helley 1977; Atwater 1979).

Pre-Holocene Geologic History

Although the pre-Holocene (before 11,000 years B.P.) history of Elkhorn Slough is poorly understood, it is apparent that the slough now occupies a portion of what was once the much larger Elkhorn Valley, the on-land extension of the Monterey Submarine Canyon (Schwartz, Mullins, and Belknap 1986). Samples from oil wells drilled in the 1950s indicate that directly landward of the canyon is a deep, sediment-filled paleo-canyon or gorge (fig. 2.2). This narrow erosional canyon is incised into granitic material dating to the Cretaceous period (144–66.4 million years B.P.) and is filled with 2.5 kilometers of sediment deposited between the early Miocene and Pliocene (28–5 million years B.P.) (Starke and Howard 1968). This gorge was apparently eroded in the late Oligocene (approximately 30 million years B.P.) during a period of tectonic uplift and major sea level decline (Greene and Clark 1979; Vail and Hardenbol 1979).

By contrast, the present-day Elkhorn Valley formed relatively recently, having eroded into the Aromas Sands Formation (Dupre 1975). The Aromas Formation is a nonmarine deposit of late Pliocene–early Pleistocene age (1–3 million years B.P.) and is widespread in the central region of the Monterey embayment. Composed of interbedded aeolian (wind-deposited) sands, stream deposits, lake deposits, and nearshore marine sands, the heterogeneous Aromas Sands Formation also reflects the complex, rapidly changing set of environments in which it was deposited. Since Elkhorn Valley is incised into deposits dating from 1–3 million years B.P. , it can be no older than early Pleistocene (Dupre, Clifton, and Hunter 1980).

Based on a variety of data, including the size of the Elkhorn Valley and the fact that the Carquinez Straits did not act as the drainage for the Central Valley until just recently, researchers have proposed that large-scale drainage entering the Pacific Ocean from the Santa Clara Valley and possibly the Central Valley of California originally reached the ocean via the Elkhorn Valley. During periods of low sea level this ancestral stream slowly carved the Elkhorn Valley and probably the nearshore portions of the Monterey Submarine Canyon (Beard 1941; Martin and Emery 1967; Jenkins 1974).

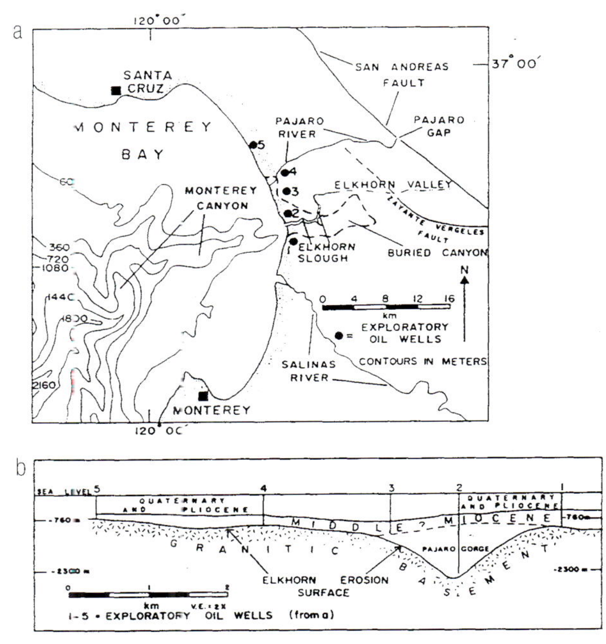

Figure 2.2. (a) The general relationship between the buried Pajaro Gorge at the head of the Monterey Submarine Canyon and Elkhorn Slough and Valley. (b) Generalized north-south geologic cross-section of the central Monterey Bay region based on exploratory oil well data. Both illustrations based on Starke and Howard (1968).

Monterey Canyon and Monterey Bay Coastline

Although the Monterey Submarine Canyon probably began to form 10 million years ago, most of its development took place over the last 5 million years. Numerous processes helped form this underwater canyon, including the mass movement of underwater sediments called turbidity currents and debris flows, fault movement along the San Gregorio Fault zone, and in the upper 100 meters, river erosion.

Today, different portions of Monterey Bay's coastline are changing in dramatically different ways. The central region of the Monterey embayment (the lowlands near Castroville) is subsiding, while the northern and southern edges of the bay are experiencing uplift. As a result, there are no rock outcrops along the subsiding coastline in the central bay area, where accelerated coastal erosion of the soft sand dunes is taking place. In contrast, slow uplift along the bay's north and south sides exposes bedrock in the surf zone, resulting in relatively slower coastal erosion rates. Image provided courtesy of Monterey Bay Sanctuary Foundation.

Although once much more extensive, Elkhorn Valley is now “beheaded” at its eastern end, its original headwaters literally cut off by the San Andreas Fault. During the mid to late Pleistocene (approximately 500,000 years B.P.), uplift associated with faulting along the San Andreas created the Northern Gabilan Range. In the process, the large paleo-drainage system of the original Elkhorn Valley was disrupted as right lateral fault movement truncated the valley, resulting in a much smaller watershed drained by Carneros Creek, which continues to drain the relic Elkhorn Valley (fig. 2.3).

Holocene Events: Sea Level Changes and Slough Development

The relative size of the central Monterey Bay region’s wetlands has changed dramatically over the last 20,000 years in response to a worldwide sea level increase linked to global glacial retreats. Prior to the onset of sea level rise, the region’s coastline looked very different from what we see today. Nonmarine sandy gravels recovered from a depth of approximately 30 meters below present *Mean Lower Low Water (MLLW)* at the mouth of Elkhorn Slough indicate that the drainage in Elkhorn Valley was sufficient to incise, or carve, its channel to this depth during the last glacial maximum, about 18,000 years B.P. (CLIMAP 1976). At that time, sea level was approximately 90–120 meters lower and the shoreline lay 5–8 kilometers to the west of its present location.

As glaciers retreated and melted, marine waters gradually invaded the incised channel in the western Elkhorn Valley. Approximately 8,000 years B.P., relative sea level was still 15–17 meters lower than present. A *high-energy tidal inlet* existed at the mouth of Elkhorn Slough, comparable to the present-day environment (Schwartz, Mullins, and Belknap 1986). As sea level rose, the mouth of Elkhorn Slough gradually filled with marine sediment and the depositional energy decreased. These events are reflected in a sequence of coarse-grain gravel and sand at the base of the slough’s stream channel, covered by sand, then silt, and finally clay. The deposition of this characteristic sequence of marine sediments or sedimentary rocks is known as a fining upward transgressive sequence, with particle size becoming finer as you move upward through the layers.

Between 4,000 and 5,000 years B.P., the continuous invasion of marine waters created a large estuary that extended into McClusky and Moro Cojo Sloughs as well as into the northern Salinas Valley. Energy within the depositional environment continued to decrease over a period of 2,000 to 3,000 years, until Elkhorn Slough became an extremely quiet-water estuary, as indicated by estuarine clays recovered at the top of marine sediments deposited at the slough mouth. These clays hold fossils of microorganisms typical of brackish waters, offering evidence that the slough was cut off from marine waters 2,000–3,000 years B.P. Salinity may have dropped to as low as 5 parts per thousand during this period (R. Laws and R.

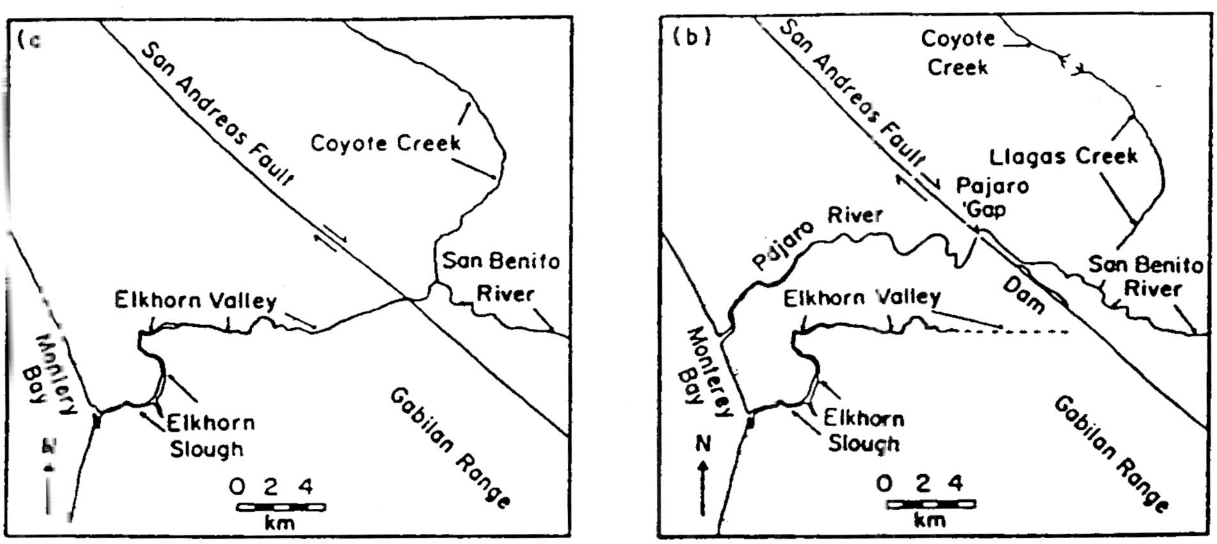

Figure 2.3. Highly generalized, interpretive, paleogeographic reconstruction of late Pleistocene drainage in the central Monterey Bay region (a); possible effects of right-lateral offset and tectonic truncation of Elkhorn Valley from the San Benito River and present development of the Pajaro River system (b). Modified from Martin and Emery (1967) and Jenkins (1974).

Dave Schwartz (far right) and Moss Landing Marine Laboratories graduate student Michelle Hornberger drill for estuarine clays, forams, and Holocene bivalve communities beneath Moro Cojo Slough. Photo credit: Mark Silberstein.

Forester, pers. comm.). In comparison, normal marine salinity is 35 parts per thousand. Although it is unclear exactly why the energy of the environment gradually decreased and the water became brackish, a beach or sand dune may have temporarily blocked the slough's direct connection with the open ocean.

Similar estuarine clay deposits and microfossils have also been recovered from the subsurface of nearby Moro Cojo and McClusky Sloughs and the Pajaro and Salinas River valleys (Dupre 1975; Tinsley 1978; Hornberger 1991; see photo, above). Thus, a large interconnecting estuarine system appears to have existed during the early to middle Holocene in the central Monterey Bay region. However, in the Salinas Valley, river- and windborne sediment accumulated at a rate at least two to three times greater than in Elkhorn Slough (Tinsley 1978). This higher sedimentation rate led to the complete infilling of the Salinas Estuary.

Higher sediment accumulation rates in the Salinas Valley probably resulted from a combination of its larger drainage basin area (Tinsley 1978) and the fact that the Salinas River has a much greater year-round discharge than Elkhorn Valley's Carneros Creek. Therefore, even though global sea level has continued to increase slowly during the last 5,000 years, the large estuarine system that once occupied the central Monterey Bay region began to shrink as sedimentation rates in the northern Salinas Valley outpaced the relative sea level rise (Schwartz, Mullins, and Belknap 1986). However, the "beheading" of Elkhorn Valley by movement on the San Andreas Fault cut off the valley's original drainage system, leaving behind a low-discharge stream (Carneros Creek) to drain a relatively large area. Because sedimentation in Elkhorn Valley has not kept up with the rise in sea level, Elkhorn Slough remains an open marine system.

Dave Schwartz examined microfossil *assemblages*, sediment grain size, and the percentage of *organic material* present in Elkhorn Slough sediments, and found that they were deposited in three different microenvironments: subtidal channels, intertidal mudflats, and salt marshes (Schwartz, Mullins, and Belknap 1986). The salt marsh environment was rich in organic material mixed with sand, silt, and clay. In contrast, the intertidal mudflats consisted of pure blue-gray clay and no organic material. The subtidal channels also lacked organic material, but differed from the other two in that the sediments consisted of sands and gravel. Each microenvironment contained a different assemblage of benthic, or bottom-dwelling, single-celled marine protozoans called Foraminifera.

Evidence from hand auger cores showed that as Elkhorn Slough filled with sediments over the last 5,000 years, the salt marsh environment developed along its landward margins and the mudflat developed along the axis of the main subtidal channel. As more sediment was introduced to Elkhorn Slough, the salt marsh broadened and advanced toward the main subtidal channel, gradually burying the mudflats (fig. 2.4). Several cores taken in the slough's salt marshes showed that the salt marsh environment is 1 to 2 meters thick and sits on top of older mudflats.

Cores collected from McClusky and Moro Cojo Sloughs indicate that these now-freshwater wetlands were once marine (Schwartz, Mullins, and Belknap 1986). The cores consist of salt marsh clays and peat with typical microfossil and macrofossil assemblages that are overlain by stream deposits. Thus McClusky and Moro Cojo Sloughs, now separate from Elkhorn Slough, appear to have evolved further into small freshwater systems and alluvial valleys, while Elkhorn Slough has remained a marine environment.

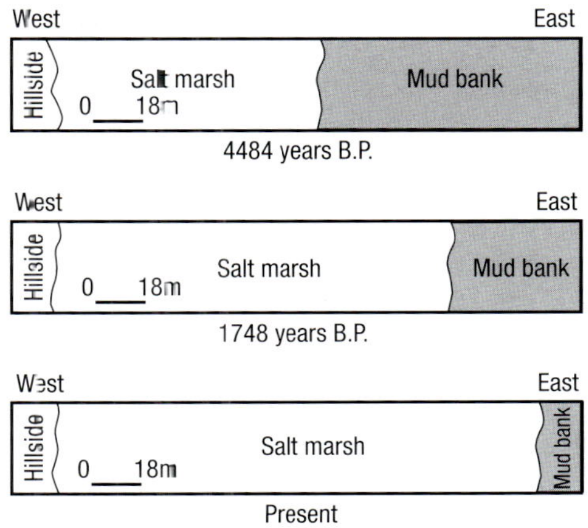

Figure 2.4. Schematic summary of salt-marsh development in Elkhorn Slough during the past 5,000 years. As more sediment was deposited the salt marsh advanced toward the main channel and gradually buried the mudflats. Adapted from D.L. Schwartz et al.

River Channel Changes and Human Impacts

Historical records show that numerous natural hydrologic changes have taken place in the Elkhorn Slough region (see chapter 4 "Hydrography," for further discussion). In the middle 1850s, Elkhorn Slough was a minor tributary to the much larger Pajaro-Salinas River system. These two rivers may have shared a common entrance to the Pacific Ocean north of the present Moss Landing Harbor (Blake 1853). However, several Coast and Geodetic Survey maps show that by the late 1800s the Salinas and Pajaro River mouths were separated. Until 1903, the Salinas River continued to discharge into the Pacific Ocean north of Moss Landing, with Elkhorn Slough as a tributary (fig. 2.5). In 1908, winter storms modified the course of the Salinas River, shifting the mouth to its present location south of Moss Landing, while Elkhorn Slough persisted as a tributary to the old Salinas River channel. Construction of the jetties at the Moss Landing Harbor in 1946 provided a direct link between the Pacific Ocean and Elkhorn Slough.

After the jetties were completed, salt marshes began to retreat from the axis of Elkhorn Slough as it evolved into its present form. Comparison of air photos taken in 1937 and in the early 1980s shows that the salt marsh environs are rapidly eroding and the subtidal channels are growing wider (Malzone 1998). This is probably the result of the direct marine tidal incursion into the slough through the harbor entrance created by the construction of the Moss Landing Harbor (see chapter 4). As discussed above, for the past 5,000 years the slough has been filling in with fine-grained marine sediments; until the jetties were built, the rate of filling was gradually outpacing sea level rise. Had the harbor jetties not been built, Elkhorn Slough would have eventually filled with sediment, slowly evolving into a freshwater wetland and finally into a dry alluvial valley, probably within 3,000 years (Schwartz, Mullins, Belknap 1986). Its fate would have been similar to that of the Salinas Estuary.

Localized tectonic subsidence of the earth's crust that occurs in response to an earthquake may also play a role in marine tidal incursion. Since the 1989 Loma Prieta earthquake, saltwater during high tides is reaching areas that have not been wet by seawater for thousands of years. In addition, marsh restoration efforts have increased the tidal prism and accelerated the erosion process (see chapter 4).

Management Issues and Research Recommendations

Elkhorn Slough is part of a coastal estuarine system that has occupied the central portions of Monterey Bay for thousands of years. The relative size of central Monterey Bay's wetlands have changed dramatically over the last 20,000 years in response to worldwide sea level fluctuations linked to glacial advances and retreats, tectonic adjustments of the crust, and human activities. Further research on Elkhorn Slough's geologic history will help reveal the extent of these changes.

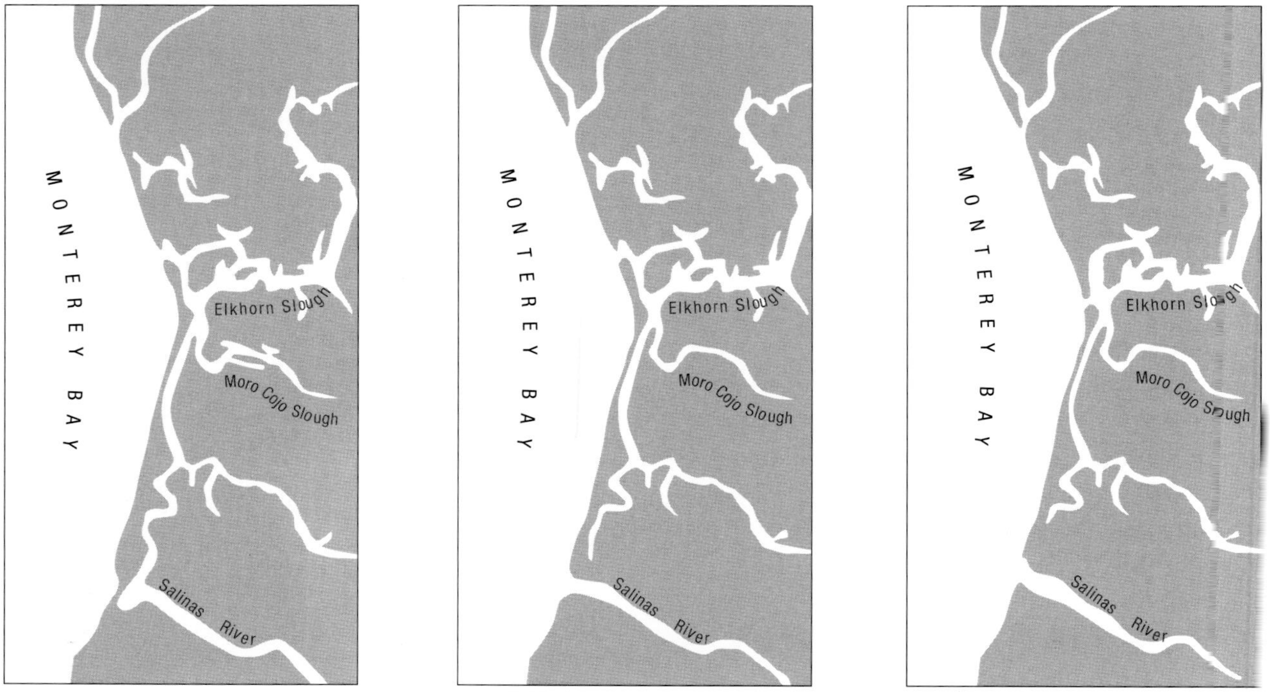

Before 1908 **1908-1946** **1946-present**

Figure 2.5. Changes in location of the Salinas River and mouth of Elkhorn Slough caused by winter storms and the construction of Moss Landing Harbor. Graphics provided courtesy of Monterey Bay Aquarium.

Late Pleistocene History

Studies from deep-sediment boreholes in San Francisco Bay have determined that the underlying structural basin (an area of the earth's crust that subsided, or sank, due to faulting and/or folding) has gradually alternated between nonmarine stream (alluvial) valley environments and marine saltwater embayments several times during the middle to late Pleistocene, or last one million years. The cyclic flooding and receding of marine waters in San Francisco Bay has occurred in response to sea level changes caused by the world's glaciers and polar caps retreating over thousands of years (Atwater 1979).

The late Pleistocene history (before 20,000 years B.P.) of central Monterey Bay and Elkhorn Slough is still unclear. To help resolve a more detailed geologic history, future studies should involve drilling several boreholes, at least 100 meters deep, in Elkhorn Slough and in the Pajaro River Valley. Information from subsurface sediments can help resolve how Elkhorn Slough and central Monterey Bay responded to late Pleistocene glacioeustatic (worldwide) sea level changes and tectonic adjustments of the crust. Such an effort would also provide a more detailed picture of how fresh or marine different areas were before humans made major hydrologic changes. This sort of study could help guide marsh restoration in the region, particularly the decision about whether an area should be fresh or brackish.

Effects of Tidal Scouring

Installation of the jetties at the mouth of Elkhorn Slough has altered the slough's natural evolution toward a freshwater wetland. Increased tidal scouring and saltwater incursion are rapidly eroding the slough's mudflats and salt marshes, habitats that serve as home and breeding ground to hundreds of plant and animal species. Future research efforts should address whether the tidal currents in Elkhorn Slough can be reduced to help retard erosion of this sensitive wetland environment.

Geologic Timeline of Elkhorn Valley and Elkhorn Slough

30 to 29 million years B.P. Elkhorn erosion surface exposed. Pajaro Gorge is formed by fluvial (river) systems discharging toward the southwest and dissecting granitic basement rock of the Salinian Block.

25 to 5 million years B.P. Pajaro Gorge fills with Tertiary marine sediments. Slow subsidence of central Monterey Bay begins, which continues to present day. San Andreas Fault develops and begins to displace central Monterey Bay and the Salinian Block to the northwest, a process that likewise continues today.

10 to 5 million years B.P. Begin formation of Monterey Submarine Canyon?

5 to 3 million years B.P. Formation of modern Monterey Submarine Canyon; submarine erosional and depositional processes responsible continue to present day.

1.5 million to 500,000 years B.P. Large paleo-drainage system drains Santa Clara Valley (and possibly Central Valley of California) to central Monterey Bay. Elkhorn Valley forms, probably at the same time as the Aromas Sands Formation is deposited.

Approximately 500,000 years B.P. Uplift and faulting of the Northern Gabilan Range and the destruction of eastern Elkhorn Valley. The large Pleistocene paleo-drainage system of ancient Elkhorn Valley is disrupted. A much smaller watershed area, Carneros Creek, is established, which continues to drain Elkhorn Valley today.

500,000 to 18,000 years B.P. Unknown. Indirect evidence indicates glacioeustatic sea level fluctuations in the late Pleistocene caused erosional/fluvial systems to alternate with depositional/estuarine systems in the central Monterey Bay region. Much scientific evidence suggests that this occurred in San Francisco Bay and other estuaries around the world.

18,000 years B.P. Last maximum glaciation. Global sea level is approximately 120 meters lower than today. Larger fluvial system drains to Monterey Bay via Elkhorn Valley.

16,000 to 11,000 years B.P. Global sea level rises rapidly. Climate is moister, all rivers and streams have higher discharge than today and are capable of downcutting and erosion. Erosion of Elkhorn and Salinas Valleys occurs.

11,000 to 8,000 years B.P. Initial flooding of fluvial systems in the central Monterey Bay lowlands begins approximately 11,000 years B.P. This includes Salinas Valley, Moro Cojo, Elkhorn, Bennett, and McClusky Sloughs. (Pajaro Valley also probably experienced flooding by rising sea level; however, more research is needed to confirm this.) Continued rise of sea level creates large, quiet-water estuarine system in central Monterey Bay.

8,000 to 5,000 years B.P. Maximum extent of estuarine system in central Monterey Bay. Salinas Estuary extends from northwest Salinas to the Pacific Ocean. Moro Cojo and McClusky Sloughs are part of a much larger Elkhorn Slough Estuary. Global sea level rise begins to slow at approximately 5,000 years B.P.

5,000 years B.P. to 1946 Filling in of estuarine environment in central Monterey Bay with mostly fluvial and some aeolian (windborne) sediments. Sea level rises slowly as estuarine system decreases in size as a result of faster sedimentation rates. Salinas Estuary completely infills; Moro Cojo, Elkhorn, and McClusky Sloughs are left as relicts. Salinas River changes course in 1908 to present location; such course changes likely have happened numerous times in last several 100,000 years.

1946–1948 Construction of jetties for the Moss Landing Harbor. Northern end of the abandoned Salinas River channel fills in with sediments. Regional water table continues to be drawn down for irrigation of agricultural lands adjacent to Elkhorn Slough.

1946 to Present Elkhorn Slough changes from a brackish, quiet-water lagoon environment to a high-energy tidal inlet with normal marine salinity as a result of the jetties constructed for the Moss Landing Harbor. High-energy tidal currents erode all wetland environments, salt marsh deteriorates, and intertidal channels and main channel widen and deepen—processes that continue today. Marsh restoration activities increase the area subject to tidal flooding, thus increasing current speed and erosion rates. Local subsidence resulting from seismic activity may also contribute to accelerated erosion.

References

Atwater, B. F. 1979. Ancient processes at the site of southern San Francisco Bay: Movement of the crust and changes in sea level. In T. J. Conomos, ed., *San Francisco Bay: The Urbanized Estuary. Investigations into the Natural History of San Francisco Bay and Delta with Reference to the Influence of Man*, 31–45. 58th Annual Meeting Pacific Division AAAS, June 1997, San Francisco State University. San Francisco, CA: The Division.

Atwater, B. F., C. W. Hedel, and E. J. Helley. 1977. Late Quaternary depositional history, Holocene sea-level changes, and vertical crustal movement, southern San Francisco Bay, California. United States Geological Survey Professional Paper 1014.

Beard, C. N. 1941. Drainage development in the vicinity of Monterey Bay, California. Ph.D. diss., University of Illinois, Urbana.

Blake, W. P. 1853. Geologic map of California. In *Explorations and Surveys for a Railroad Route from the Mississippi River to the Pacific Ocean.* Washington, D.C.: U.S. War Department.

CLIMAP. 1976. The surface of the ice-age Earth. *Science* 191: 1131–1137.

Dupre, W. C. 1975. Quaternary history of the Watsonville Lowlands, north central Monterey Bay region, California. Ph.D. diss., Stanford University.

Dupre, W. R., H. E. Clifton, and R. E. Hunter. 1980. Modern sedimentary facies of the open Pacific coast and Pleistocene analogs from Monterey Bay, California. In M. E. Field, A. H. Bouma, I. P. Colburn, R. G. Douglas, and J. C. Ingle, eds., *Society of Economic Paleontologists and Mineralogists Pacific Coast Paleogeography Symposium* 4:105–119.

Gordon, B. L. 1979. *Monterey Bay Area: Natural History and Cultural Imprints.* Pacific Grove, CA: Boxwood Press.

Greene, H. G., and J. G. Clark. 1979. Neogene paleogeography of the Monterey Bay area, California. In J. M. Armentrout, M. R. Cole, and H. TerBest Jr., eds., *Society Economic Paleontologists and Mineralogists Pacific Coast Paleogeography Symposium*, 3:277–298.

Hornberger, M. I. 1991. Paleoenvironment of Elkhorn Slough and surrounding wetland habitats: A geological study using an ecological approach. M.S. thesis, Moss Landing Marine Laboratories. San Jose State University,

Inman, D. L., and C. E. Norstrom. 1971. On the tectonic and morphologic classification of coasts. *Journal of Geology* 79: 1–21.

Jenkins, O. P. 1974. Pleistocene lake San Benito. *California Geology*, 151–163.

Malzone, C. M. 1999. Tidal scour and its relation to erosion and sediment transport in Elkhorn Slough. M.S. thesis, Moss Landing Marine Laboratories. San Jose State University.

Martin, B. D., and K. O. Emery. 1967. Geology of Monterey Canyon, California. *Bulletin American Association Petroleum Geologists* 51: 2281–2304.

Schwartz, D. L., H. T. Mullins, and D. F. Belknap. 1986. Holocene geologic history of a transform margin estuary: Elkhorn Slough, central California. *Estuarine, Coastal and Shelf Science* 22: 285–302.

Starke, G. W., and A. D. Howard. 1968. Polygenetic origin of Monterey Submarine Canyon. *Geological Society of America Bulletin* 79: 813–826.

Tinsley, J. 1978. Quaternary geology of the north Salinas Valley, California. Ph.D. diss., Stanford University.

Vail, P. R., and J. Hardenbol. 1979. Sea-level changes during the Tertiary. *Oceanus* 22: 71–79.

Vedder, J. G., D. G. Howell, and H. McLean. 1983. Stratigraphy, sedimentation, and tectonic accretion of exotic terranes, southern Coast Ranges, California. *American Association Petroleum Geologists Memoir* 34: 471–498.

3

Chapter Three

Climate

Jane Caffrey

Elkhorn Slough enjoys a mild, Mediterranean climate, thanks to the Pacific Ocean's moderating effects. Because ocean temperatures fluctuate very little, the adjacent land experiences a fairly narrow range of air temperatures. Although air temperatures at a weather station on the Elkhorn Slough National Estuarine Research Reserve (ESNERR; fig. 3.1) range between 5° and 35°C (41° and 95°F) on a seasonal basis, the monthly means range from only 11.1° to 15.4°C (52° to 60°F; figs. 3.2 and 3.3). These moderate temperatures combine with distinct seasonal patterns of wind, fog, and rain to create a maritime climate regime limited to a handful of coastal areas around the world.

In this chapter I describe seasonal conditions at Elkhorn Slough, including the rainfall patterns, cyclic droughts, and El Niño conditions that affect the slough's salinity and circulation patterns. The chapter concludes with suggestions for additional research to expand our knowledge of climatic influence on the slough ecosystem.

Elkhorn Slough's Three Seasons

Oceanic conditions dictate three distinct seasons along California's central coast. In spring and summer, prevailing winds from the northwest blow along the coast, combining with the earth's rotation to move surface water offshore. Cold water from below replaces the surface water in a phenomenon known as upwelling. These prevailing winds commonly are greatest in the afternoon, ranging between 2 and 5 m s^{-1} (4.5–11 mph[1]), while evening and morning winds are

Upwelling of cold, deep water in Monterey Bay generates summer fog that blankets Elkhorn Slough. Photo credit: ESNERR

[1] 1m s^{-1} = 2.23 mph. See page 278 for a table of English and metric equivalents.

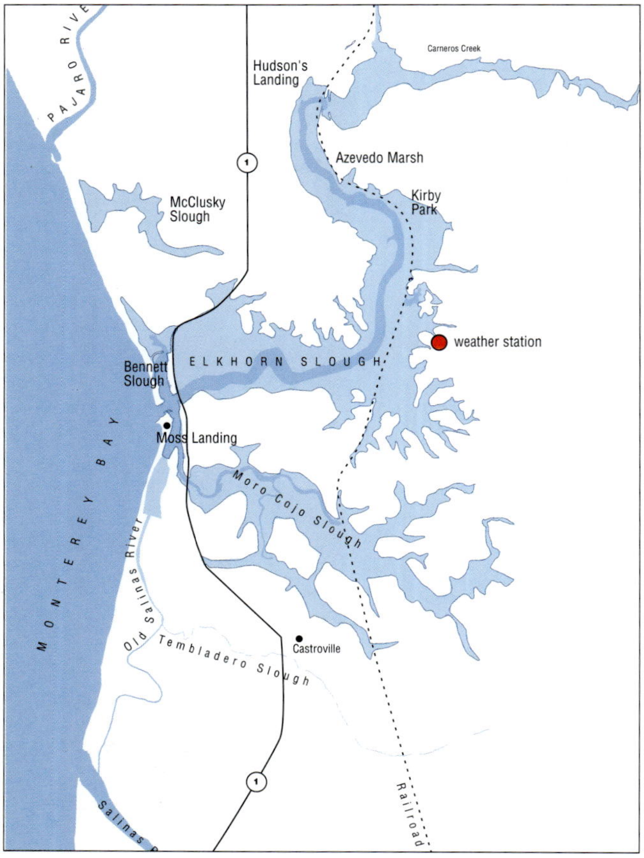

Figure 3.1 Location of the weather station on the Elkhorn Slough National Estuarine Research Reserve.

generally less than 2 m s^{-1}. Compounding the prevailing winds are strong afternoon winds (up to 8 m s^{-1}), like those seen at San Francisco Bay, which occur between April and September as a result of solar heating in the Central Valley (Conomos, Smith, and Gartner 1985; Nuss 1996). These strong winds increase water column mixing and resuspension of bottom sediments.

Upwelled water also generates fog, as relatively warm air comes in contact with the cold water at the sea surface and condenses. Blankets of fog cool the Elkhorn Slough watershed in late spring and summer. In July, for example, average afternoon temperatures peak at 15°C (59°F), only 2°C warmer than temperatures at midnight (fig. 3.2). In contrast, daily temperature ranges are greater during April and October (before and after upwelling season), when midday temperatures often exceed 25°C (77°F; fig. 3.2). In summer, when upwelling is more intense, relative humidity is generally high near the slough, averaging 80% and often climbing to 95% in the midafternoon when fog rolls in. Fog is an important source of moisture for plants (Holton, Barbour, and Martens 1991) and probably has a significant effect on fire frequency. Average daily solar radiation is highest in June, for although days are longer in the summer, fog can reduce light availability.

Rainfall averages are less than 0.6 cm per month between June and September (fig. 3.4), when a persistent high pressure cell becomes established offshore of central or northern California and drives most storms northward (Conomos, Smith, and Gartner 1985). Solar insolation (the amount of solar radiation striking the earth's surface), an important determinant of plant growth, peaks at 14.75 hours at the summer solstice.

In early fall, northwest winds decline, warmer water returns to the coast, and Elkhorn Slough experiences its warmest period (fig. 3.3). Seas are relatively calm during this season between upwelling and the beginning of winter storms. This is also the driest time of the year, with lowest relative humidities, averaging 65%, occurring in late fall (primarily October and November).

After October, the wet season begins as the high pressure system moves south and winter storms originating in the North Pacific track into the central coast. Storms lasting 2 to 5 days can occur every 7 to 14 days (Conomos, Smith, and Gartner 1985), causing intense but brief wind events. The rainiest months are December and January, with average rainfall of 10.9 cm per month at Watsonville and 7.6 cm per month at the ESNERR (fig. 3.4). In general, rainfall at the ESNERR is about 25% lower than at the Watsonville station. This is consistent with the pattern of average rainfall totals in the Monterey Bay region, which decline as one moves south and east between Santa Cruz and Salinas/Monterey (fig. 3.4). Minimum winter temperatures occur in December and January (fig. 3.3). Daylength drops to 9.5 hours at the winter solstice, and average daily solar radiation falls to 1900 watts m^{-2} d^{-1} in January.

Interannual Rainfall

Rainfall in the Elkhorn Slough region is highly variable, not only on a seasonal basis but on annual and interannual time scales. Totals collected at Watsonville since 1879 and at the ESNERR weather station since 1992 show that most of the rainfall occurs between October and May, with an average annual rainfall of 55.2 cm. Over the 119-year period, annual

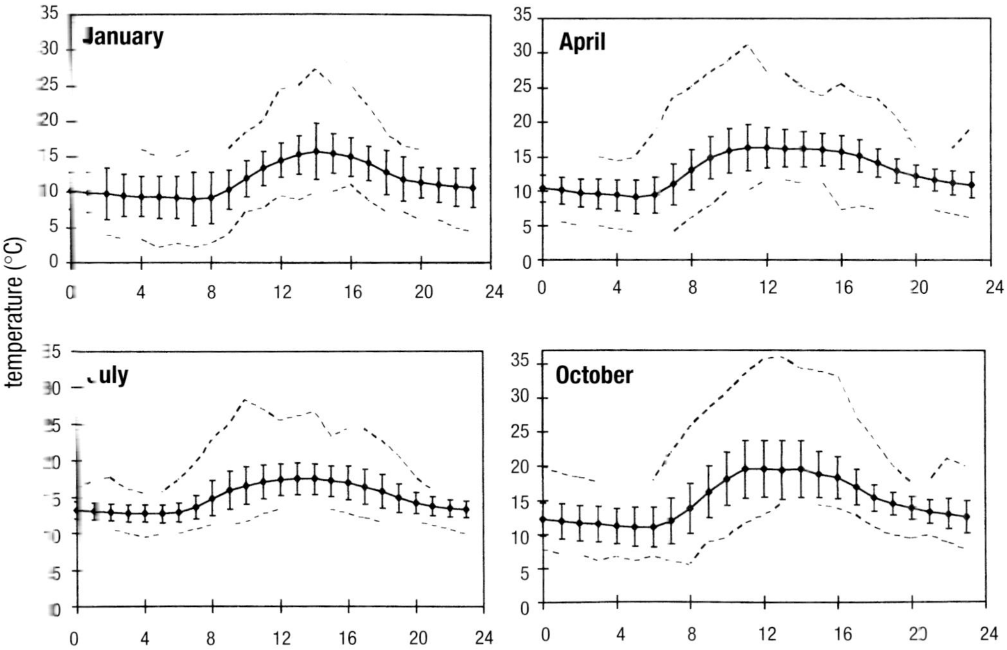

Figure 3.2 Average diurnal air temperature (°C) at Elkhorn Slough National Estuarine Research Reserve weather station in January, April, July, and October. Data averaged from years 1994–1997. Error bars are S.D. Dashed lines are minimum and maximum temperature.

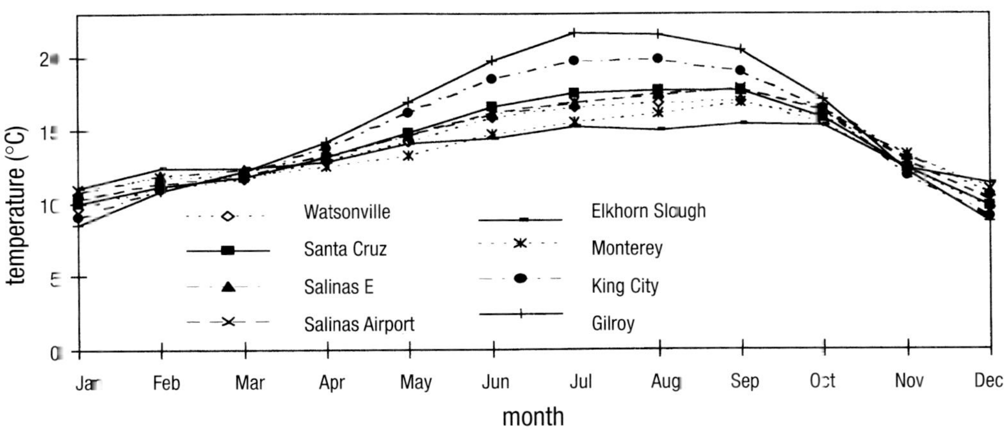

Figure 3.3. Average seasonal temperatures (°C) in the Monterey Bay region (National Weather Service data).

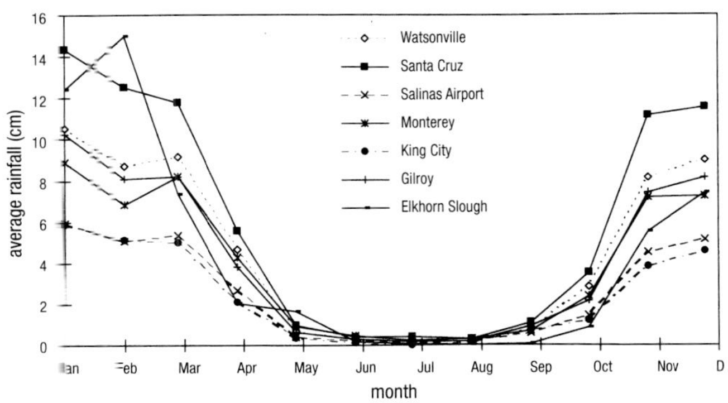

Figure 3.4. Average rainfall (cm) in the Monterey Bay region (NWS data, ESNERR weather station).

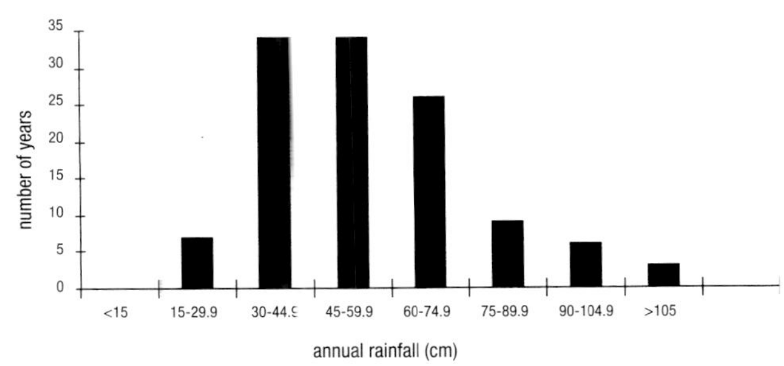

Figure 3.5. Frequency distribution of rainfall at the Watsonville Waterworks.

rainfall varied widely (fig. 3.5), with drought periods lasting two to six years and wet years where rainfall was significantly above average. Five distinct droughts stand out in this record: 1927–1933, 1943–1949, 1958–1961, 1976–77, 1986–1993 (fig. 3.6). The average rainfall for these years was 42.2 cm. Although the pattern of El Niños may have a significant impact on interannual variation in rainfall, the effects of El Niño are not consistent throughout California (Cayan and Peterson 1989).

Management Issues and Research Recommendations

The Elkhorn Slough National Estuarine Research Reserve currently maintains a weather station to monitor rainfall, wind speed and direction, air temperature, relative humidity, and solar photosynthetically active radiation. This monitoring provides critical information for managers and researchers in the area. Expanding these monitoring efforts could help managers correlate climatic factors with large-scale ecosystem changes.

Climate change has profound effects on coastal environments, from the relatively short-term (3- to 7-year) cycle of El Niño events to the anticipated rise in sea level due to global warming over the next decades. Exactly how these changes will affect the slough is unknown, but they need to be considered as we develop restoration plans and manage wetland areas within Elkhorn Slough.

One of the most significant changes predicted by global warming models is a rise in sea level, which would have the greatest impact on coastal marshes and other coastal areas Reference stations in Elkhorn Slough and throughout the region could be used to track sea level and monitor the slough's response to these changes, as well as to evaluate the way that geological processes such as uplift and subsidence can affect sea level.

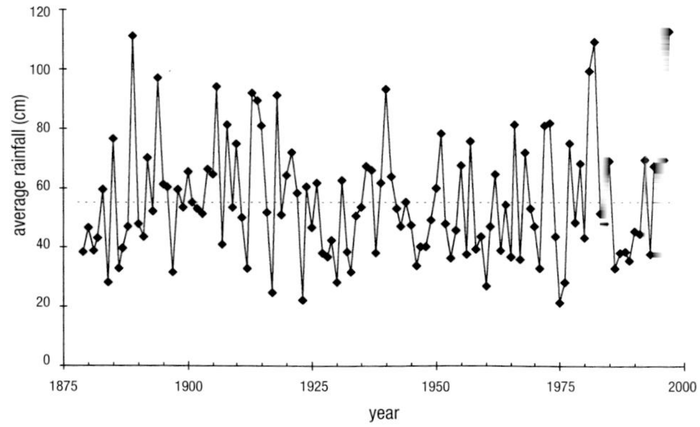

Figure 3.6. Annual rainfall (cm) at the Watsonville Waterworks between 1877 and 1997.

References

Cayan, D. R., and D. H. Peterson. 1989.The influence of North Pacific atmospheric circulation on streamflow in the West. In D. H. Peterson, ed., *Aspects of Climate Variability in the Pacific and the Western Americas*, pp. 375–397. Geophysical Monograph 55. Washington, D.C.: American Geophysical Union.

Conomos, T. J., R. E. Smith, and J. W. Gartner. 1985. Environmental setting of San Francisco Bay. *Hydrobiologia* 129: 1–12.

Holton, B., Jr., M. G. Barbour, and S. N. Martens. 1991. Some aspects of the nitrogen cycle in a California Strand ecosystem. *Madroño* 38: 170–184.

Nuss, W. 1996. Climate and meteorology. Monterey Bay National Marine Sanctuary site characterization. http://bonita.mbnms.nos.noaa.gov/sitechar/clim.html.

4

CHAPTER FOUR

Hydrography

Jane Caffrey, William Broenkow

Present-day Elkhorn Slough is a shallow tidal embayment undergoing dynamic, rapid changes: its channel is getting deeper, tidal currents are getting faster, and salt marsh habitat is disappearing as erosion increases. Tidal currents in the main channel approach 150 cm s^{-1}, representing a hazard for unwary recreational boaters. Maximum ebb currents measured in 1995 had doubled since the first measurements were made in 1970. Increasing tidal currents created by opening the harbor and restoring diked areas to tidal exchange have led to an increasing erosion rate in a positive feedback loop: as the slough is enlarged by erosion, the tidal prism, or the volume of saltwater exchanged by the tides also enlarges and the erosional capacity of the tidal currents increases.

Elkhorn Slough has been described as a "seasonal estuary" or "tidal embayment." Both terms are appropriate, but the term "slough," meaning a stagnant backwater, no longer applies. Before the Moss Landing Harbor entrance was created in 1946, this waterway was in fact sloughlike (MacGinitie 1935), as are nearby Bennett, Moro Cojo, McClusky, and Tembladero Sloughs today (fig. 4.1). Although the harbor's opening generated the most significant impacts to the slough's hydrology, human alterations dating as far back as the 1880s have altered the slough's fundamental character.

In this chapter we describe what is known about water transport within the slough and the way that it has changed as a result of human and natural impacts. We discuss the seasonal nature of slough circulation and salinity patterns; examine the shifting morphometry, or shape of the slough; and summarize the research that has documented changes in tidal currents and erosional impacts that continue to shape the slough's physical character and affect its plant and animal communities. Because water transports sediments, nutrients, contaminants, and organisms, Elkhorn Slough's dynamic hydrography has implications for many aspects of slough management. The chapter closes with a discussion of management issues related to slough hydrography and recommendations for future research.

Human and Natural Changes to Elkhorn Slough's Hydrography

Beginning in the late 1800s, a series of human activities removed marshes from tidal exchange and altered the slough's basic circulation. In 1872, rail beds installed for the Southern Pacific Railroad isolated marshes and restricted tidal exchange along the slough's eastern edge. Between the 1880s and 1940s, ranchers and farmers ditched and diked the lands surrounding the slough to remove tidal flow over low-lying marshes and create pasture for cattle grazing. Around the turn of the century, a 150-hectare[1] area near the slough mouth was

[1] 1 hectare=2.47 acres. See page 278 for a table of English and metric equivalents.

Aerial view of Moss Landing Harbor construction on August 30, 1946. A channel is being cut through the dunes in line with the axis of Elkhorn Slough. Note the natural, sinuous channel opening to the bay north of the dredged harbor entrance. Photo credit: Army Corps of Engineers.

blocked off to create evaporative salt ponds, removing this area from natural tidal exchange.

In 1908, winter storms diverted the Salinas River mouth from north of the present-day Moss Landing Harbor entrance to its current location 7 kilometers south of the harbor entrance (see fig. 2.5 in Chapter 2, "Geology"). During this period, farmers along the lower Salinas River opened a direct channel with Monterey Bay to reduce flooding. This new channel bypassed Elkhorn Slough, cutting it off from its major source of freshwater and sediments and depriving the slough of its tidal character.

In 1946, the slough's hydrography underwent its most dramatic change when the U.S. Army Corps of Engineers dredged a channel across the sand spit at Moss Landing to form the present-day entrance to Moss Landing Harbor. A portion of the old Salinas River channel was dredged to form the harbor, and tide gates were constructed to block tidal flow into upper Elkhorn Slough above Hudson's Landing, Moro Cojo Slough, and the old Salinas River channel (Browning 1972). The channel not only opened the slough to direct tidal incursion, but its breadth and depth are now such that periodic dredging is necessary to maintain it for the commercial and research vessels using the harbor.

Restoration activities were initiated in the 1980s to try to reverse the loss of Elkhorn Slough's marshes to agriculture. In 1983 the California State Department of Fish and Game returned 160 hectares of diked dairy pasture to tidal flooding by digging a channel across former dikes and excavating about 10% of the former marsh to a depth of 2 meters below *Mean Lower Low Water (MLLW)*. Now called the South Marsh (fig. 4.1), this wetland forms part of the Elkhorn Slough National Estuarine Research Reserve (ESNERR). At the time, no one recognized the significant effect restoration would have on the slough's hydrography. Returning this area to tidal exchange increased the tidal prism and thus the tidal current's capacity for erosion (see section on morphometry, below).

Tidal exchange remains restricted between the main channel and many marshes along the slough's eastern edge. In the North and Estrada Marshes, which are part of the ESNERR, the tide gate is set to allow exchange with the main channel during the highest high tides. On the Azevedo Ranch, culverts in the Upper Pond allow exchange with the main channel when the tide level exceeds approximately 1.2 meters. Culverts

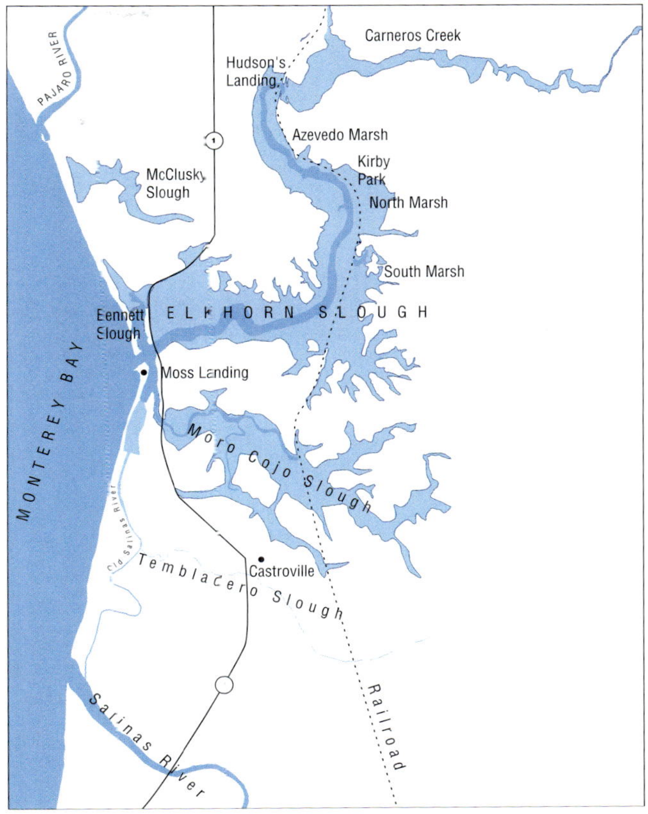

Figure 4.1. Elkhorn Slough and surrounding sloughs.

between the slough and the ranch's middle and lower ponds are completely blocked. Restoration of Moro Cojo Slough began with the removal of the tide gate in 1988, reopening that area to tidal exchange through culverts under Moss Landing Road.

In 1984 the California State Department of Fish and Game acquired the salt ponds near the mouth of Elkhorn Slough and began managing them for wildlife, particularly nesting birds such as the Snowy Plover. Over time the dikes and levees for the ponds began to fail, so in the early 1990s a new levee and water control structures were built, splitting the salt ponds in half. The water levels in the inner half of the salt ponds could be controlled, while the outer half next to the slough was increasingly exposed to tidal inundation as the old levee system failed.

While saltwater flow into Elkhorn Slough has increased since 1946, freshwater flow has declined. Groundwater was probably a significant source of freshwater before the advent of widespread irrigated farming. Although no accurate records of groundwater flow exist, the artesian wells and seeps that were

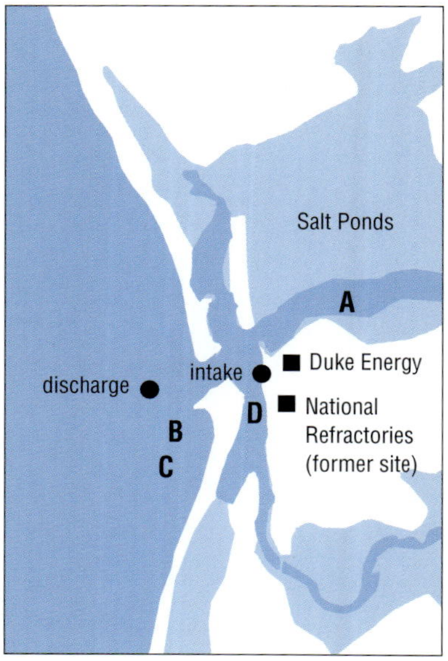

Figure 4.2. Lower Elkhorn Slough and Moss Landing Harbor. Duke Energy (and previously, National Refractories and Minerals) pumps seawater from intakes located near D. National Refractories spent seawater was discharged at C. Between 1950 and the 1990s, spent seawater from the power plant was discharged into the slough at A; discharge continues at the head of the Monterey Submarine Canyon at B.

once common throughout the watershed are nonexistent today. Construction of several dams on the Salinas River and the heavy demands for water by agriculture in the Salinas Valley have cut the Salinas River flow in half since 1941 (USGS stream flow data). Due to the diversion of the Salinas River, rainfall and agricultural runoff are now the slough's major sources of freshwater, although some freshwater occasionally reaches the slough via the old Salinas River channel.

Natural events have also affected slough hydrography. In 1989, following the Loma Prieta earthquake, culverts that restricted tidal flow into Bennett Slough and upper Elkhorn Slough at Hudson's Landing collapsed. Replacement culverts in Bennett Slough allow for complete tidal exchange. In 1995 culverts in the upper slough were replaced with tide gates to restrict saltwater intrusion into this area. Cracks in the marsh and loss of pickleweed in the upper reaches of the slough above Kirby Park suggest that subsidence occurred following the earthquake, although a survey of this area is required to confirm the anecdotal reports.

A variety of agricultural and industrial activities affect water flow into Moss Landing Harbor itself. Agricultural runoff enters the harbor from Elkhorn, Bennett, and Moro Cojo Sloughs and from Tembladero Slough via the old Salinas River channel. Until 2000, National Refractories and Minerals operated a magnesium extraction plant just to the east of the harbor, where approximately 100,000 m^3 of seawater were pumped from the harbor each day. Effluent from the magnesium plant was discharged offshore at the head of Monterey Submarine Canyon (fig. 4.2). Duke Energy operates a natural gas-fired power generation plant on the east side of the harbor (formerly operated by Pacific Gas & Electric). Between 1950 and the 1990s, seawater coolant was pumped from the harbor and was discharged into Elkhorn Slough (about 1 million m^3 day^{-1} at 5° C above ambient temperature) and into the offshore waters at the head of Monterey Submarine Canyon (about 3 million m^3 day^{-1} at 7–10° C above ambient). Discharge into the canyon continues. These industrial uses have not appeared to impact the slough's hydrography (Broenkow 1971; Nybakken et al. 1975).

Estuarine Classification and Salinity Patterns

In an idealized estuary, a surface layer of freshwater overlies a denser layer of seawater: freshwater flows seaward from the head of the estuary and saltwater moves up the estuary in a two-layer flow. The principal factors controlling water circulation within estuaries are the mixing that occurs when fresh and saltwater mix (density-driven mixing), tidal currents or tidal prism (controlling the strength of mixing), and winds (Kjerfve 1989).

Estuaries can be classified into one of four general types according to the relative importance of freshwater flow and mixing (Pritchard 1955). Salt wedge estuaries occur when river flow dominates over tidal mixing, so that little mixing occurs between fresh surface and saline bottom waters. In partially mixed estuaries, there is some mixing between surface and bottom layers. Well-mixed estuaries are characterized by strong mixing between surface and bottom water and little difference between surface and bottom salinities, although there is usually a horizontal (along axis of estuary) gradient in salinity, with fresher water at the head and saltier water near the mouth. Negative estuaries occur in regions where evaporation is greater than freshwater flow, so that salinity at the head of the estuary may be greater (i.e., hypersaline) than seawater salinities offshore.

Elkhorn Slough is a *seasonally* negative estuary due to the Mediterranean climate of the region (Largier, Hollibaugh, and Smith 1997). During the dry season (summer and fall), when freshwater flow is low, water may remain in the upper reaches of the slough for as long as fifty days. During this long *residency time*, hypersaline conditions are created as evaporation exceeds freshwater input. The rest of the year Elkhorn Slough is not hypersaline and is categorized as well mixed. Even during high runoff periods following winter rains, there is little evidence of salinity stratification (Smith 1974).

Carneros Creek, at the head of the estuary, is the slough's only significant source of freshwater. Discharge from the creek is generally nil from July to October, and ranges between 0.2 and 38 $m^3 s^{-1}$ from December to April; flow in the creek is highest during February and March (Monterey County, unpublished data). The Salinas River has significantly greater flows, ranging from 0 to 9 $m^3 s^{-1}$ in July and August up to 300 $m^3 s^{-1}$ in March (USGS stream flow data). During high discharge, the sandbar at the Salinas Lagoon is breached and water flows directly into Monterey Bay. During low discharge, water enters the old Salinas River channel and flows out into Moss Landing Harbor at the mouth of Elkhorn Slough. When the Salinas Lagoon is closed off, flow in the Salinas River ranges between 0 and 10 $m^3 s^{-1}$ with a median of 0.05 $m^3 s^{-1}$ (USGS stream flow data).

Seasonal changes in rainfall affect discharges from Carneros Creek, Moro Cojo, and the Salinas River that are reflected in the slough's seasonal salinity pattern. Since 1988, salinity has been monitored monthly in the Elkhorn Slough watershed and lower Salinas River (Caffrey et al. 1997). At Carneros Creek, surface salinity varies between 0 and 35.7 annually (fig. 4.3a). Tide gates installed at Hudson's Landing following the creation of Moss Landing Harbor have restricted tidal exchange between Carneros Creek and the rest of the slough. These gates were destroyed by the 1989 Loma Prieta earthquake and not replaced until the summer of 1995. Between 1989 and 1995 salinity at this station was often high during the dry season, but summer values declined following the replacement of the gates, except for periodic intrusions of saltwater during the late summer and fall. At Hudson's Landing, downstream of the tide gates, salinity ranges between 0 and 35.8. The average salinity at this station is 25.9, with a regular decline in the winter months during high runoff periods (fig. 4.3b). In the mid region of Elkhorn Slough at Kirby Park, salinity ranges between 0 and 34.7, averaging 27.8 (fig. 4.3c). At the mouth (Skippers Restaurant site), salinity ranges between 0 and 34.0, and average salinity is 28.7 (fig. 4.3d). Periodic inputs of freshwater from the old Salinas River channel may explain why salinity is occasionally lower at the harbor mouth in winter than it is at Kirby Park.

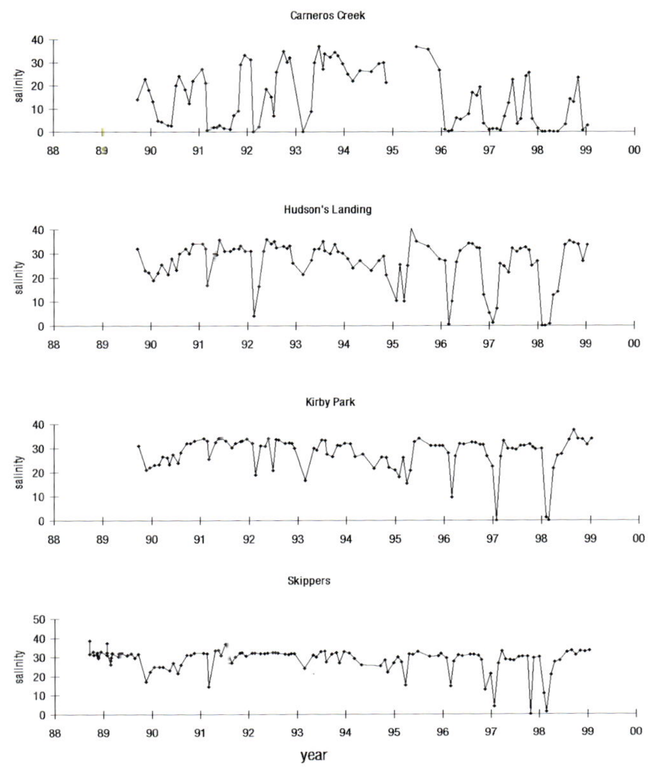

Figure 4.3. Salinity (in PSU) in Elkhorn Slough between 1988 and 1998 at Carneros Creek, Hudson's Landing, Kirby Park, and the mouth (Skippers).

Salinity and water temperature are also variable over periods of weeks, days, and hours. These parameters, along with dissolved oxygen (the amount of oxygen in the water), pH, and water depth, have been monitored continuously since July 1995 at a tidal channel in the South Marsh restoration area of ESNERR. Two examples of the twice-daily (semidiurnal) and fortnightly variation in temperature, salinity, and water depth are presented in figure 4.4.

In September (during the dry season), both temperature and salinity have a strong semidiurnal signal as warm, saline water (22.9° C, 33.8 Practical Salinity Units [PSU]) in the restoration area mixed with cooler, less saline water (15.9° C, 31.6 PSU) from the main channel of the slough on high tides (fig. 4.4a). In January, a similar semidiurnal pattern was evident in the temperature and salinity record, although the temperature of water mixing in from the main channel was fairly constant. As a result, temperatures at high tide were fairly constant at about 15° C, while water in the restoration area warmed up to nearly 17° C during the day and cooled to about 12° C at night (fig. 4.4b). Salinity was also highly variable and affected by freshwater runoff from ephemeral creeks on the reserve.

Figure 4.4. Salinity, temperature (°C), water depth (m), and rainfall (cm) in the South Marsh tidal channel in (a) September 1997 and (b) January 1998.

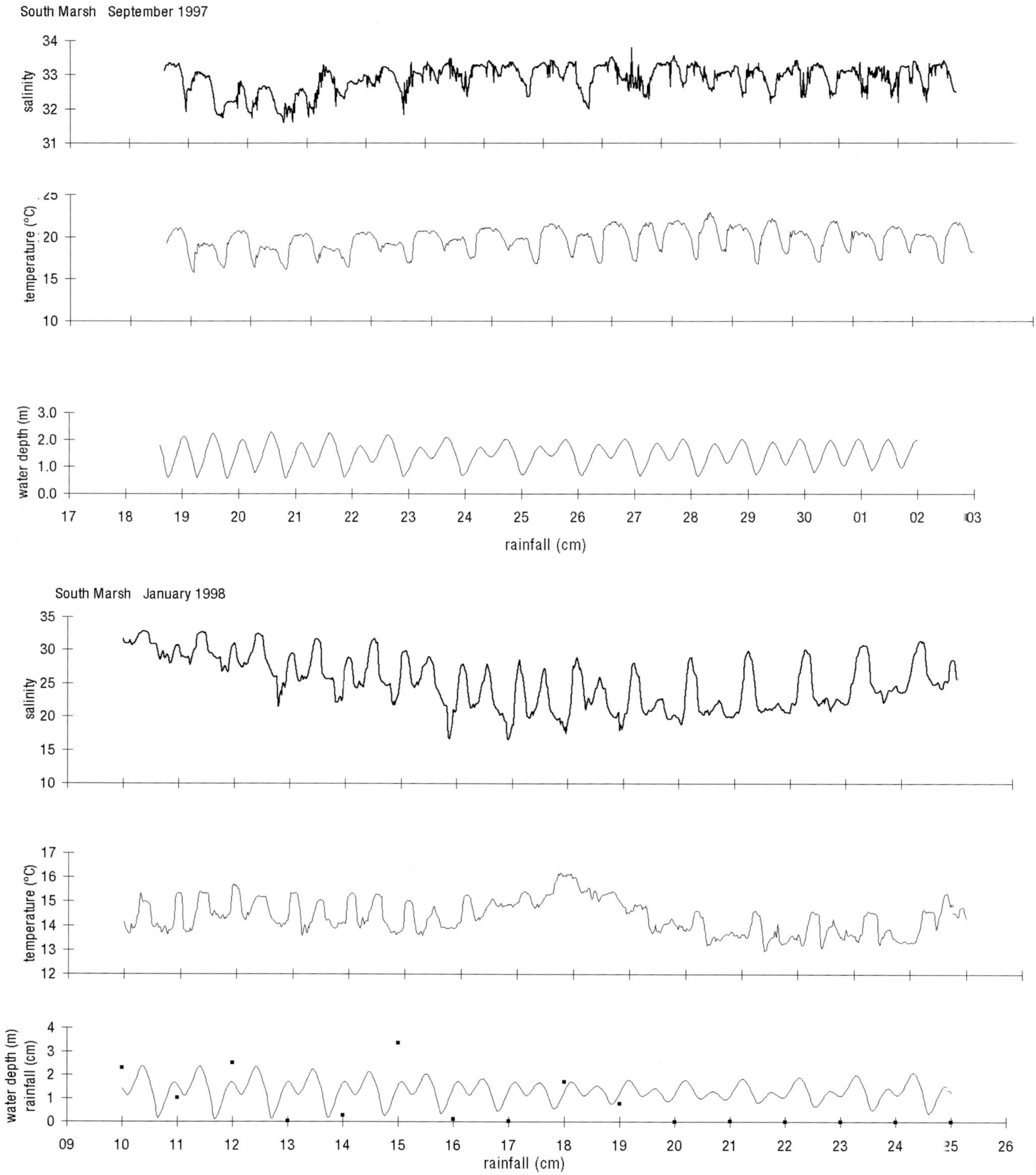

Overall, salinities declined to 16.5 between January 10–17 as storms dropped 13 cm of rain. In addition, salinity in the restoration area showed a spring-neap (see sidebar) tidal cycle signal where equal bimodal salinity peaks occurred during spring tides, but single peaks in salinity dominated during neap tides (fig. 4.4). This is probably the result of reduced tidal energy and mixing during the neap tides compared to the greater tidal energy associated with spring tides, which would promote greater mixing between the restoration area and main channel.

Tidal Characteristics of Elkhorn Slough

Tidal characteristics of Elkhorn Slough were first investigated in detail in the 1970s. In 1970–1971, Clark (1972) obtained several short series of tide height records. In 1976, the National Ocean Survey (NOS) made a yearlong continuous recording of tide levels at four slough locations and obtained a 29-day record from Sandholdt Pier at the head of Monterey Submarine Canyon. Based on these records at the slough and the long-term tide height record in San Francisco, NOS was able to make tidal predictions for six locations within the slough. Clark's results were generally consistent with the 1976 NOS survey.

This study revealed that Elkhorn Slough has a mixed semidiurnal tide, with a principal tidal range of 1.2 meters (based on the four most important tidal components). The mean diurnal tide height is 1.7 meters. This is the annual average difference between high and low tides and is calculated as the average of two daily high minus the average of two daily low tides for a year. The strongest spring tides occur in December and June, while minimal spring tides occur in March and September.

At any given moment, tide level is not identical throughout the slough. There is a lag time as tidal effects progress up and down the slough, to and from the mouth. The NOS study measured a 25-minute delay in tide level between Sandholdt Pier at Moss Landing and Kirby Park. This means that high tide occurs 25 minutes later at Kirby Park than at Moss Landing (table 4.1).

The delays within the ESNERR, which is removed from the slough's main channel, are much greater than those along the main channel. Tide heights in the ESNERR lag behind those at the Highway 1 bridge by about 60 minutes for high tides and 12 minutes for low tides (Wong 1989). This phase lag is

Table 4.1. National Ocean Survey corrections to the daily tidal predictions based on San Francisco for Elkhorn Slough and Monterey tide stations.

Station	Time Correction (minutes)		Height Correction (feet)	
	high	low	high	low
Santa Cruz	-:05	-:13	-0.6	0.0
Monterey	:02	-:02	-0.5	0.0
Moss Landing Ocean Pier	:00	:00	-0.7	-0.1
General Fish Company	:02	-:01	-0.6	-0.1
Elkhorn Yacht Club	:01	:00	-0.6	-0.1
Elkhorn Slough Hwy 1 Bridge	:04	-:04	-0.7	-0.1
Elkhorn, Elkhorn Slough	:22	:03	-0.6	0.0
Kirby Park, Elkhorn Slough	:27	:06	-0.4	-0.1
Elkhorn Slough Railroad Bridge	:34	:06	-0.4	-0.1

probably caused by the highly restricted water flow into and out of the ESNERR at the Parsons Slough railway bridge.

More recent data suggest that the slough's tidal characteristics have changed dramatically since the 1976 NOS study. Between 1993 and 1996, tide gauges were installed at Kirby Park and the Highway 1 bridge. Tidal records showed a 1-hour phase lag between these two stations (Malzone 1999), over twice as long as the lags measured by the NOS study. This suggests that the current official tide predictions are incorrect. The slough's tidal volume has been changed significantly by restoration and erosion in the last two decades and a new set of measurements is needed to update the tide tables.

Morphometry

Many estuaries, including Elkhorn Slough, are made up of a complex mix of channels, mudflats, marshes, and small tidal creeks, making it difficult to estimate the area and volume associated with each of these habitats. However, such estimates are critical to determining the different types and extent of habitats available for different species and the changes in each habitat. In addition, a detailed understanding of the slough's physical geography and geometry—surface area, elevation,

Tides: A Primer

Tides are the rise and fall of sea level caused by the gravitational pull of the moon and sun. The interactions among the various phenomena that create tides are very complex, involving the rotation of the earth, the alignment of the sun and moon relative to the earth, and the declination (height relative to the celestial equator) of the moon and sun. Tides are also influenced by the shape of the body of water in which they occur. In basins, tides act as standing waves that oscillate or slosh back and forth, like water in a bathtub. Every basin has a natural rate or period of oscillation based mostly on its size and shape. When the oscillation period in a basin is the same length as the tidal period, tides are enhanced. Tidal patterns vary significantly from day to day, month to month, and region to region as a result of the combination of these tidal components or harmonic constituents.

Daily or diurnal tide cycles result from changes in the position of the moon as the earth completes one full rotation every 24 hours: high tides occur when the moon is aligned with a site, low tides occur when the moon is at right angles. Mixed semidiurnal tides occur along the California coast. The main semidiurnal tidal period is 12 hours 25 minutes, so there are two high tides

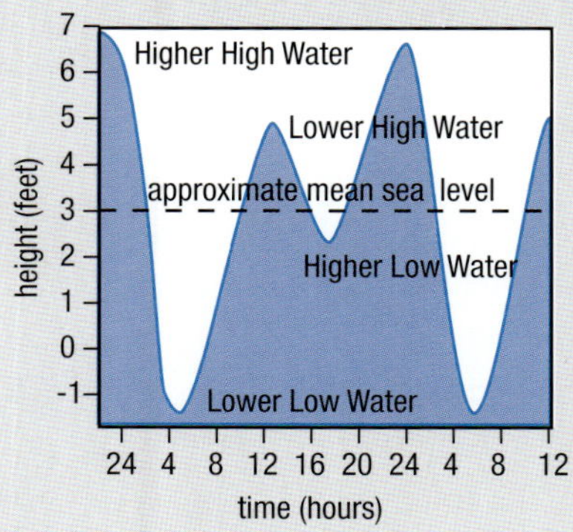

Characteristics of tide at San Francisco. Adapted from J. W. Hedgpeth, Introduction to Seashore Life of the San Francisco Bay Region and the Coast of Northern California. *Berkeley: University of California Press, 1962.*

and two low tides each day, with the new cycle starting about an hour later each day. Thus it takes about 6 hours for the tide to rise (this is termed flooding) and another 6 hours for it to recede, or ebb. The tide is termed "mixed" because one of the two tide cycles is more extreme (higher high and lower low) than the other. The highest tide level, Higher High Water (HHW), is followed by the lowest tide level, Lower Low Water (LLW), then by a moderately high tide (Lower High Water or LHW), and finally by a moderately low tide (Higher Low Water or HLW).

Daily tide levels are also the result of a lunar or monthly tidal cycle. When the sun, moon, and earth are all aligned, as happens during full and new moons, the tide range is enhanced. These are called spring tides and occur about every 14.7 days. Tidal range, the difference between the highest and lowest tide levels, is greatest at this time. During the quarter moon, the gravitational forces of the moon and sun counteract one another, so tides are diminished. These lower highs and higher lows are termed the neap tides.

Measurement and prediction of tide levels is very complex, but for most of us a local tide table is sufficiently precise. Tide levels presented in tide tables are based not on actual sea level but on mean low water (MLW) at a nearby reference point called a datum. Thus a tide of 3.0 feet is 3 feet higher than the average low tide level (0.0 feet) for that area.

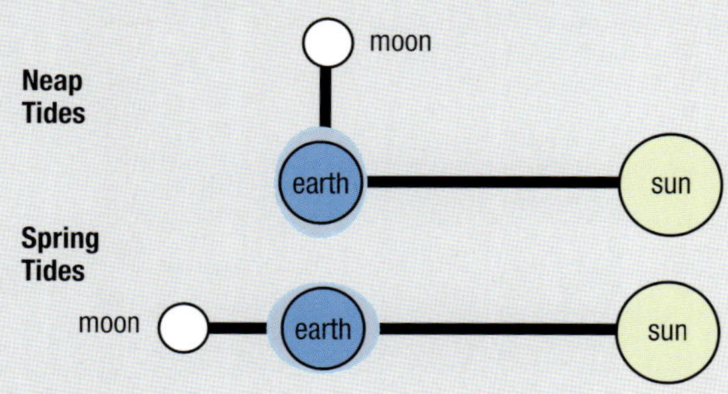

Spatial arrangements of earth, sun and moon at neap tide and spring tide. Darkened area indicates relative depth of seawater at various points on the earth. The point of view is directly above one of the earth's poles. Adapted from S. Hinton. Seashore Life of Southern California. *Berkeley: University of California Press, 1987.*

cross-sectional area, etc.—is essential for calculating the volume of the tidal prism and modeling tidal currents and circulation. Often physical circulation models can then be used as the basis for sediment transport or water quality models.

Surveys show that Elkhorn Slough is about 10 kilometers long and has a main channel width of about 200 meters, with channel depths decreasing from 6.5 meters at the slough's mouth (Highway 1 bridge) to 3 meters near the entrance to Parsons Slough and 1.6 meters at Hudson's Landing (fig. 4.5a; Malzone 1999; by 2002 the depth had increased to 7.5 meters at the Highway 1 bridge–K. Wasson, pers.comm.). These measurements contrast markedly with early bathymetric surveys of the lower reaches of Elkhorn Slough. Surveys made in 1909 and 1940 found that the mouth was between 30 and 60 meters wide with a maximum depth in the channel of about 2 meters (Phillip Williams and Associates 1992). A subsequent survey by the U.S. Army Corps of Engineers made in 1947, after opening of the harbor mouth, showed a deepening of the channel at the Highway 1 bridge to 4.3 meters and a fivefold increase in the width of the slough mouth (Phillip Williams and Associates 1992).

To determine changes in the slough's morphometry since the harbor opening, R. E. Smith (1974), a graduate student at Moss Landing Marine Laboratories, developed an elegant cross-sectional model to estimate the slough's tidal prism and residence time. Using a 1973 map made from infrared aerial photographs, he determined the surface area and volume of marshes, mudflats, and channel, including Parsons Slough and North Marsh but not the South Marsh restoration area or the salt ponds. According to Smith's estimates, the total volume (tidal prism) of the slough in 1973 was 8.7 million m^3.

The slough's tidal prism increased significantly in the 1980s as a result of projects to restore tidal marshes to provide additional habitat for birds, fish, and invertebrates. For example, in 1983 the California Department of Fish and Game directed the restoration of tidal action to 1.6 km^2 of diked salt marsh in the ESNERR next to Parsons Slough. Dikes were breached and channels were dredged to allow pastures to flood. The potential effects of this work on slough hydrography received little consideration at the time, but because the restored marshland flooded and drained completely during each tidal cycle, it significantly increased the slough's volume.

Because of the restoration activities, Elkhorn Slough's surface areas and prism volume were recalculated using a 1983 aerial photograph (Broenkow unpublished data) and were found to be generally similar, although somewhat larger than Smith's (1974) for comparable areas of the slough. This suggests there had been some change due to erosion, although there are uncertainties in the calculations. When the area of the restored habitat was included, the total volume of the slough was 11.3 million m^3; this represented a 30–40% increase in the slough's tidal volume (Smith 1974; Broenkow unpublished data).

Elkhorn Slough's tidal prism has continued to rise. A detailed study of slough bathymetry in 1994 revealed a deepening and widening of the channel and tidal creeks (Malzone 1999). Tidal volume increased 55% between 1972 and 1994 because of erosion and restoration in the South Marsh, North Marsh, Dolan Marsh, Porter Marsh, and additions of the salt ponds (fig. 4.5b). The current estimate of slough volume is 13.1 million m^3 (Malzone 1999)—an increase of more than 50% since 1973. Since 1972, the maximum water depth at the Highway 1 bridge increased from 5 meters to 6.5 meters (fig. 4.6) as a result of the increased tidal prism as well as dredging activities and the 1985 bridge replacement (Malzone 1999).

Circulation Patterns

Tidal current flow or velocity is a function of the volume of water in the slough's tidal prism and the cross-sectional area of the channel through which it passes. Increases in the prism cause faster currents because the greater volume of water must pass through the same spot (the channel mouth) in the same time period (one tidal cycle). The only way this can happen is for the water to flow faster. If the channel is made larger by dredging or erosion, the water has more room to flow and the current speed will decrease.

Tidal currents in Elkhorn Slough have increased significantly during the last few decades, though some would argue that currents have been increasing ever since Moss Landing Harbor was opened to tidal exchange in 1946. These changes have been difficult to quantify, however, because of the technical challenges in measuring currents in shallow estuaries like Elkhorn Slough. Most studies have used a single current meter to measure the volume of water transported through a specific section of the slough. The placement of the meter can influence the results and make interstudy comparisons

John Mason and Kai Parker illustrate tidal changes and amplitude (and great patience) during a flooding spring tide in Elkhorn Slough. Photo credits: Scott Hartley.

problematic. This is because in tidal embayments like Elkhorn Slough, water flow accelerates at the outside of bends; thus, ebbing flow is not necessarily exactly 180° different in direction from flooding flow. In this case, a current meter oriented properly for one flow will not be oriented properly for the other. In addition, friction slows flow near the banks and bottom of the channel.

The first measurements of tidal currents in Elkhorn Slough were made near the Highway 1 bridge in May 1970 and March–May 1971 (Clark 1972; fig. 4.7). Maximum ebb and flood current speeds measured during this study were about 60 cm s^{-1} and 40 cm s^{-1}, respectively. Clark observed that maximum currents occurred an hour or more following the greatest change in tide level. He noted that the tidal current periodicity was asymmetric: that is, flooding currents rapidly reverse to ebbing currents, but ebbing currents slowly change to flooding flow. These observations are consistent with well-known shallow-water tide effects. The differences between flood and ebb velocities are due to the fact that the maximum change

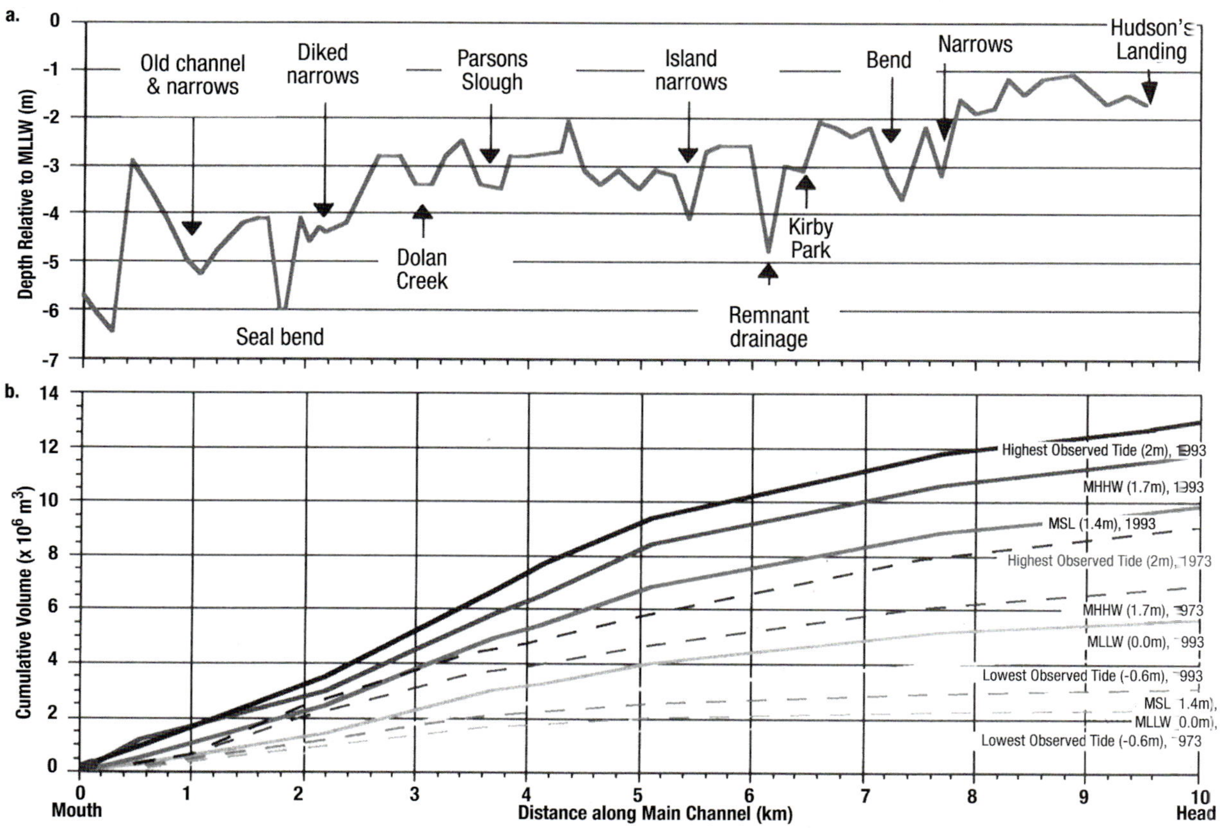

Figure 4.5. (a) Profile of Elkhorn Slough main channel with important landmarks and (b) cumulative volume of water in Elkhorn Slough with distance inland (from Malzone 1999). MHHW: mean higher high water; MSL: mean sea level; MLLW: mean lower low water.

in water level occurs on the ebb tide. The change from HHW to LLW produces the greatest current velocities. Clark (1972) estimated that current velocities in the main channel could reach 72 cm s−1 during ebbing December spring tides, when the tidal range is at its annual maximum. These velocities are about one third of the very strong ebb currents that occur at the Golden Gate Bridge in San Francisco Bay (NOAA tide tables).

Tidal currents were not measured again until the mid-1980s, by which time restoration projects had significantly increased the slough's tidal volume. In September 1986, Wong (1989) measured current flows at the Highway 1 bridge, at the entrance to the restored Parsons Slough/South Marsh area, and just upslough of the restoration area (fig. 4.7). The record from the Highway 1 bridge showed that ebb currents had increased to nearly 80 cm s−1. Wong estimated that maximum current flows at the Highway 1 bridge during December spring tides were about 113 cm s−1 for ebbing tides and 75 cm s−1 during flood tides. Interestingly, maximum current velocities at the entrance to the South Marsh/Parsons Slough were higher than those at the bridge, 171 and 153 cm s−1 for ebb and flood tides, respectively.

Elkhorn Slough's tidal current flows have continued to rise as the tidal prism continues to increase. During the most recent study of water and sediment transport, maximum ebb current velocity at the Highway 1 bridge was 147 cm s−1, twice the flow maximum observed by Clark in 1970 (Malzone 1999). Flood current velocities also increased, with a maximum of 130 cm s−1.

Sediment Transport and Tidal Scour

One effect of these increased tidal currents is the scouring of the channels and the reshaping of Elkhorn Slough's physical appearance. For example, in the early 1970s, low tides exposed an island about 1 kilometer inland of the Highway 1 bridge. Boaters had to be careful to steer north of this feature when it was covered at intermediate tidal stages. This island has slowly eroded away, and by 1989 was no longer visible even at low tide. Erosion continues to alter the slough's shape, or morphology, as muddy effluent ebbs into Monterey Bay and clear Bay waters flood in on the following tide.

Although it is clear that erosion is taking place, quantifying the erosion rate is difficult. Various methods have been used, but none are simple.

1. A series of aerial photographs could be taken at various tidal stages over a several-year period. Measurements of the flooded areas at 30 centimeter intervals would allow the volume of the slough to be determined, with increases representing the volume eroded over that time period.

2. Similarly, precision bathymetric measurements at extreme high tides over a several-year period could be used to determine a history of volume changes over the years.

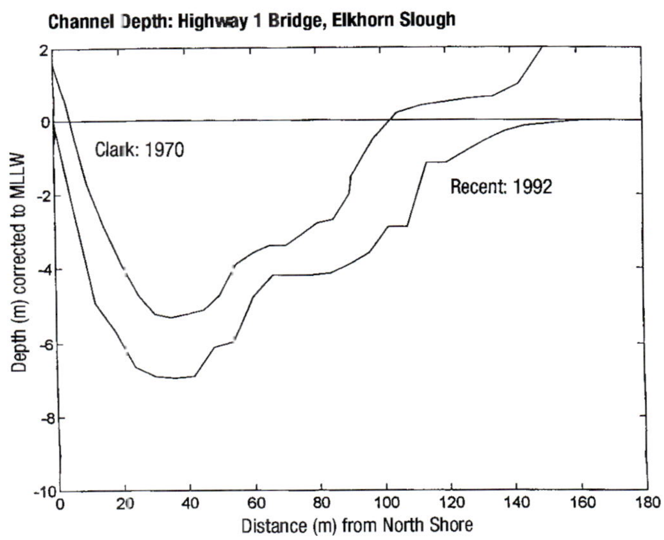

Figure 4.6. Measured cross-section at the Highway 1 bridge in 1970 (Clark 1972) and 1992 (Malzone 1999).

3. Direct measurements of the advective transport of suspended sediment could be made. While this is conceptually simple, it may not be accurate because it depends on accurately measuring the net transport of water out of Elkhorn Slough.

These techniques have been tried with varying success. Historical aerial photographs have been analyzed to determine changes in the three major habitats: salt marsh, mudflats, and channel. Changes in the salt marsh, specifically the reduction in pickleweed, have been attributed to tidal scour. Lowe (1999) analyzed a series of aerial photographs taken between 1931 and 1997 for percent cover of pickleweed marsh along the main channel of Elkhorn Slough. Pickleweed cover throughout the main stem of the slough decreased by 20% following the opening of the harbor and by 30% in the early 1990s. One explanation for the declines in the early 1990s is marsh subsidence following the 1989 Loma Prieta earthquake. Another possible explanation is the restoration of South Marsh in 1983. Either or both of these mechanisms may be responsible for the observed changes and further research is necessary to determine the cause for the decline.

Bathymetric surveys made along the main stem channel in 1940, 1987, and 1994 were used to estimate erosion rates in the slough. The erosion of sediment between 1940 and 1987 was estimated to be approximately 22,000 m^3 y^{-1}, assuming that most of the erosion was occurring along the channel and banks in the lower reaches of the slough (Phillip Williams and Associates 1992). A more thorough study was done between 1993 and 1996, examining changes along the channel and in the tidal creeks (Malzone 1999). A series of bathymetric transects across the channel showed that it had widened since surveys done in 1987 (Oliver and Schwartz 1988). Bank erosion of the lower intertidal zone was greatest in the mid-slough region, and most of this material was deposited in the subtidal zone. Malzone (1999) estimated that the sediment loss rate for Elkhorn Slough as a whole was 80,000 m^3 yr^{-1}, or almost four times the previous estimate, probably because Malzone's estimate includes changes in the tidal creeks and in the upper reaches of the slough.

Although a study of sediment transport using advective transport modeling has not been conducted in Elkhorn Slough, a study of phosphate transport (Reilly 1978) illustrates the challenges of this approach. The advective transport of any material is the current speed **x** concentration **x** cross-section area.

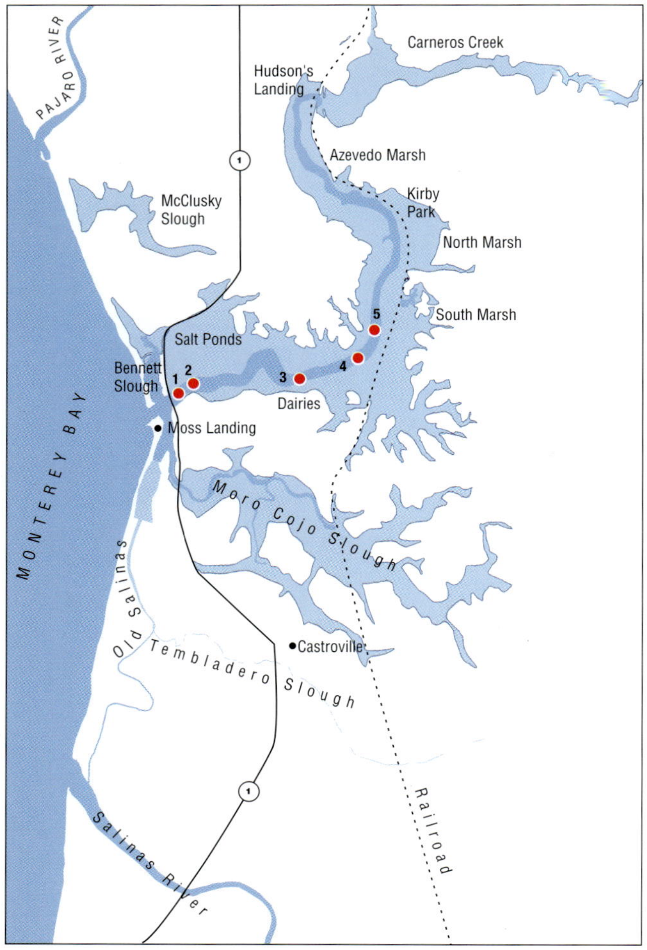

Figure 4.7. Red circles mark metering stations used to measure currents in Elkhorn Slough: 1 is the Highway 1 bridge station used by Clark in 1970 and 1971, and for more recent measurements (Malzone 1999); 2, 4, and 5 are locations of Wong's 1986 measurements; 3 is location of Reilly's measurements.

The fundamental assumption of this method is that one can very accurately measure both flood and ebb currents through a cross-sectional area. The difference between the material transported on the flood and ebb tides is the net transport or net flux. If the net flux is out the slough, that means that the flux of materials out on ebb was greater than the flux of materials in during flood. Although Reilly obtained a figure for the phosphorus flux from the slough that agreed with other such estimates from bays and lagoons, his numbers were dependent on an accurate estimate of the net transport of water. While this technique could be used to estimate sediment transport in the slough, the result could be seriously biased because of the difficulty in determining net transport through a cross-sectional area.

Despite the challenges in accurately measuring currents and sediment transport, there is sufficient evidence to conclude that the volume of water exchanged on each tide has nearly doubled since 1970. It is not known whether restoring the ESNERR marsh to tidal flooding caused all of this increase in the tidal prism, but that certainly contributed directly to the situation. Even without this large and sudden impact, Elkhorn Slough would have enlarged itself due to increased tidal scour caused by the creation of Moss Landing Harbor.

Management Issues and Research Recommendations

In this section we highlight some of the most critical management issues within Elkhorn Slough and point out the need for further research. A firm grasp of the physical processes is key to our understanding of changes in water quality, the distributions of various species, changes in the extent of slough habitats, and other factors influenced by circulation, erosion, and tidal currents.

Water Transport

Although various investigators have studied circulation within Elkhorn Slough since 1970, questions still remain about the transport of water throughout the slough. Past studies made clear the difficulty of characterizing tidal currents with just one or two current meters, even in a channel that is 200 meters wide or less. Further studies of circulation are critical, not just to give us a better understanding of water transport throughout the slough, but because water transports sediments, nutrients, contaminants, and organisms, and thus affects many aspects of the slough ecosystem.

Tidal Scour

One of the most intractable management issues within Elkhorn Slough has been how to reduce tidal scour and the loss of salt marsh along the main channel. A critical research need is accurate mapping of mudflat, marsh, and tidal creeks within the slough, in order to document changes in these habitats. The bathymetry of the main channel was recently mapped by Malzone (1999), but the area and topography of these other habitats needs to be measured. Monitoring of creek bank erosion stations, channel bathymetry, and pickleweed cover should be continued. This information could be useful in guiding restoration efforts within the slough to ensure that sufficient marsh and mudflat habitats are retained.

Groundwater Resources

Groundwater resources have been only briefly mentioned in this chapter. Farmers and residents within this region depend on groundwater as their main source of freshwater. In fact, groundwater is being pumped for irrigation faster than it is recharged, so saltwater intrusion into the aquifers is a serious problem. There is a need for studies examining the links between surface waters and groundwater within the Elkhorn Slough watershed.

References

Broenkow, W. 1971. *Chemical and Physical Effects of Discharging Kaiser Waste Effluent into Monterey Bay and Moro Cojo Slough.* Moss Landing Marine Laboratories Technical Publication 71-3, Moss Landing, CA: Moss Landing Marine Laboratories.

Browning, B. M. 1972. *The Natural Resources of Elkhorn Slough: Their Present and Future Use. Coastal Wetland Series #4.* Sacramento, CA.: California Department of Fish and Game.

Caffrey, J. M., S. Shaw, M. Silberstein, A. De Vogelaere, and M. White. 1997. Water quality monitoring in Elkhorn Slough: A summary of results, 1988–1996. Unpublished report. Elkhorn Slough Foundation and Elkhorn Slough National Estuarine Research Reserve.

Clark, L. R. 1972. Long period wave characteristics in Moss Landing Harbor and Elkhorn Slough. M.A. thesis, Moss Landing Marine Laboratories. San Jose State University.

Conomos, T. J., R. E. Smith, and J. W. Gartner. 1985. Environmental setting of San Francisco Bay. *Hydrobiologia* 129: 1–12.

Kjerfve, B. 1989. Estuarine geomorphology and physical oceanography. In J. W. Day, Jr., C. A. S. Hall, W. M. Kemp, and A. Yañez-Arancibia, eds., *Estuarine Ecology*, pp. 47–78. New York: John Wiley.

Largier, J. L., J. T. Hollibaugh, and S. V. Smith. 1997. Seasonally hypersaline estuaries in Mediterranean-climate regions. *Estuarine Coastal and Shelf Science* 45: 789–797.

Lowe, P. 1999. Marsh loss in Elkhorn Slough, CA: Patterns, mechanisms, and impact on shorebirds. M.S. thesis, Moss Landing Marine Laboratories. San Jose State University.

MacGinitie, G. 1935. Ecological aspects of a California marine estuary. *American Midland Naturalist* 16: 629–765.

Malzone, C. M. 1999. Tidal scour and its relation to erosion and sediment transport in Elkhorn Slough. M.S. thesis, Moss Landing Marine Laboratories. San Jose State University.

Nybakken, J. G., G. Cailliet, and W. W. Broenkow. 1975. A baseline study of the Moss Landing/Elkhorn Slough environment: Progress report to PG&E. Moss Landing, CA: Moss Landing Marine Laboratories.

Oliver, J. S., and D. L. Schwartz. 1988. Photograph and map library for Elkhorn Slough Reserve: Erosion and sedimentation in Elkhorn Slough. Elkhorn Slough National Estuarine Research Reserve. Final Report. Watsonville, CA.: Elkhorn Slough National Estuarine Research Reserve.

Phillip Williams and Associates. 1992. Elkhorn Slough tidal hydraulics erosion study. Report to the U.S. Army Corps of Engineers, San Francisco District. No. 830.

Pritchard, D. W. 1955. Estuarine circulation patterns. Proceedings American Society of Civil Engineering. 81:717/1-717/11.

Reilly, P. 1978. Phosphate fluctuation from Elkhorn Slough, California. M.A. thesis, Moss Landing Marine Laboratories. San Jose State University.

Smith, R. E. 1974. The hydrography of Elkhorn Slough, a shallow California coastal embayment. M.A. thesis, Moss Landing Marine Laboratories. San Jose State University.

Wong, C. R. 1989. Observations of tides and tidal currents in Elkhorn Slough. M.A. thesis, Moss Landing Marine Laboratories. San Jose State University.

5

Chapter Five

Soils

Marc Los Huertos and Carol Shennan

Soils in the Elkhorn Slough watershed reflect a relatively young, dynamic setting, with the majority of soils formed from recently stabilized sand dunes and others typical of grasslands that have experienced a fluctuating water table. Soil types range from clay soils in the wetland areas to beach sands near Moss Landing and McCluskey Slough. This variation in soil characteristics influences the distribution of plants and animals, as well as nutrient cycling, water quality, slope stability, and land use, throughout the watershed.

Soils play a vital role in the function of both the upland and wetland Elkhorn Slough ecosystems. They provide the physical substrate for plant roots and habitat for a myriad of organisms, from microorganisms and invertebrates to reptiles, birds, and mammals. Slough and watershed soils also mediate nutrient and moisture availability for plant growth, as well as the processes of organic matter decomposition and accumulation. The watershed's soils control rainwater infiltration, reducing the volume of surface runoff during storms, and store water, releasing it to the atmosphere by evaporation or plant evapotranspiration, to the groundwater by downward movement or to surface waters such as creeks and wetlands by lateral movement.

Land use, including agricultural and residential development, is also dictated in part by the watershed's soils. For example, the well-drained soils of the Arnold series that dominate much of the watershed are ideal for strawberry cultivation. In turn, land use practices have a profound impact on soil resources. Human activities ranging from tillage and fertilization for agriculture to road building and vegetation clearing for development affect the soil's ability to modulate nutrient cycles, store water, and maintain slope stability.

Despite their key role in the ecosystem and their importance to land use in the watershed, little research has been done on Elkhorn Slough's soils beyond basic mapping of soil type distribution. In this chapter we discuss what is known about the development of soil in the watershed, review the area's soil types, and examine the issue of soil erosion and contamination by agricultural chemicals. We close with suggestions for research to enhance the existing knowledge of soils, reduce erosion, and improve fertility management.

Soil Formation in the Elkhorn Slough Watershed

As with all soils, a variety of factors, most notably parent material, climate, organisms associated with the soil, topography, and time (Jenny 1980), have influenced soil development in the Elkhorn Slough watershed. Parent material—the substrate from which soils form—provides the mineral portion of soils. Parent material may come from the breakdown of the underlying rock, such as granite in soils of the Sierra Nevada Mountains or uplifted sedimentary rock in Rocky Mountain soils. It may be deposited by colluvial (water transported) or aeolian (wind transported) particles. The soils

The Aromas Sands Formation of the Elkhorn Slough watershed is highly erodible, as seen in this feature above Werner Lake. Photo credit: Mark Silberstein.

of the Great Plains are derived from aeolian dust deposits; some of the soils around the Monterey Bay, including many at Elkhorn Slough, are derived from aeolian sand deposits.

Aromas sands are the most familiar aeolian sandy parent material in the Monterey Bay region. Stabilized by plant growth, these sands have formed most of the soils in the Elkhorn Slough watershed. The Aromas sands were deposited in the Pleistocene (within the last 100,000 years) and have been subject to pedogenesis—the process that leads to the development of soil horizons with distinct chemical and physical characteristics—since that time (Buol 1997). Uplifted marine terraces, some granitic outcrops, and some metamorphic materials in the upper part of the watershed have also provided parent material for Elkhorn Slough's soils (Cook and Buetler 1978; Isgrig 1969).

As the name implies, the Aromas Formation is composed largely of sands. However, fine-textured silt and clay particles (see sidebar) also occur in these soils. Soil development in Aromas sands includes the breakdown of the crystalline structure of the mineral particles. These processes are especially important in generating the clay-sized particles, although it is a relatively slow process.

Depending on their composition, these clays are subject to chemical changes in their crystalline structure, including the substitution of cations (positively charged ions) in the lattice structure or the layers of the crystals themselves. These chemical substitutions increase clays' ability to provide negatively charged sites for cations to bind, giving soils derived from Aromas sands some *cation exchange capacity*. Thus the clays provide otherwise sandy soils with both a characteristic soil structure that makes them well suited to agriculture and cation exchange sites for plant uptake of important nutrients. The source of the clays in the Elkhorn Slough watershed has not been studied, but they are likely to be composed of a combination of the breakdown of particles in situ, along with wind- and water-deposited material, perhaps from outside the watershed.

Climatic conditions act on parent material as soils develop. Infiltrating water reacts with soil particles as it moves through the soil, carrying dissolved chemicals into lower parts of the soil profile. As water is removed by plant roots and evaporation, many of these dissolved chemicals precipitate or accumulate in particular depths in the soil profile. This transport and deposition of chemicals creates distinct soil horizons, resulting in

Soil Texture

Soil texture is the relative proportions of different size classes that make up a soil. The U.S. Department of Agriculture classifies particles between 2 and 0.05 mm as sands, those between 0.05 and 0.002 mm as silts, and those less than 0.002 mm as clays (United States Natural Resources Conservation Service 1998).

changes in color and characteristics in the soil so noticeable that the soil can appear striated with depth. Although in some cases soil layers were created by different materials being deposited over time, horizons are often due to pedogenic development.

Vegetation and soil organisms have added additional material to the watershed's soils. Vegetation contributes organic matter both to the soil surface in the form of litter and below ground via root turnover. Invertebrates, including worms, nematodes, and arthropods, break down vegetation and animal tissues into smaller particles, which are further decomposed by microbes, such as fungi and bacteria. The resulting organic compounds dissolve in the soil and are carried by infiltrating water to lower soil horizons, where they often become stabilized in the soil into *humus*, giving the soil a characteristic brown color in a process known as melanization. This interaction among soil particles, plant root decomposition, and animal biota is best developed in the grassland soils classified as Mollisols, of which there are several series in the watershed (see section on soil classification below).

Topography and aspect (orientation) also influence soil development and diversity throughout the watershed. For example, tree growth is limited on the soils of dryer, south-facing slopes, increasing the importance of chaparral on these aspects. These chaparral-dominated soils receive both a lesser amount and different type of organic matter input annually. Over time, these differences create different soil characteristics. Soils at the top, middle, and bottom of a sloped hillside also show different characteristics. Water transports dissolved chemicals from the top to the bottom of the slope, where they accumulate and give the slope different soil properties.

Soil Types in the Elkhorn Slough Watershed

The Monterey County Soil Survey (Cook and Buetler 1978) and the San Benito County Soil Survey (Isgrig 1969) provide important resources for understanding the soils in the Elkhorn Slough watershed. Of the twelve soil orders recognized in the United States (see sidebar on soil taxonomy), six occur in the watershed, with Entisols and Mollisols making up the dominant orders. At the time of writing, only the Monterey County surveys were available digitally (United States Department of Agriculture 1998). Digital surveys make it easy to calculate the extent of soils in the watershed using a geographical information system (GIS). From these data, we have calculated the contribution of each soil type only in the

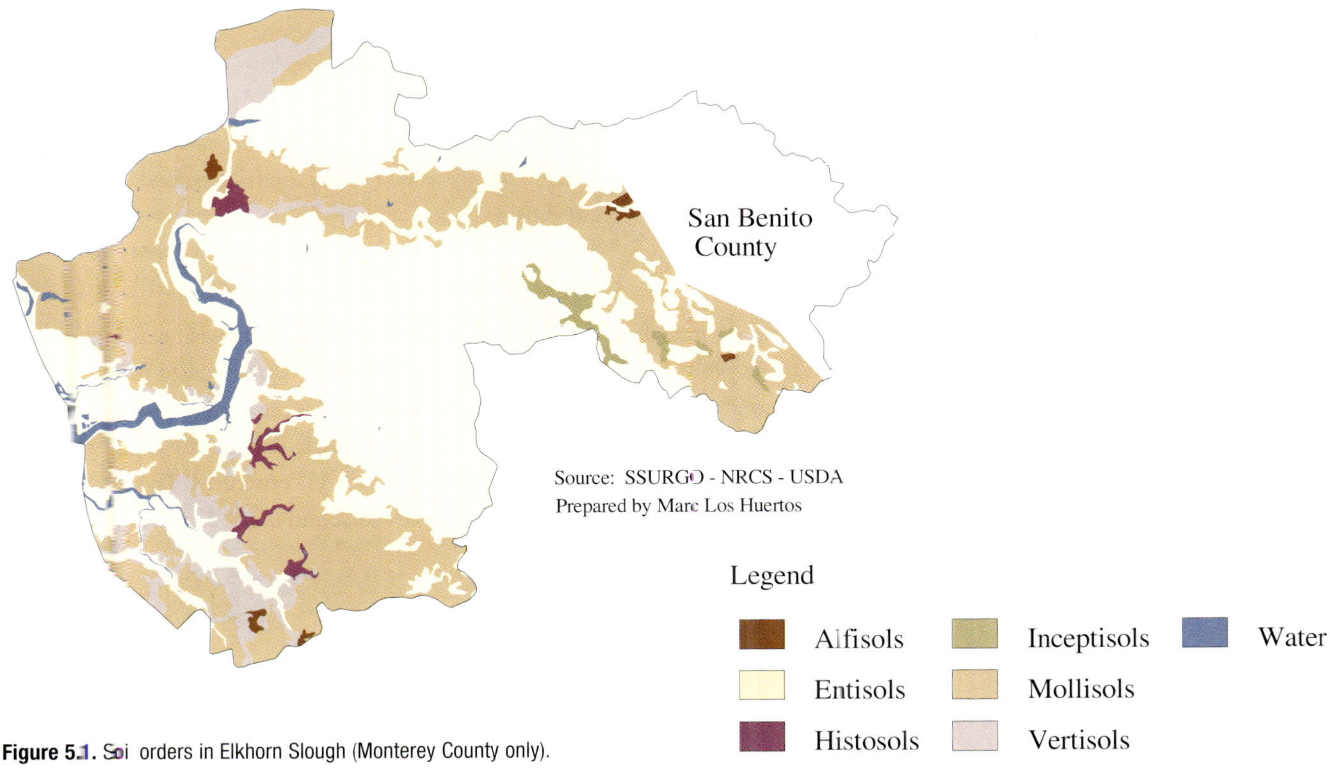

Figure 5.1. Soil orders in Elkhorn Slough (Monterey County only).

Elkhorn Series

Elkhorn sandy loam is a common soil found on marine terraces and stabilized sand dunes north of Castroville and west of Prunedale to the Pajaro Valley. This series is also common on marine terraces in Santa Cruz and San Mateo Counties. The Elkhorn series is similar to the Santa Ynez series; both are in the Xeroll suborder.

The Elkhorn Series is a Mollisol (a grassland soil) formed on weakly consolidated sandy sediments or sandstone. In contrast to the Santa Ynez soil series, the Elkhorn series is classified in the Argixeroll great group, which means that it has an argillic horizon but does not show the same kind of extensive weathering as a Palexeroll. Finally, the Elkhorn series is classified in the Pachic Argixeroll subgroup. The Pachic subgroup name means that the mollic epipedon—i.e., the top layer of soil that is dark in color—is relatively thick: greater than 50 centimeters. The relatively high concentration of nitrogen and carbon has encouraged agricultural development on these soils, which have commonly been used for specialty crops such as artichokes, strawberries, and brussels sprouts.

Soil Erosion

The Elkhorn Slough watershed has a reputation for its high erosion rates, often quoted as being the highest rates measured west of the Mississippi. These unusually high rates are generated in part by agriculture and other human activities acting on soil types that are particularly vulnerable to erosion.

In the winter of 1982–1983 the Soil Conservation Service (SCS; now the Natural Resources Conservation Service, NRCS) carried out an extensive study of erosion in Elkhorn Slough (United States Department of Agriculture Soil Conservation Service River Basin Planning Staff Target Area Team 1984). While natural erosion rates occur at an approximate rate of 1–5 tons per acre per year, depending on soil type, strawberry farming combined with natural erosion can generate erosion rates estimated at 33 tons per acre per year. During the winter of 1982–1983, erosion rates on newly formed farm access roads were estimated at 1,870 tons per road acre.

There are some clear limitations to the SCS study. First, 1982–1983 was an El Niño year, with high rainfall totals occurring over short periods of time. In the years prior to this study, the region had relatively mild winters during which growers did not develop good erosion control practices. Although these points do not invalidate the SCS study, its results should be seen as a worst-case scenario.

In addition, the study predicted that soil loss would reduce agricultural productivity and that 15% of the strawberry land would go out of production by 1994. These predictions have not yet been borne out, in part because the study considered erosion as a measure of lost soil. However, because eroded soil is often deposited lower on the slope and growers often transport this soil back upslope, soil erosion measurements can overestimate soil lost from the farm.

In response to these high erosion rates, growers in the watershed—with the assistance of the NRCS and the Resource Conservation District of Monterey County—have worked hard since the 1982–1983 study to improve runoff management practices and reduce erosion.

Types of Soil Erosion

Although soil erosion is a natural process, soil formation processes usually outpace erosional processes. When soils are disturbed, soil erosion is greatly accelerated—the surface horizon can lose its structural integrity and be washed away. Soils saturated by rainfall or irrigation water are particularly vulnerable. Under these conditions, soil air spaces fill with water and surface runoff (the flow of water downslope at the soil surface) increases dramatically.

Several forms of erosion are common to disturbed soils in the Elkhorn Slough watershed. When raindrops land with an impact high enough to disrupt soil particles, they loosen the particles from each other in a process called splash erosion. Soil particles can then be transported with surface runoff. Sheet erosion is the removal of a fairly uniform layer of soil particles with a thin, even flow of water across a surface. Although sheet erosion does not usually occur on well-drained soils, which are common to the uplands of the watershed, it can take place when soils are saturated by large amounts of continuous rainfall and is exacerbated when tractor operations compact soils. When surface runoff becomes concentrated due to an uneven soil surface, it forms small rivulets (small linear depressions on the soil surface) in a process termed rill erosion. Sheet and rill erosion from strawberry farms in the watershed was estimated at 10 tons per acre per year (United States Department of Agriculture Soil Conservation Service River Basin Planning

Heavy rains can trigger gully erosion, as seen on this strawberry farm near Elkhorn Slough. Photo credit: Mark Silberstein.

Staff Target Area Team 1984). Gully erosion, the biggest contributor to erosion in the Elkhorn Slough watershed, generates an estimated 23 tons per acre per year of soil lost from land planted in strawberries. Like rills (but much larger), gullies form when water flow is concentrated in a particular area and they deepen and expand as runoff increases.

Access roadways on strawberry farms are prime sources for the types of concentrated flows that create gullies, and the NRCS has worked hard to promote techniques that protect roadways from this type of damage. Concentrated flows on roadways can be reduced through improved furrow alignment (with slopes less than 2%) and by seeding furrow bottoms with grasses to increase infiltration and attenuate peak runoff. Other techniques include seeding roadbeds with grasses to improve water infiltration and sediment trapping, and installing buried drainage pipes. Although these methods are not foolproof, with proper maintenance they have successfully reduced erosion rates.

Surface crusts can exacerbate all these types of erosion. Sandy soils with a clay component, such as those found in much of the watershed, form surface crusts when rain or overhead irrigation impact bare soil. The impact rearranges clay particles so that a hard crust forms when the soil dries. Soil crust formation impedes water infiltration down into the soil profile, increasing runoff and the potential for erosion. Even during summer it appears that fog drip and sun shining directly on the soil can create surface crusts. In Elkhorn Slough, soil crusts often occur in agricultural fields or areas that have been cleared of vegetation. In most cases, these crusts are destroyed with the first rainfalls, but crusts may contribute to erosion in early rainfall events.

The Effects of Farming on Soil Structure

Although management improvements can decrease erosion rates, farm practices by their nature will continue to cause erosion. Clearing and cultivating land automatically increase the potential for erosion: removing the protection of permanent plant cover and surface litter (fallen leaves, dead grasses, etc.) and replacing deep-rooted perennials with shallow-rooted annual crops result in decreased water infiltration and reduce the detention time of water before surface runoff and erosion begin.

Farming practices can also accelerate erosion by altering soil structure. Tillage and soil disturbance disrupt soil aggregates (clusters of soil particles), the semi-stable structures in the soil that form pore spaces for water and air. This disruption is especially severe when soil is wet. Decreased water infiltration and increased runoff occur on soils where pore spaces have been compacted. Compaction also limits oxygen diffusion to the rooting zone and makes it more difficult for plants to penetrate the soil to access water and nutrients.

Farmers use tillage to make it easier for crop roots to grow in the soil. But by breaking up soil aggregates, tillage also exposes more soil aggregates to microbial attack. Although this microbial activity can increase the available nitrogen for crop plants in the short term, the soil's organic matter levels eventually decline as microbes digest the available carbon (Jackson 1998). This loss of soil organic matter occurs rapidly when soils are first tilled; the rate declines after several decades, depending on soil type and cultural practices. In addition to causing a decline in organic matter, disruption of individual soil aggregates eventually decreases populations of structure-stabilizing fungi and earthworms.

The loss of soil organic matter and resultant loss of soil structure also decrease soils' resistance to slaking (aggregates that break apart in water), surface sealing (crusting), and accelerated erosion by wind and water. Disrupted soils are especially vulnerable immediately after bed preparation in strawberry fields, when soils may be powdery and thus easily eroded by early rains. Intense early rains have caused extensive erosion in the Elkhorn Slough watershed.

Contaminated Soils

The use of nitrogen fertilizers and pesticides throughout the Elkhorn Slough watershed has led to problems with soil contamination. Nitrate, which is an important nutrient for plant growth, is easily leached by rainfall and carried down the soil profile into the groundwater. Although it may take many years for nitrate to reach the water table, nitrate contamination of groundwater wells is common along the central coast and is a serious problem in wells that supply drinking water throughout the Elkhorn Slough watershed. It is not clear whether nitrate contamination is due to historical or current agricultural practices or the use of septic systems, but it is likely a combination of these factors. Nitrate pollution also affects plant and animal communities, a topic discussed in chapter 12, "Biogeochemical Cycling."

Pesticides used in agriculture and residential settings have also contributed to soil contamination in the watershed. For example, DDT, an organochlorine pesticide, continues to be found in soils today because it is a "persistent" pesticide (Werner et al. 1997). Although its use was banned in 1972, DDT and its breakdown products (DDD and DDE) remain toxic for decades.

When erosion occurs, DDT and other pesticides that adhere to soil particles enter the water column and wetlands. There they can bioaccumulate in invertebrates, fish, and their prey leading to toxic effects on the biota. However, to date few studies of the distribution of DDT and other pesticides in upland soils have been carried out, so we cannot state the extent of soil contamination in the watershed. (Pesticide pollution, transport routes, and effects on slough biota are discussed in chapter 13, "Land Use and Contaminants.")

Management Issues and Research Recommendations

Seismic activity, sea level fluctuations, climate changes, and shifts in vegetation communities have helped create a complex mosaic of soil types along California's central coast. Human uses of these various soil types have added another layer of complexity, both for soil scientists who study the region and for managers working to minimize the impacts of human activity. Although soils are a critical resource that provide both ecosystem and economic benefits, they are sensitive to disturbance and can become a source of contamination.

Careful land use planning and management have already reduced many of the impacts of soil disturbance in the Elkhorn Slough watershed. However, as residential development increases in the next decades, it will be a challenge to maintain agricultural production and limit the impact of urbanization while protecting Elkhorn Slough. Here we suggest a number of research and management areas that should be considered as resource managers work to better understand the region's soils, improve soil protection, and minimize erosion.

Soil Surveys

Although basic soil surveys of the Elkhorn Slough watershed exist, we need to know how soil types change following decades of cultivation and the preferential loss of silt and clay particles. In addition, surveys of soils to quantify concentrations of DDT and other persistent pesticides will determine the relative importance of these compounds as potential toxins to slough biota. A survey of soil contaminants will allow landowners, growers, and resource agencies to gauge the potential liabilities of farming particular pieces of land.

Erosion Management

The issues of soil erosion are paramount in the Elkhorn Slough watershed. At some level, controlling erosion is a relatively simple engineering problem, but in the context of farming and its socioeconomic considerations, the problem requires a multidisciplinary approach, for which the NRCS has demonstrated an outstanding aptitude.

Because gully erosion is the most visible and damaging form of erosion in the area, many growers work closely with the NRCS to install underground pipes to prevent surface runoff along farm access roads, carefully improve row arrangements (to reduce slope and runoff), and plant grass in furrow bottoms. However, full-bed mulching (plastic covering the beds) has become an increasingly popular technique, such that bed erosion may be minimal but runoff rates will increase. This is likely to increase bottom furrow erosion and gully formation on roadways. It will take several years for resource agencies to understand the effects of full-bed mulching and develop ways to reduce its impact on farms and the watershed as a whole. We believe that when gully erosion is solved, more emphasis will be placed on management of soil structure to improve infiltration and reduce erosion.

Developing and promoting methods to increase soil infiltration rates, reduce runoff volumes, and prevent erosion are key steps to long-term soil and water management in the Elkhorn Slough watershed. Growers, resource agency staff (NRCS and Resource Conservation District), extension agents, and university researchers will continue to investigate the potential for farming practices such as compost additions, cover crops, vegetative buffer strips, and alternative tillage techniques that improve water infiltration rates. In addition, a survey of soils to quantify DDT concentrations would help pinpoint areas where stepped-up erosion control measures should be implemented.

Nutrient Management and Demonstration Sites

Relatively high soil fertility is one of the characteristics of Mollisol (grassland) soils in Elkhorn Slough. Farmers and researchers are increasingly interested in managing this fertility and in the continued sustainability of farming in the region. University of California (UC) Cooperative Extension, UC Davis, and UC Santa Cruz have the opportunity to develop research and extension programs that will help growers protect the slough even as they face the potential of increasing costs and declining revenues. Understanding the role of compost and cover crops in soil structure and fertility is a critical area for development in the Elkhorn Slough watershed and the surrounding area. This work should link soil management and how it directly affects water quality, e.g., its effects on nitrate leaching. The establishment of a demonstration farm will encourage researchers to establish field trials. A site where experiments can be made in a replicated and randomized fashion is important to researchers, who are often judged by their ability to publish results from these types of studies.

Elkhorn Slough is atypical of the prime agricultural lands found in the valley floors of the region, for example in the Pajaro and Salinas Valleys. For this reason, soil and agricultural scientists may be hard-pressed to justify their work in the watershed. However, as urban growth continues to convert prime farmland to residential uses, agriculture is increasingly looking to the marginal soils of sloping uplands. The Elkhorn Slough area provides an opportunity to study these farming systems and anticipate potential soil problems in new upland farms. Involving scientists and growers in this type of research will require continued outreach and encouragement from NRCS, the Resource Conservation District of Monterey County, and the Elkhorn Slough Foundation.

Acknowledgments

The authors would like to thank Marc Buchanan and Heidi Simonson for their comments on earlier drafts of this chapter.

References

Brady, N. C., and R. R. Weil. 1999. *The Nature and Properties of Soils.* 12th ed. Upper Saddle River, N.J.: Prentice Hall.

Buol, S. W. 1997. *Soil Genesis and Classification.* 4th ed. Ames: Iowa State University Press.

Cook, T. D., and C. S. Buetler. 1978. *Soil Survey of Monterey County, California.* Washington, D.C.: U.S. Soil Conservation Service.

Isgrig, D. 1969. *Soil Survey: San Benito County, California.* Washington, D.C.: U.S. Soil Conservation Service.

Jackson, L. 1998. Carbon and nitrogen dynamics after tillage in California soils. In *Proceedings, Emerging Soil Management Options for California.* Kearney, CA: UC Cooperative Extension, Kearney Agricultural Center.

Jenny, H. 1980. *The Soil Resource: Origin and Behavior.* New York: Springer-Verlag.

United States. Department of Agriculture. 1998. *SSTNOG Monterey County.* Washington, D.C.: USDA.

———. Soil Conservation Service River Basin Planning Staff Target Area Team. 1984. *Strawberry Hills Target Area, Monterey County, California: Watershed Area Study Report.* 2 vols. Davis, Calif.: USDA Soil Conservation Service Forest Service.

United States. Natural Resources Conservation Service. 1998. *Keys to Soil Taxonomy*. 7th ed. Washington, D.C.: USDA.

Werner, M., M. Silberstein, M. L. Huertos, S. Gliessman, J. Newman, R. Tjeerdema, J. Caffrey, S. Shaw, A. D. Vogelaere, and J. Mason. 1997. Agriculture and wetlands: Examining the interface at the Azevedo Ranch and the Elkhorn Slough National Research Reserve in Coastal Central California. Final report submitted to the US Department of Commerce, National Oceanic and Atmospheric Administration. Ocean and Coastal Management Award No. NA 370R0162.

6

CHAPTER SIX

Archaeology & Prehistory

Terry L. Jones

As recently as the 1980s, the distinctive role that estuaries such as Elkhorn Slough played in the prehistory of California was poorly understood and underestimated. Early perceptions of estuarine prehistory in California were heavily influenced by findings from the San Francisco Bay shell mounds, which were a focus of archaeological investigations beginning in the early 1900s.

Because radiocarbon dates from these mounds indicate little evidence for occupation earlier than 5,000 years ago, human use of estuaries was thought to be relatively recent in California. This notion supported ideas advanced by some prehistorians (e.g., Osborn 1977; Beaton 1985; Yesner 1980) that for hunter-gatherers, marine habitats are inferior to terrestrial ones, and that they were used only after inland areas had been settled.

As data accumulated in the 1970s and 1980s from smaller estuaries such as Batiquitos Lagoon (San Diego County), Morro Bay (San Luis Obispo County), and Elkhorn Slough, it became apparent that San Francisco Bay was chronologically anomalous and that humans had been using coastal wetlands for much longer than was previously suspected. Because of their rich plant and animal resources, estuaries may actually have been among the first habitats selected by California's initial colonists (T. Jones 1991). Based on radiocarbon dating and other data, we now know that humans used estuarine embayments at Bodega Bay (Schwaderer 1992), San Dieguito Lagoon, and Agua Hedionda (Gallegos 1992) between 10,000 and 9,000 years before present (B.P.). Most other estuaries, including Elkhorn Slough (T. Jones and Jones 1992; Breschini and Haversat 1995; T. Jones et al. 1996), Morro Bay (T. Jones et al. 1994), Goleta Slough, and Batiquitos Lagoon (Gallegos 1992), have yielded evidence for occupations dating to between 9,000 and 8,000 years B.P., and in many regions estuary use seems to predate heavy use of exposed rocky and sandy coasts. Recent excavations on the south central coast show exploitation of an estuary near present-day Pismo Beach as early as 10,300 years ago; referred to as Halcyon Bay, this estuary is now extinct (Fitzgerald 1998). Similarly extinct estuaries dating to 9000–8000 B.P. are represented at archaeological sites along the coast of Santa Barbara (Erlandson 1994) (fig. 6.1).

Estuaries provide a unique marine and terrestrial resource base with dynamic environmental histories. Unlike many other habitats exploited by Native Californians, Elkhorn Slough provided resources year-round, and the area's resource diversity and richness raise interesting questions about the relative mobility or sedentism of the people adapted to this environment.

Estuaries in California also show evidence of constant change. At the peak of the Last Glacial Maximum approximately 18,000 years ago, most of what are now estuaries were the backwaters of deeply incised drainages. With the onset of the Flandrian Transgression (the interval of melting glacial ice and rising seas that following the glacial peak, dating ca. 18,000–7000 years B.P.), the mouths of these drainages

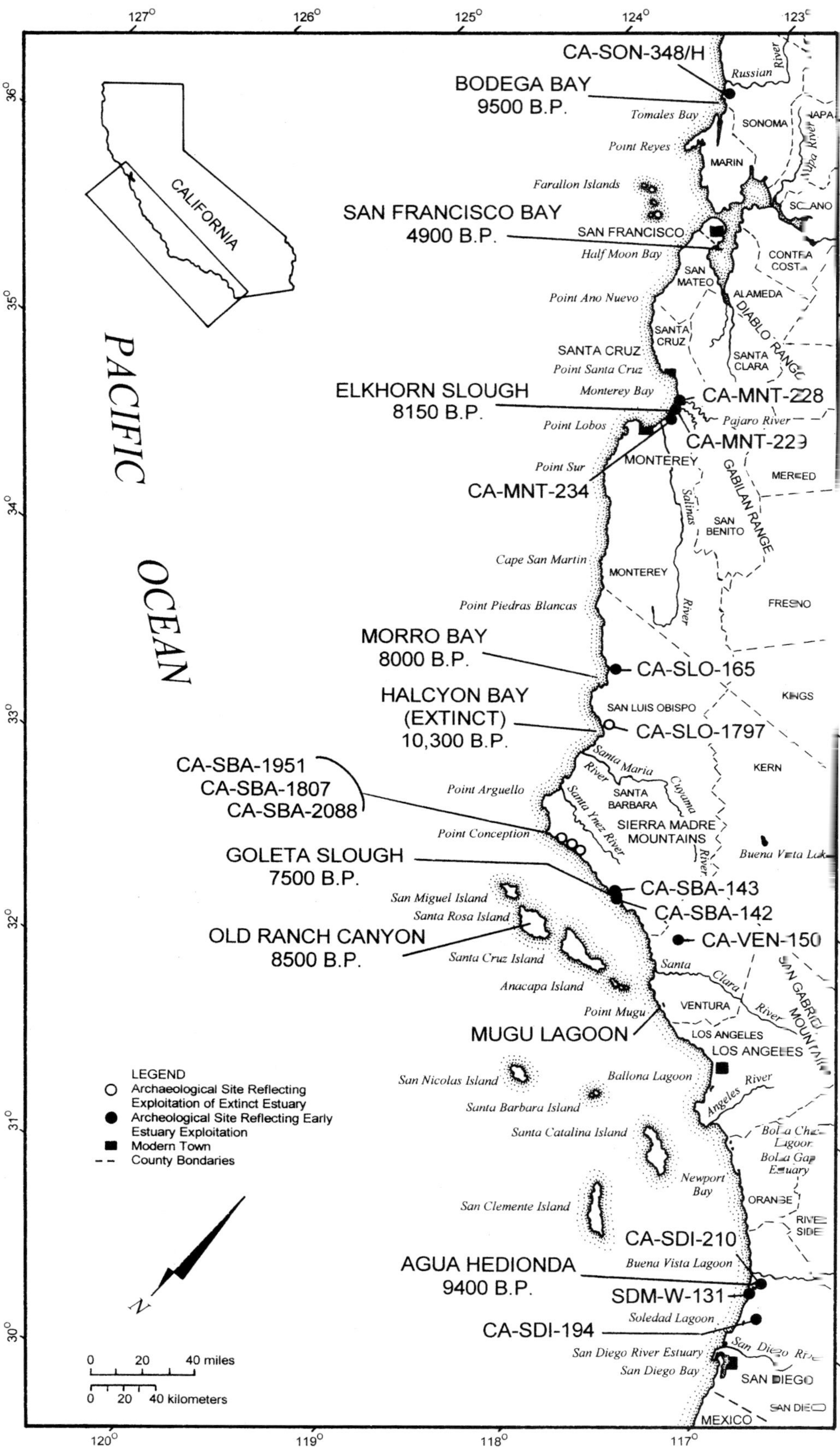

Figure 6.1. Major estuaries of central and northern California and dates of their initial settlement by humans.

flooded and embayments formed. As sea level rose during this 11,000-year period, these embayments progressed inland, but when the rate of sea-level rise slowed at mid-Holocene, sediments began to accumulate and the systems started to fill in (Atwater, Helley, and Hedel 1977; Bickel 1978). By 2,000 years ago, some estuaries, including Batiquitos Lagoon and Agua Hedionda in San Diego County, were cut off from tidal waters, and their marine habitats deteriorated rapidly. At Elkhorn Slough, this process was complicated by the convergence of three discrete drainages—Elkhorn Slough and the Pajaro and Salinas Rivers—that occasionally shifted course over time, altering the location of tidal inflows. This dynamic hydrographic landscape presented challenges and opportunities for the resident hunter-gatherer populations of the central Monterey Bay area.

This chapter describes prehistoric human settlement of the central Monterey Bay area beginning with the first evidence for human occupation during the early Holocene and extending up to the eighteenth century when Spanish explorers first crossed Elkhorn Slough. The discussion begins with an examination of changes in Holocene climate, freshwater systems, and sea level that affected the prehistoric environment of the Elkhorn Slough region. It reviews impressions of the area's native inhabitants by the first European visitors and what these impressions tell us about native lifeways, settlement, subsistence, and resources, including the still-unanswered question of whether one or two major groups occupied the Elkhorn Slough region. Archaeological efforts at Elkhorn Slough are detailed and their findings interpreted to give a picture of settlement patterns, diet, population movements, and use of the slough over time. The chapter closes with management recommendations for preserving important archaeological sites and suggestions for research projects that will help fill in the picture of the physical environment and human uses of the Elkhorn Slough region prior to European contact.

Paleoenvironment

In contrast to the more environmentally stable outer shores of the central coast, the Elkhorn Slough region shows evidence of a dynamic past, typical of small California estuaries. Climate shifts influenced vegetation distribution, and sea level rise altered the hydrographic landscape, creating an ever-changing resource base for the region's human inhabitants.

Holocene Climate

The North American continent experienced a series of large-scale, low-intensity environmental fluctuations during the Holocene Epoch (Antevs 1948, 1952). Although we don't know exactly how these fluctuations affected humans living along the California coast, the impact of climate change in the coastal zone seems to have been less extreme than in the interior due to the tempering influence of ocean water (Johnson 1977).

Based on pollen samples from the Santa Barbara Channel, we know that the climate of the California coast during the early Holocene (9000–7000 B.P.) was cool and wet, with a higher incidence of pine and fern than is seen today. The mid-Holocene (7000–5000 B.P.) was a warm and dry period, and drought-tolerant species (oak, sagebrush, and sunflower) became more prominent in vegetation bordering the channel, with the warmest period occurring 5400–4400 B.P. The climate became cooler and moister after 4400 B.P., although modern conditions of temperature, rainfall, and vegetation were not established until after approximately 2300 B.P.

Closer to Elkhorn Slough, a pine-dominated coniferous forest covered the Santa Cruz Mountains (Adam, Byrne, and Luther 1981) during the terminal Pleistocene (12,000–10,000 B.P.), much like that which is now found 800 kilometers to the north. Warmer, drier climates apparently progressed coastward and northward as the glacial epoch ended (Axelrod 1981), and pine-dominated forests retreated northward from their former extension well south of Big Sur. *Disjunct* stands of ponderosa pine in the Santa Lucia and Santa Cruz Mountains are vestiges of this earlier vegetation distribution.

Changes in the central Monterey Bay's vegetation through the mid-Holocene can be interpreted from a pollen profile collected at Elkhorn Slough (West 1988). T. Jones and Waugh (1997) determined that the oldest pollen in the 6.9-meter-deep core dates to 6300 B.P. (see also Dietz, Hildebrandt, and Jones 1988; T. Jones 1992, 14; T. Jones and Jones 1992). The pollen composition reveals that a combination of large-scale Holocene climate changes combined with local events to create distinct shifts in both the terrestrial and aquatic landscapes. At the lowermost levels of the profile, moderate levels of pine, redwood, oak, and grass pollen occur, but a significant shift takes place between 400 and 356 centimeters, where pine reaches its lowest frequency and oak declines (fig. 6.2).

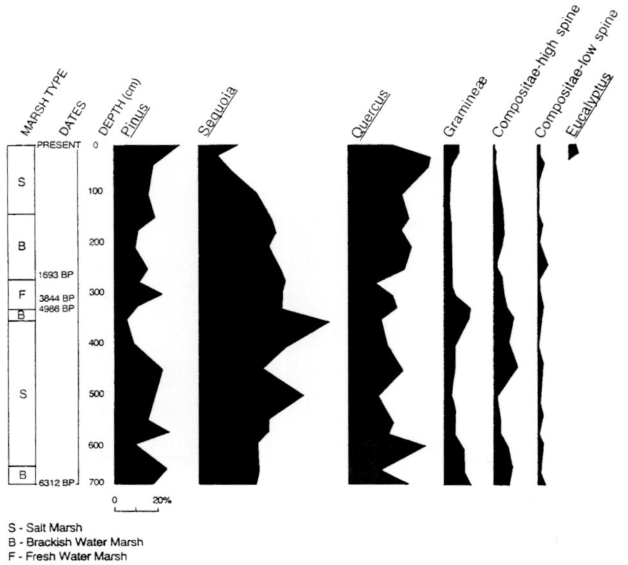

Figure 6.2. Terrestrial taxa from the Elkhorn Slough pollen core.

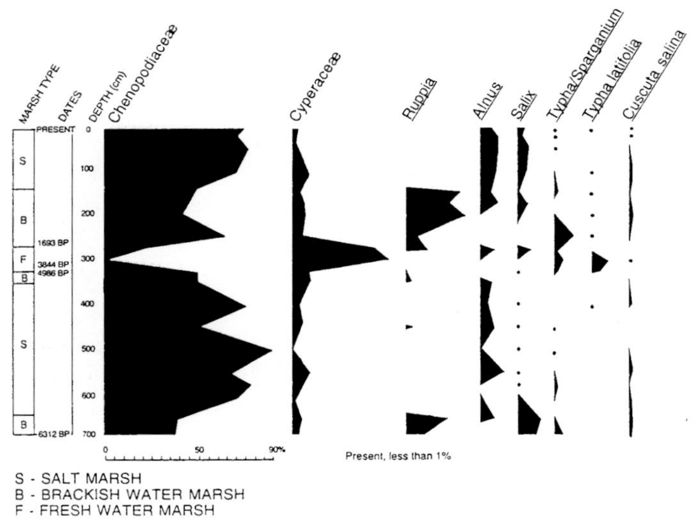

Figure 6.3. Hydrophytic taxa from the Elkhorn Slough pollen core.

At the same time, higher proportions of grasses, high-spine composites, and redwood all appear. Dating to ca. 5200 B.P., this pattern conforms to the decrease in pine and increase in redwood associated with the mid-Holocene in the San Francisco Bay area (Adam, Byrne, and Luther 1981), and likewise seems to correlate with peak warming in the Santa Barbara Channel (Heusser 1978). As in San Francisco Bay, high frequencies of redwood pollen suggest this was not a period of extended drought.

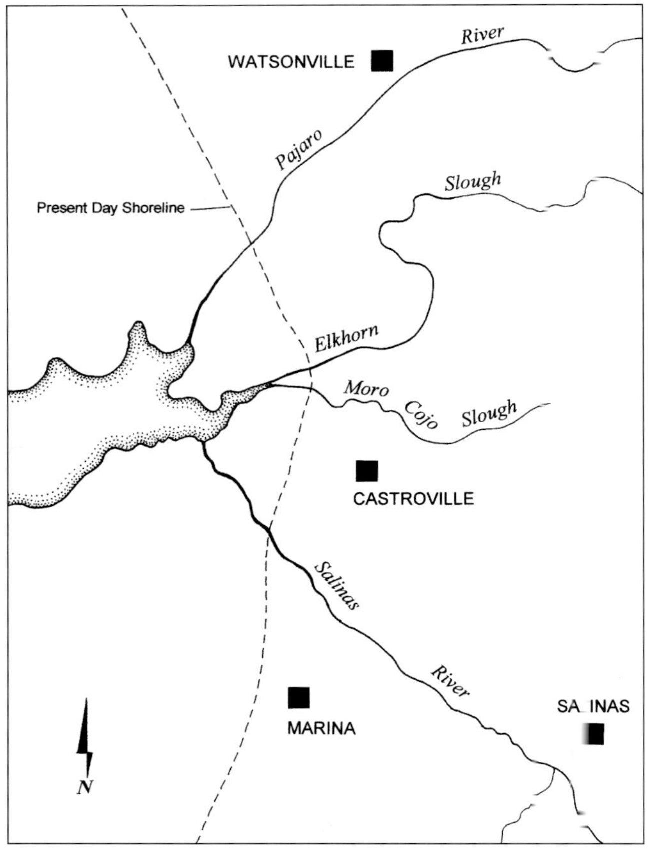

Figure 6.4. Central Monterey Bay shoreline during the Last Glacial Maximum (18,000 years B.P.) when sea level was 120 meters below its present level (adapted from Milliken et al. 1999).

Fluctuations in both freshwater and salt-tolerant plants (*Salicornia* spp., Cyperaceae, and *Ruppia* spp.; fig. 6.3) represented in the Elkhorn Slough pollen sequence probably reflect a localized event during which freshwater dominated the slough area, rather than widespread climatic change. During this apparent infusion of freshwater, pollen of salt-tolerant species (e.g., *Salicornia* [pickleweed], of the Chenopodiacae family) disappear and freshwater species of the cypress family increase sharply. Based on radiocarbon dating, this event began approximately 5000 B.P. and continued until sometime after approximately 3780 B.P., when the slough reverted to saline conditions similar to those that prevail at present. Sedimentation rates within the slough were noticeably lower during the freshwater reversal. From approximately 5000 to 3790 B.P., Elkhorn Slough was probably cut off from the open ocean, with no freshwater drainage cutting through the sand barrier to the sea, and few sediments entering the slough channel.

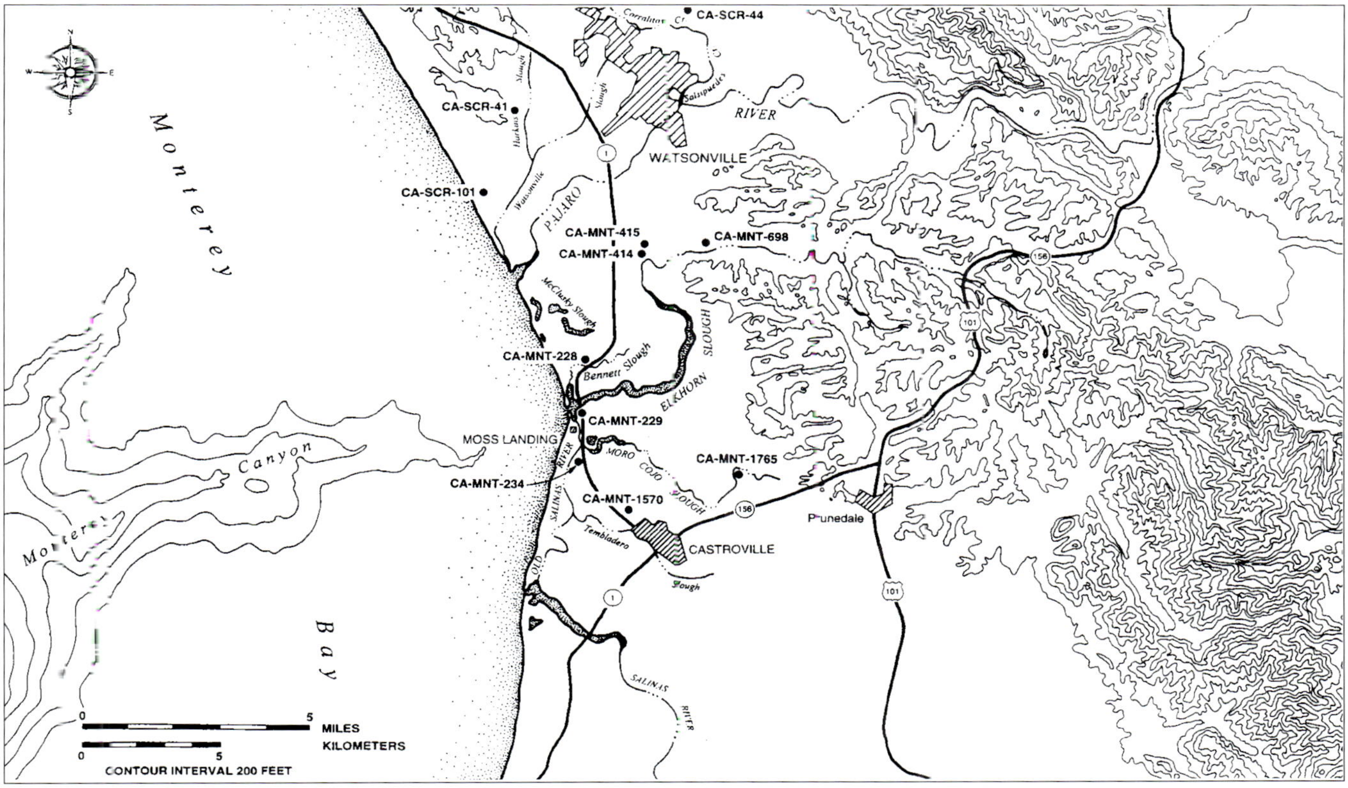

Figure 6.5. Investigated archaeological sites in the Elkhorn Slough locality.

California's climate also experienced significant fluctuations during the most recent millennium in its prehistory. In particular, coastal habitats were probably affected by the period of widespread drought approximately 1,150 and 650 years ago during the Little Climatic Optimum or Medieval Climatic Anomaly, as well as the ensuing Little Ice Age, 500–300 years ago (Boerner, Killingley, and Taylor 1985; Fagan 2000). Tree rings from the southern Sierra Nevada show signs of prolonged drought between A.D. 1100 and 1375 (Graumlich 1993). Droughts triggered declines in the level of Mono Lake in the eastern Sierra between A.D. 890 and 1110, and again between 1210 and 1350 (Stine 1990; 1994, 549). Stine (1994, 549) further suggests that these extended dry intervals were caused by an unusual climate shift, such as has not recurred in the contemporary weather cycle. This drought interval further corresponds with an interglacial period in the Sierra Nevada (Curry 1969) and a warm dry period reflected in tree rings in the White Mountains (LaMarche 1974). It was followed by glacial advances during the Little Ice Age.

Hydrographic Changes and Slough Occupation

Changes in both sea level and freshwater systems have dramatically altered the central Monterey Bay landscape over time. Tectonic movements along the San Andreas Fault periodically dammed the course of the Pajaro River, forming lakes in the San Juan and Santa Clara Valleys (Jenkins 1973) that affected drainage into Elkhorn Slough and the Pajaro River. More significant in terms of direct effects on the landscape of the central Monterey Bay has been sea level rise. At the time of the last glacial maximum (18,000 years B.P.), when sea level was 120 meters lower than it is today, the Monterey Bay shoreline in some areas was as much 15 kilometers west of its current location. At Elkhorn Slough, the head of the Monterey Submarine Canyon brought the Pacific Ocean to within 2 kilometers of its current location (fig. 6.4). Following the last glacial maximum, rapid sea level rise brought Pacific waters to within 10 meters of their current level by approximately 7,000 years B.P. (Inman 1983, 8–9), flooding the deeply incised channels of Elkhorn Slough and the Pajaro and Salinas Rivers and forming large embayments (see also chapter 2, "Geology"). These changes affected the human use and occupation of the region.

Evidence for large embayments occurs in both the archaeological record and present-day features. Invertebrate remains from an archaeological site northeast of present-day Castroville on the Tembladero Slough (CA-MNT-1570*; see fig. 6.5), most of which date to 5800–4800 B.C., suggest that a substantial estuary once existed there. Wetland-derived soils (Clear Lake clay) in the Castroville area apparently mark the boundaries of the estuary, which was associated with a former channel of the Salinas River. The site near Castroville was located on a marine terrace on the shoreline of this embayment. Similar estuary features were present in the lower courses of Elkhorn Valley and the Pajaro River, as marked by peat deposits (California Division of Mines 1956, 2), wetland soils, existing stands of marshland, and open water. Open water in present-day Harkins Slough, several kilometers inland from the mouth of the Pajaro River, is probably a remnant of a more

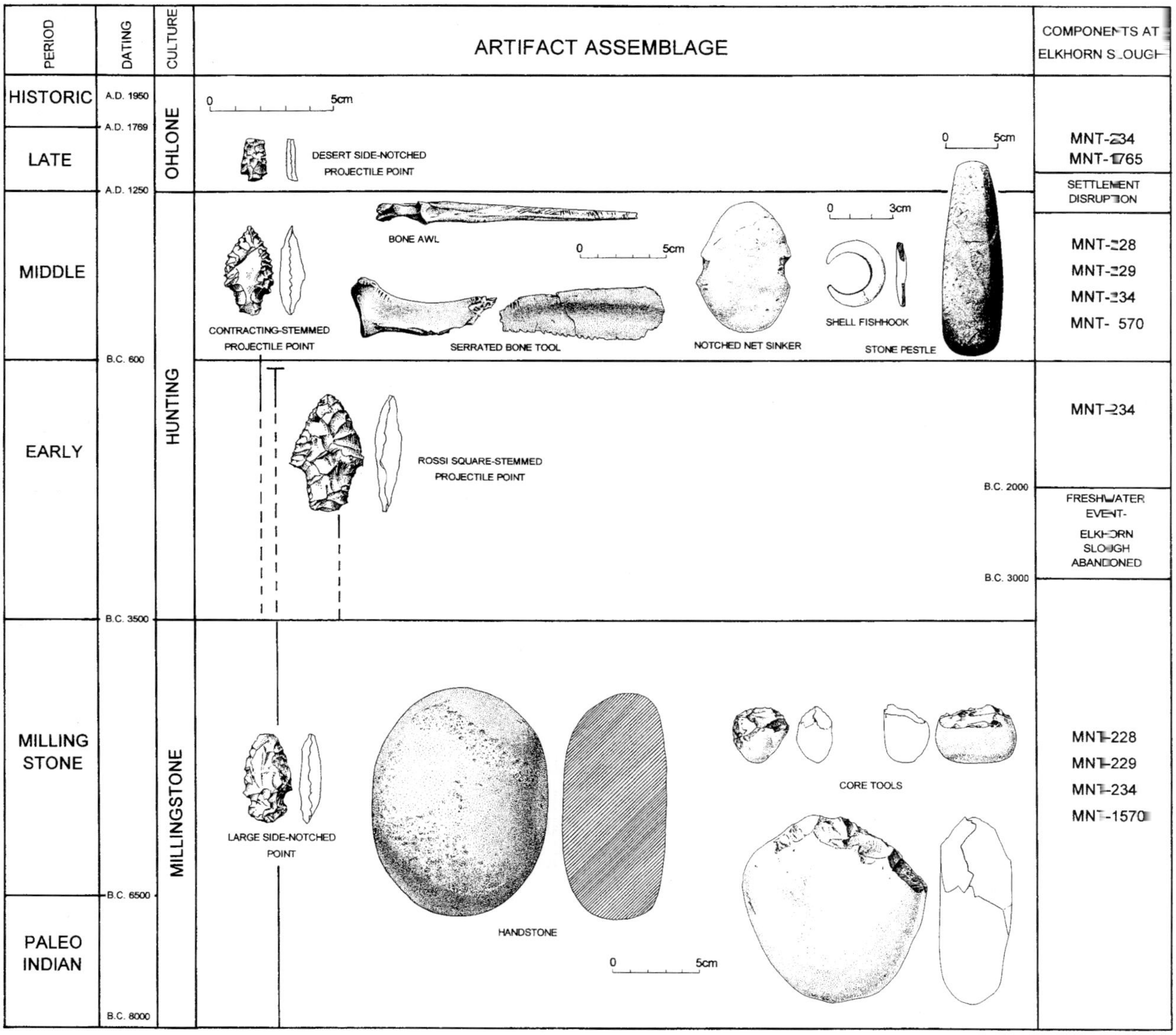

Figure 6.6. Summary of Elkhorn Slough culture and settlement history.

* As they are discovered, archaeological sites in California are assigned sequential, three-part numbers (trinomials) at regional archaeological information centers (CA = California; MNT = Monterey). Information on Monterey County archaeological sites is housed at the Historical Resources Information System of the California Archaeological Inventory at Sonoma State University, Rohnert Park.

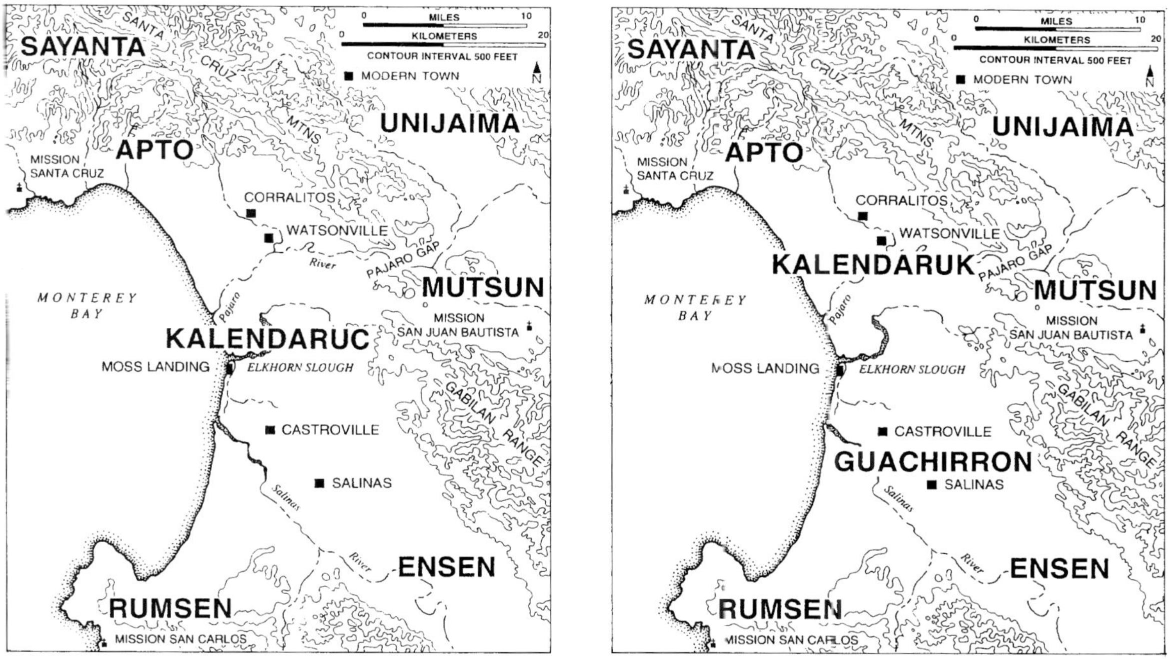

Figure 6.7. Alternative conceptualizations of Monterey Bay political boundaries based on Spanish Mission records.

expansive mid-Holocene embayment. Estuarine shell remains at a site on the coastal terrace north of the Pajaro River adjacent to Harkins Slough (CA-SCR-41) indicate that humans used this environment. The expanse of open water in these features was probably greatest immediately prior to the mid-Holocene (approximately 7,000 to 5,000 years ago), when sea level rise was still rapid enough to outpace sediment accumulation. After the rate of sea level rise declined at mid-Holocene, these systems began to fill with sediments. Dating of at least one site (CA-SCR-41) suggests that the estuary persisted until at least 2,500 years ago.

Shellfish *assemblages* from sites near Bennett, Elkhorn, Moro Cojo, and Tembladero Sloughs (CA-MNT-228, -229, -234, and -1570; fig. 6.5) show that mudflat habitats suitable for clams and cockles were well established in these embayments by 6200 B.C. This pattern contrasts markedly with that found in San Francisco Bay, where clam-dominated middens are not common until after A.D. 500 (Elsasser1978:39). More recent sites at Elkhorn Slough show fewer clams and more mussels (table A-6.1), a trend also very different from that through time in San Francisco Bay.

The co-occurrence of estuarine shellfish and freshwater fish remains at three Elkhorn Slough sites (CA-MNT-228, -229, and -234) marks two intervals during which the Salinas and/or Pajaro Rivers joined the slough with the Pacific Ocean, the first spanning from approximately 8200 to 6000 B.P., and the second beginning approximately 5000 B.P. The 1,000-year period between these two episodes is intriguing, in that none of the sites excavated in the Elkhorn Slough area (CA-MNT-228, -229, -234, -414, or -1570) show any evidence of human occupation during this time, and some sites (e.g., CA-MNT-228, -229, and -234) reveal a distinct gap in occupation (fig. 6.6). This gap coincides with the period when freshwater plant species appear in the slough's pollen profile (fig. 6.3). With no connection to the ocean, the slough would have been a freshwater lagoon with no shellfish, marine fish, or mammals. This decidedly inferior habitat would have been very unattractive to humans. We know from radiocarbon-dated shells from CA-MNT-234 that the slough had reverted to marine conditions by approximately 4000 B.P. Dating of *Mytilus* shells from two sites in the Gilroy area (CA-SCL-577 and -698) likewise indicate that marine conditions were reestablished by 4000 B.P. (Hildebrandt and Mikkelsen 1993, 75)—which is consistent with dating from the pollen core. Following reestablishment of estuarine habitat, humans again settled in the slough area, as indicated by settlement debris and radiocarbon dates from CA-MNT-228, -229, and -234.

Infusion of freshwater into Elkhorn Slough could reflect discharge from one of several lakes that were intermittently present east of the San Andreas Fault in the southern Santa Clara Valley. Alternatively, the change in salinity could have been caused by a meander of either the Pajaro or Salinas River, closing the slough's connection to the ocean. This seems the most likely cause, since many other lagoons and estuaries in California experienced similar closings at one time or another (see Gallegos 1987, 1992). Such a closure might have been related to mid-Holocene climatic warming—if river flows were reduced and sand barriers established, the flow of ocean water into the slough would have been blocked. However, this scenario is made more complicated by the fact that many other estuaries (e.g., Goleta Slough, Morro Bay, Batiquitos Lagoon) retained their connections to the ocean during this time along with their saline environments (T. Jones and Waugh 1997). Furthermore, the slough's pollen record does not show signs of extensive mid-Holocene drought.

The slough area's hydrographic landscape and human settlement also seem to have been affected by the Medieval Climatic Anomaly. As seen elsewhere in central California (see Jones et al. 1999), the slough shows signs of a disruption in settlement during the centuries of medieval drought. Sites like CA-MNT-228 and -229 that were occupied prior to the Medieval Climatic Anomaly were abandoned at the same time that other sites, such as CA-MNT-1765, were initially occupied.

Ethnohistory

Only the earliest Euro-American explorers and colonists of the Monterey Bay area observed the Costanoan-speaking natives of the Elkhorn Slough area in an unacculturated state. Diaries of the early Spanish explorers, soldiers, and missionaries provide limited descriptions of native life before the mission period; mission records from Carmel, San Juan Bautista, and Santa Cruz provide clues to village locations and affiliations of specific individuals. Most of what has been gleaned from these early historic accounts was summarized by anthropologists in the 1970s (e.g., Broadbent 1972; Heizer 1974; Kroeber 1925; Levy 1978). Also relevant is the work completed by Milliken (1988) for a Caltrans archaeological project at the mouth of Elkhorn Slough (CA-MNT-229). Milliken's extensive ethnographic and ethnohistoric research on Elkhorn Slough and the Pajaro and Salinas River system represents the most concise treatment of local ethnohistory to date. The following discussion incorporates general observations on the Costanoan with locally specific conclusions developed by Milliken.

The term Costanoan, first used by Latham in 1856, designates dialects spoken by five groups at Mission Dolores in San Francisco, whose collective territories ranged from Soledad to the San Francisco Bay (Milliken 1991). An alternative term, Ohlone, was applied to these people by Merriam (1967) and has subsequently been used by many contemporary descendants of Costanoan speakers. Attempts to assign political boundaries based on language differences have been summarized by Milliken (1993).

Sociopolitical Organization and Ethnogeography

Costanoan social order was marked by distinct political groups, with the "tribelet" forming the basic political unit. According to Kroeber (1962, 33), a tribelet would

> contain several settlements. These several settlements—there might be three or four or five of them—sometimes more or less the same size, but more often one was dominant or permanent, the other more like suburbs of it. They might be situated some miles away. The smaller settlements were likely to be inhabited seasonally, or by certain families only perhaps for a stretch of years, after which their population might drift back to the main settlement.

A chief acquired his power through patrilineal descent and conducted tribelet affairs, although this role was more advisory and ceremonial in nature than peremptory (Levy 1978, 487). According to Harrington (in Levy 1978, 488), Costanoan tribelets were organized into clans, which were subdivided into deer and bear *moieties*. In apparent allusion to tribelet organization, Fages (1937, 66) notes that the natives in the Carmel area rarely traveled more than 4 or 5 leagues (approximately 22.5–28 km, 14–17.5 mi), so as to avoid conflict with their neighbors and enemies.

The location and organizational structure of tribelets in the Elkhorn Slough/Castroville area has been a topic of uncertainty and ongoing debate (cf. Kroeber 1925; C. King 1974; Levy 1978; Milliken 1988), largely as a consequence of an observation made by early Spanish explorers that conflicts with records at Missions San Carlos and San Juan Bautista. Milliken (1988) made the most authoritative attempt to reconcile this discrepancy.

First European contact with the natives of the Watsonville-Castroville area occurred on October 6, 1769, when scouts of the Portolá overland expedition observed a settlement near present-day Watsonville:

> They had seen…a numerous village of heathens living camped in grass-covered huts, which must by what they said have been over 500 souls. These Indians had had no notice of our coming to their lands, as was seen by the consternation and terror our presence caused among them: for some, amazed and confounded, scarce knowing what they did, ran to their weapons; others shouted and cried out; the women dissolved into tears. Our people did all they could to quiet them, and the sergeant of Loreto Presidio, who was in charge of the party, managed it with great difficulty by getting down from his mount and approaching them with signs of peace.… The heathens became very happy, applauding our men's behavior, who then, the better to assure them their intent was not to do them hurt, but rather that they wished their friendship, asked them by signs for food. At this the Indians grew better pleased, and at once their women set to grinding seeds of which they made some dough balls, and made them a present of them. The sergeant gave them some beads, and the Indians were left very well satisfied and pleased. (Costansó 1911:245)

When the rest of the Portolá expedition passed along this route two days later, they found that this village had been "burned and abandoned," with "not a heathen in sight anywhere" (Fray Juan Crespí, in Stanger and Brown 1969). On November 26, 1769, on their way back to San Diego, this entourage passed through the same area and observed a group of natives building a village near present-day Castroville. In an observation that has proven critical to reconstruction of contact-era sociopolitical organization, scouts claimed that these were the same people who had been seen near Watsonville the month before (Crespí 1927, 240). The only other account of natives in this area was by Crespí; setting out from the presidio at Monterey with Pedro Fages in 1772, he traveled along the Salinas River toward Salinas and San Juan Bautista. Upon rounding a marshy area northeast of Salinas, Crespí (1927, 280) noted the presence of "two huts of the heathen, made of branches and covered with grass." Subsequent Spanish explorers bypassed the Pajaro-Salinas area, taking inland routes to San Francisco Bay.

The presidio of Monterey and Mission San Carlos Borromeo de Carmelo were established in the summer of 1770. Mission founder Junípero Serra soon discovered that the nearby presidio and its soldiers frightened away potential neophytes, and in 1771 he moved Mission San Carlos to its present location in Carmel. Missions San Carlos, Santa Cruz (est. 1790), and San Juan Bautista (est. 1797) all drew converts from villages near the mouth of both the Pajaro and Salinas Rivers. Mission records on baptisms, marriages, and deaths include information on personal genealogies and villages affiliations. Occasionally, these affiliations included location descriptions, providing clues to settlement locations. (For additional detail from mission records, see appendix 6.1.) However, these records present conflicting pictures of tribelet organization.

Baptismal, marriage, and death records from Mission San Carlos and observations by the Portolá expedition suggest that one large tribelet, Calenda Ruc, covered the central Monterey Bay area. At least seven villages are associated with that area in mission records: Kalenda Ruc, Mustac, Culul, Locuyusta, Tusquesta, Chalicta, and Tiuvta (Milliken 1988, 67–68). However, records from Mission San Juan Bautista suggest that there were there two tribelets: Calenda Ruc, in the vicinity of Watsonville (with main villages of Calenda Ruc and Tiuvta); and Guachirron, in the vicinity of Castroville (including the village Locuyusta; see fig. 6.7). This discrepancy represents a vexing, ongoing problem in the study of local ethnohistory and archaeology.

This question will not likely be resolved based on the existing archival record, but archaeological data and a more detailed consideration of local geography may provide some new insights. C. King (1974) and Milliken (1981; 1988, 85) attempted to overlay cultural boundaries on to the Monterey Bay area in a fairly regular pattern, with two tribelets of roughly equal size and shape. Such a configuration belies the complexity of the precontact Monterey Bay landscape, which included vast tracts of land that were largely uninhabitable (e.g., extensive wetlands of the lower Salinas and Pajaro Rivers, and the dry dunes in the Fort Ord area). Villages could be established only in scattered pockets, and tribelets probably conformed more to the distribution and configuration of habitable tracts than with a regular template. Archaeological site locations define the areas

The Ohlone constructed boats of tule reed to use on central coast sloughs, bays, and wetlands. Louis Choris, official artist for Kotzebue's expedition on the Rurik, drew this view from the Rurik's anchorage near the San Francisco Presidio in 1816. (Courtesy of the Bancroft Library)

Table 6.1. Mammal species found in Elkhorn Slough region archaeological sites (CA-MNT-228, -229, and -234).

Taxon	Common Name
Terrestrial	
Antilocapra americana	pronghorn
Canis latrans	coyote/dog
Cervus elaphus	tule elk
Lepus californicus	jackrabbit
Lynx rufus	bobcat
Mephitis mephitis	skunk
Mustela frenata	weasel
Neotoma fuscipes	dusky-footed woodrat
Odocoileus hemionus	deer
Procyon lotor	raccoon
Scapanus latimanus	mole
Sylvilagus audubonii	cottontail rabbit
Taxa taxidea	badger
Urocyon cinereoargentus	gray fox
Ursus americanus	black bear
Marine	
Arctocephalus townsendi	Guadalupe fur seal
Callorhinus ursinus	northern fur seal
Enhydra lutris	sea otter
Eumetopias jubatus	Steller sea lion
Phoca vitulina	harbor seal
Phocoena phocoena	harbor porpoise
Zalophus californianus	California sea lion

Europeans exploring the central coast region described the Ohlone hunting small game with bows and arrows. Bowstrings were made of sinew or vegetable fibers; arrows were tipped with chipped stone or bone arrowheads. (By Louis Choris, Courtesy of the Bancroft Library)

where settlement was possible; dense clusters of sites suggest areas that were consistently attractive for habitation.

Although archaeological survey is far from complete in this area, some preliminary locational patterns are apparent. Site density is very high along the edges of Elkhorn Slough, where deposits are found near the slough's current mouth, and 4–5 kilometers inland at its terminus. Very few sites have been identified on the lower floor of the Pajaro Valley, but many have been recorded on the slopes and bluffs above the river. Very few sites have been found in either the Watsonville area or in the lower Salinas Valley. Sites may occur in Castroville, which prehistorically was an upland amid extensive marshes, but survey of that community is incomplete. Very few sites have been found in the Santa Cruz Mountains or in the Fort Ord area. Milliken (1988, 64) suggested that Elkhorn Slough would have been a logical boundary between northern and southern tribelets, with members of each using the slough seasonally. However, the high site density in the slough suggests otherwise. Furthermore, the slough's inland segment harbors a full complement of potential food-bearing habitats, including oak grassland, and there is no reason to assume that the slough's use was seasonally restricted.

Shelter and Clothing

Spanish accounts described dwellings built in the Watsonville-Castroville area as being "spherical-shaped houses of poles and tule" (Crespí 1927, 240). Tule was also used to build balsa rafts for hunting and fishing. In warm weather, men usually were unclothed. Women wore short woven aprons of red and white twisted cords or dry green tule, and rear aprons of deerskin (Fages 1937, 66). In cold weather, both men and women wore robes of rabbit skin, sea otter skin, duck feathers, or buckskin fastened with a cord under the chin (Levy 1978, 493).

Settlement and Subsistence

Precisely how the natives used the rich resource base of the central Monterey Bay is not entirely clear, partly due to the uncertainty surrounding tribelet boundaries. Early historical accounts describe natives as eating a wide variety of foods, though the reports conflict as to whether they lived in permanent villages or moved about, taking advantage of seasonally available resources.

In 1602, during his exploration of the California coast, Sebastian Vizcaíno recorded the following observations in the Monterey Bay area:

> The land is well populated with Indians without number, many of whom came on different occasions to our camp. They seem to be gentle and peaceful people; they say with signs that there are many villages inland. The sustenance which these Indians eat most daily, besides fish and shellfish, is acorns and another fruit larger than a chestnut; this is what we could understand of them. (Broadbent 1972, 47)

Pedro Fages described natives gathering food in the Monterey Bay area in his 1769 expedition journal. Near Point Nuevo he noted that they were "very clever at going out to fish embarked on rafts of reeds, and they succeed, during good weather, in getting their provisions from the sea...the land also provides them with an abundance of seeds and fruits" (Fages 1937, 70). He also noted heavy exploitation of a summer run of sardines (Fages 1937, 69), and remarked that local inhabitants "do not have fixed places for their villages, but wander here and there wherever they can find provisions at hand" (Fages 1937, 67). In the Salinas Valley he observed: "Many antelope were seen going by, and the place was named Real de los Cazadores, for there were then round about it some Indians who were so absorbed and occupied in hunting game that they did not notice us" (Fages 1911, 67). At Mission San Carlos, he observed:

> Those who are in this mission and nearby obtain few acorns, the lack of which they supplement in part with blackberries and strawberries, which abound around the point of the Monte de Pinos; there are many boletes or mushrooms, and another wild fruit about the size of an ordinary pear which is eaten roasted and boiled though it is somewhat bitter. The tree which bears it is rather whitish, like a fig tree, but not very tall. When it bears fruit it sheds its leaves entirely. (Fages 1937, 68)

Archibald Menzies, the Scottish naturalist for the George Vancouver expedition, recorded the following at Mission San Carlos in 1792: "Their food at this time was chiefly shellfish which the women collected along the shore, while the men lounged about the country with their bows and arrows, killing rabbits and quails" (Menzies 1924, 293–294).

These descriptions make clear that the native inhabitants of the Monterey Bay area relied on a broad spectrum of terrestrial and aquatic foods. However, researchers disagree as to how the collection and consumptions of these resources corresponded to settlement patterns. Dietz and Jackson (1981) presented a settlement model for the Monterey Peninsula based on a

Table 6.2. Bird species found in Elkhorn Slough region archaelogical sites (CA-MNT-228, -229, and -234).

Taxon	Common Name
Accipiter/*Buteo* spp.	hawk
Aechmophorous occidentalis	Western Grebe
Anas sp.	teal
Ardea herodias	Great Blue Heron
Asio sp.	owl
Aytha/Melanitta spp.	duck
Branta/Anser/Chen spp.	goose
Branta bernicla	Brant
B. canadensis	Canadian Goose
Bubo viginianus	Great Horned Owl
Callipepla californica	California Quail
Corvus brachyrhynchos	American Crow
Cygnus sp.	swan
Diomedea albatrus	Short-tailed Albatross
Falco spp.	falcon
Fulica americana	American Coot
Fulmarus glacialis	Northern Fulmar
Gavia sp.	loon
Larus spp.	gull
Pelecanus occidentalis	Brown Pelican
Phalacrocorax auritus	Double-crested Cormorant
P. penicillatus	Brandt's Cormorant
Podiceps spp.	grebe
Podilymbus podiceps	Pied-billed Grebe
Puffinus griseus	Sooty Shearwater
Tyto alba	Barn Owl
Uria aalge	Common Murre

combination of archaeological findings and historic accounts. Noting an apparent absence of references to shoreline villages among mission records, they postulated that late prehistoric (since 1,000 years B.P.) and postcontact residents were largely inland-based "collectors" (Binford 1980), who maintained permanent villages in the interior and visited the coast only to harvest specific resources. Dense deposits or "pavements" of whole abalone shells recovered from archaeological sites on the Monterey Peninsula are thought to represent one type of specialized resource collection, as abalone were apparently harvested en masse, shelled, dried, and transported inland. Subsequent researchers suggest that the natives may have preferred inland sites because of their access to acorns (T. Jones et al. 1989; Hildebrandt and Jones 1992). D. Jones (1992) challenged labeling the Costanoan as "collectors," while Rivers (n.d.), in light of the abundant historic accounts emphasizing shellfish and other marine foods, questioned the proposed focus on inland resources.

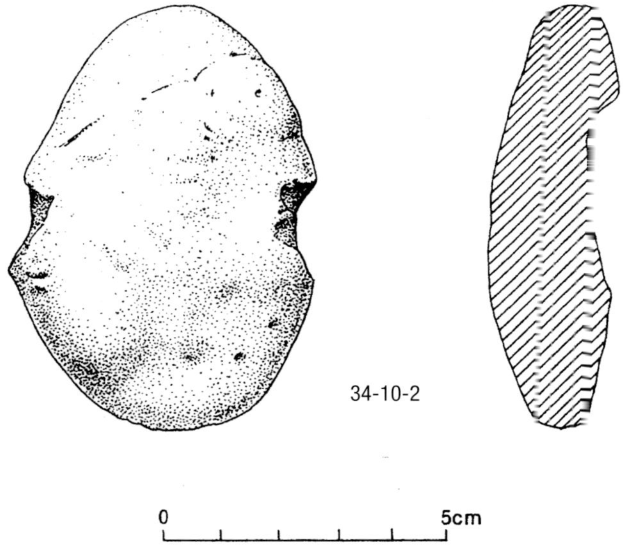

Figure 6.8. Notched stone netsinker from the Castroville area.

In the central Monterey Bay area, settlement systems are even less understood, as Elkhorn Slough's resource base contrasts markedly with that of the Monterey Peninsula's coastline. Marine habitats and acorns are both accessible at the rear of the slough. However, the slough's clams and mussels are considerably smaller than abalone, and less profitably dried; no dense collection of clam shells, equivalent to the abalone pavements of the Monterey Peninsula, have been identified in the slough region's Late Prehistoric or Historic archaeological record. In point of fact, archaeological evidence for human use of the slough during the centuries immediately preceding arrival of the Spanish is sorely lacking. No archaeological site in the area has been firmly linked to a named ethnohistoric village. Most sites predate the Late Prehistoric and Historic era. Late occupation is known only from three locations: CA-MNT-1765, a Late Period residential base 5 kilometers inland near Moro Cojo Slough (fig. 6.5), CA-MNT-234 in Moss Landing, and CA-SCR-44 near Watsonville (Breschini and Haversat 2000). The tools and faunal remains from CA-MNT-1765 suggest the site functioned as a residential base, while its inland location suggests a terrestrial emphasis in subsistence (Fitzgerald et al. 1995). CA-MNT-234 produced radiocarbon evidence for occupation between A.D. 1200 and 1750 (Milliken et al. 1999). Occupation at this location seems to have been so light and intermittent that it resulted in deposition of food refuse only (clam shells and deer bones) and virtually no artifacts. The adaptation of people inhabiting the slough area

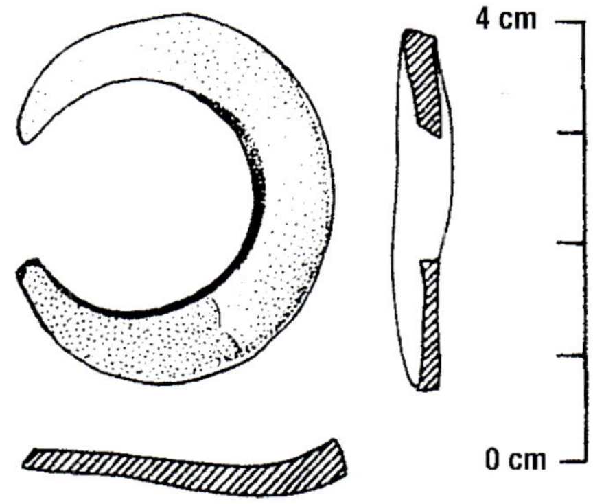

Figure 6.9. Shell fishhook from the southern Monterey County coast.

during the centuries immediately preceding historic contact seems to have been different from that of earlier peoples, in that site locations and faunal remains suggest a greater emphasis on terrestrial resources and habitats.

Subsistence Resources

Historic accounts indicate that native peoples harvested a wide variety of mammals, birds, fish, shellfish, and plant foods, but do not provide much detail on the exact species exploited; most of this information can be obtained only from the archaeological record. The following is a summary of the resources used by natives of the Elkhorn Slough region, based on archaeological findings from sites near Bennett, Elkhorn, and Moro Cojo Sloughs (CA-MNT-228, -229, and -234). These sites contain evidence from the full time range represented locally (8000 B.C.–contact).

Mammals Native inhabitants of Elkhorn Slough hunted a variety of mammals for meat, and used seal, otter, and deer skins for clothing. Long bones from deer, antelope, and elk were fashioned into awls, punches, sweatscrapers, and other tools. Sections of rib from large marine mammals were occasionally used for prying tools. Table 6.1 lists evidence of 18 terrestrial and 7 marine mammal species. At the time of historic contact, the primary weapon used for hunting animals was the bow and arrow. Deer were stalked by individual hunters who wore a deer's head as a disguise (Levy 1978, 491). The flesh of stranded whales and sea lions was roasted in earthen ovens (Levy 1978, 491).

The bow and arrow were adopted in prehistoric California relatively recently. This technology is represented in the archaeological record by very small, often finely made projectile points. Prior to about A.D. 1000, hunting was accomplished with spears and darts launched with spearthrowers.

Birds While the ethnohistoric accounts again lack detail, birds were probably used for food, with feathers also used to decorate clothing and baskets. The Pajaro (meaning "bird") River was named for the carcass of a large black bird (probably a California condor) discovered in an Indian village near the riverbank by Spanish explorers in 1769 (Gordon 1979, 166). While bird bones have been recovered from most local excavations, they are never as numerous as mammal or fish bones, indicating that birds were a relatively less important resource. Waterfowl seem to have been more important than other birds. They were caught with nets and decoys (Levy 1978, 491). Table 6.2 shows that, at a minimum, 27 species were exploited.

Fish In the last several years, archaeologists have used smaller mesh to sift deposits, producing much greater numbers of fish remains than in the past. These remains testify to the previously unrecognized importance of fish in prehistoric diets in the Elkhorn Slough area (table 6.3). Indeed, fish were probably the slough's most important resource, although the intensity of fishing apparently changed through time. (Discussions of Elkhorn Slough prehistoric fisheries can be found in Gobalet 1990, 1993; and Gobalet and Jones 1995.) Implements used to catch fish included dip nets, basketry fish traps, nets, and hook and line. Nets are represented in the archaeological record by notched stone sinkers, used to weight the nets down (fig. 6.8). Hook-and-line line fishing is represented archaeologically by fishhooks made from mussel, abalone, or chiton shell (fig. 6.9).

Shellfish While some historic accounts refer to natives collecting rocky coast shellfish in the Monterey Peninsula/ Mission San Carlos area, there are no historic descriptions of clams or other estuarine shellfish being collected at Elkhorn Slough. Nonetheless, mollusks are abundant in local midden sites (table 6.4).

All of the clams, cockles, the bay mussel, oysters, and the moon snail were collected from the slough for food. Pismo clams came from exposed beaches and, along with boring clam and gumboot chiton shells, are uncommon in local sites. Barnacles and unicorn snails were probably riders that arrived in the middens attached

Table 6.3. Fish species found in Elkhorn Slough region archaeological sites (CA-MNT-228, -229, and -234).

Taxon	Common name
Freshwater	
Archoplites interruptus	Sacramento perch
Catostomus occidentalis	Sacramento sucker
Cyprinidae fam.	minnows
Gila crassicauda	thicktail chub
Hysterocarpus traskii	tule perch
Lavinia exilicauda	hitch
Orthodon microlepidotus	Sacramento blackfish
Ptychocheilus grandis	Sacramento squawfish
Euryhaline	
Acipenser spp.	sturgeon, white or green
Atherinopsidae fam.	silversides
Atherinops affinis	topsmelt
Chitonotus pugetensis	roughback sculpin
Clupeidae fam.	herring and sardine
Clupea pallasi	Pacific herring
Cottidae fam.	sculpins
Cymatogaster aggregata	shiner perch
Embiotocidae fam.	surfperches
Gasterosteus aculeatus	threespine stickleback
Gobiidae fam.	gobies
Hyperprosopon argenteum	walleye surfperch
Leptocottus armatus	Pacific staghorn sculpin
Oncorhynchus mykiss	steelhead
Platichthys stellatus	starry flounder
Marine	
Amphistichus spp.	barred or calico surfperch
Amphistichus rhodoterus	redtail surfperch
Atherinopsis californiensis	jacksmelt
Atractoscion nobilis	white seabass
Carcharhinidae fam.	requiem sharks
Chondrichthyes	cartilaginous fishes
Citharichthys sordidus	Pacific sanddab
C. stigmaeus	speckled sanddab
Clinidae fam.	clinids
Elasmobranchii	sharks and rays
Embiotoca spp.	black surfperch or striped surfperch
Engraulis mordax	northern anchovy
Leuresthes tenius	California grunion
Myliobatis californica	bat ray
Paralichthys californicus	California halibut
Pleuronectidae or Bothidae fam.	righteyed or lefteyed flounder
Parophrys vetulus	English sole
Pleuronichthys verticalis	hornyhead turbot
Porichthys notatus	plainfin midshipman
Rajidae fam.	rays
Rhacochilus toxotes	rubberlip surfperch
R. vacca	pile surfperch
Sardinops sagax	Pacific sardine
Sebastes spp.	rockfish

to other species. The California horn snail is not known from Elkhorn Slough today, but its occurrence in middens indicates its former presence and subsequent extirpation. Small though this species is, it seems to have served as a food resource.

The purple olive was used for shell beads, although there is little evidence of beadmaking at Elkhorn Slough; more beads seem to have been produced on the Monterey Peninsula. Abalone shells collected on the rocky shores of either Santa Cruz or Monterey County show up in middens at Elkhorn Slough as ornaments. Washington, Pismo, and other clams were also used for beads in prehistoric California, but there is as yet no evidence that beads from these species were produced at Elkhorn Slough.

Plant Foods Plant foods are poorly represented in the archaeological record. Charred seed and nut fragments were recovered from sites near the mouth of Elkhorn Slough (CA-MNT-229 and -234), but problems with stratigraphic mixing (mixing of recent and prehistoric soils in the same levels in the deposit) make it impossible to distinguish aboriginally used items from natural pieces. Remains from a discrete hearthlike feature on Moro Cojo Slough (at CA-MNT-1765) represent the only known macrobotanical (in this case, charred seeds and nuts) data from the area that clearly reflects aboriginal use.

Fortunately, the Costanoan's use of plants is well documented in the ethnohistoric record. Bocek (1984) presents a summary of ethnobotanical information obtained by John Peabody Harrington from Rumsen and Mutsun descendants in the 1920s and 1930s. Harrington described use of 157 plants, of which 63 were collected for food and 84 were used for other

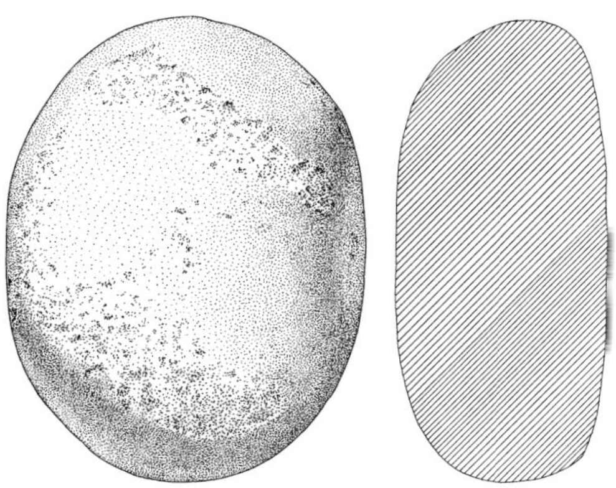

98-1-2770

Figure 6.10. Handstone from CA-MNT-234. These hand-sized shaped stones were used in conjunction with larger, flat stone (milling slabs) to grind seeds.

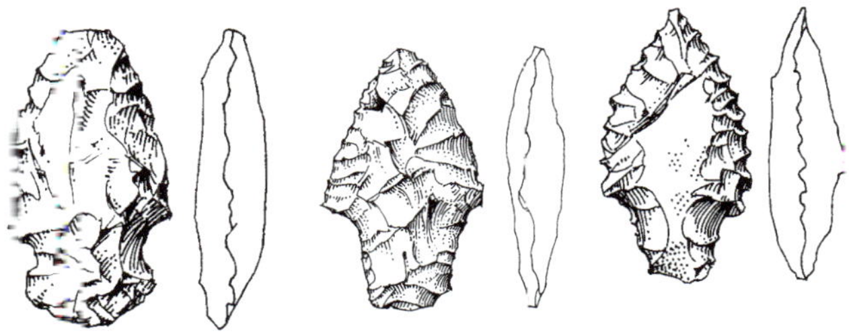

98-1-2655 98-1-258 98-1-1844

a. b. c.

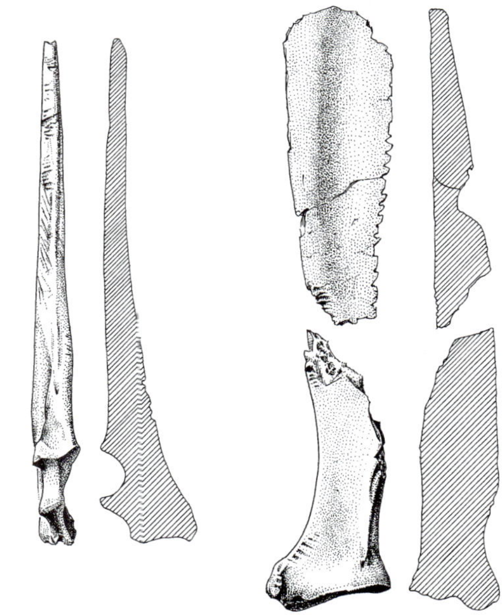

98-1-1391 98-1-2801

d. e.

Figure 6.11a. Side-notched projectile point from CA-MNT-234.

Figure 6.11b. Rossi-square-stemmed projectile point from CA-MNT-234.

Figure 6.11c. Contracting-stemmed projectile point from CA-MNT-234.

Figure 6.11d. Bone awl from CA-MNT-234.

Figure 6.11e. Serrated bone tool from CA-MNT-234.

purposes (e.g., fuel, cordage, construction materials, containers, clothing, tools, and musical instruments). Species found in and around the Elkhorn Slough area are summarized in table A-6.2.

Controlled Burning

Historic accounts indicate quite clearly that the Costanoan, like many other Native peoples around the globe, intentionally burned vegetation to enhance the growth of specific plants, aid in their collection, and make capturing game easier (see Gordon 1979, 40–48). Accounts from as early as 1769 note the extensive burning of grasslands after seeds had been harvested to encourage the growth of herbs and to make it easier to catch rabbits. Burning seems to have been done by women, since they were most commonly involved with acquisition of vegetable foods, although one account recorded by Pedro Fages suggests that men set fires to drive pronghorn near the modern-day community of Chualar. Gordon (1979, 47) believes that the long history of burning strongly influenced the character and distribution of vegetation in the Monterey Bay area, creating a more open, grassland-dominated landscape with less brush than would have been present had burning not been practiced.

Archaeology of Elkhorn Slough

Although the Monterey Bay area was generally passed over during the formative years of California archaeology (1900–1950), it has been studied more extensively in the last two decades. (For the history of archaeological research in the Elkhorn Slough region, see appendix 6.2.) This section describes what that archaeological work tells us about the slough's culture history, i.e., the basic history of local cultures represented by the distribution of artifact types through time

Table 6.4. Shellfish species found in Elkhorn Slough midden sites.

Taxon	Common name
Acanthina spirata	unicorn snail
Balanus spp.	barnacle
Cancer spp.	crab
Cerithidea californica	California horn snail
Clinocardium nuttalli	basket cockle
Cryptochiton stelleri	gumboot chiton
Haliotis cracherodii	black abalone
H. rufescens	red abalone
Macoma nasuta	bent-nosed clam
M. secta	sand clam
Mytilus californianus	sea mussel
M. trossulus	bay mussel
Nassarius fossatus	——
Olivella biplicata	purple olive
Ostrea lurida	oyster
Polinices lewissi	moon snail
Protothaca staminea	rock cockle
Saxidomus nuttalli	Washington clam
Tivela stultorum	Pismo clam
Tresus nuttalli	gaper clam
Zirfea pilsbry	boring clam

and space; about the natives' diet and settlement patterns through time; and about the movement of linguistic groups in and out of the region.

California's ethnographic record indicates that a tremendous number and variety of cultures were present in the state at the time of historic contact. Distinguishing between these cultures in the meager material record is difficult, but one of the assumptions underlying archaeological study of the past is that societies differed in terms of artifacts, diets, house types, and burial practices. Studies of present-day hunter-gatherers have corroborated links between artifact style and ethnicity (see Weisner 1982), for example, although the relationships are complex. In California as elsewhere, archaeologists classify artifacts into stylistic types and then look for types that consistently co-occur either with each other or with types of built features or mortuary practices. Clusters of artifact types, and other material traits that commonly occur together, are referred to alternately as archaeological complexes, cultures, or patterns. While these archaeological phenomena may or may not reflect cultures that were recognized by peoples themselves, they are a tool for classifying and tracking variation and patterning in the archaeological record.

Culture History

Information from Elkhorn Slough has been pivotal in developing a meaningful chronological sequence of artifact types for the Monterey Bay area, owing to the large, well-dated assemblage recovered near what is now the mouth of Elkhorn Slough (CA-MNT-229), known as the Vierra Site. Nonetheless, the Vierra Phase, defined at this site and dating to the Middle Period (see table A-6.4), remains the only well-defined phase expressed at Elkhorn Slough; other phases of Monterey Bay prehistory are so far represented only in the Monterey Peninsula area. Although testing results from Elkhorn Slough sites other than the Vierra Site show occupation before and after the Middle Period, artifact assemblages are poor, in part because of small excavations. For this reason, the following discussion of culture history draws from the entire Monterey Bay area, not just Elkhorn Slough.

Occupation of the Monterey Bay area can be divided into seven periods, outlined in table A-6.4. For the Monterey Bay area as a whole, human occupation is thought to extend back to the Pleistocene-Holocene interface (10,000 years B.P.), but formal phases are established only for occupations postdating 3500 B.C. The earliest habitation is represented in Scotts Valley (CA-SCR-177 [Cartier 1989, 1993b]); at Bennett Slough (CA-MNT-228 [T. Jones et al. 1994]); at the mouth of Elkhorn Slough (CA-MNT-229 [T. Jones and Jones 1992]); at a site along the old Salinas River channel (CA-MNT-234 [Breschini and Haversat 1995]), and possibly at sites near Gilroy (CA-SCL-178 and CA-SCL-119 [Hildebrandt and Mikkelsen 1993]). Samples from the Scotts Valley site have been dated to as far back as 11,500 B.C. (Cartier 1989, 1993b; Breschini and Haversat 1991b), but there are significant questions about the cultural origin of the samples that produced the earliest dates (T. Jones 1993, 19), and it is impossible to define the actual span of the site's use due to this uncertainty. The oldest dates from this site were obtained from samples of charcoal, but it is uncertain whether the charcoal is the result of fires ignited by humans or natural fires. The artifact assemblage found at Scotts Valley was dominated by milling slabs, handstones, and crude core tools (simple stone tools used for chopping and scraping) (fig. 6.6). This assemblage is typical of the California Milling Stone Culture, which was previously thought to be restricted to southern California but is now well represented in the north (Fitzgerald and Jones 1999). A Milling Stone Culture component was recently dated as far back as 10,300 years B.P. on the San Luis Obispo coast (Fitzgerald 1998); this is presently the oldest archaeological culture represented in central coastal California. At Elkhorn Slough, the Milling Stone Culture is best represented at CA-MNT-234, which produced a number of handstones (fig. 6.10) and a large side-notched projectile point (fig. 6.11a)—a type now recognized as one of the oldest in western North America. Crude core tools from CA-MNT-228 also represent the Milling Stone Culture.

The Early Period in Monterey Bay Area prehistory (3500–600 B.C.) is best represented by findings from a site near Pacific Grove (CA-MNT-391). Radiocarbon results firmly date use of this site between 3000–300 B.C. Artifacts include thick rectangular Olivella beads and Haliotis square beads; contracting-stemmed and Rossi square-stemmed points (fig. 6.11b) (all representing pre–bow and arrow technology—either spears or darts); mortars and pestles; and handstones and milling slabs (used to process plant foods) (Cartier 1993a). The assemblages from this location and nearby CA-MNT-387 (the Rossi Site) mark a cultural period known as the Saunders Phase in the Monterey Bay area. Two other sites on the Monterey Peninsula (CA-MNT-108 [Breschini and Haversat 1989a, 1992a] and CA-MNT-170 [Breschini and Haversat 198

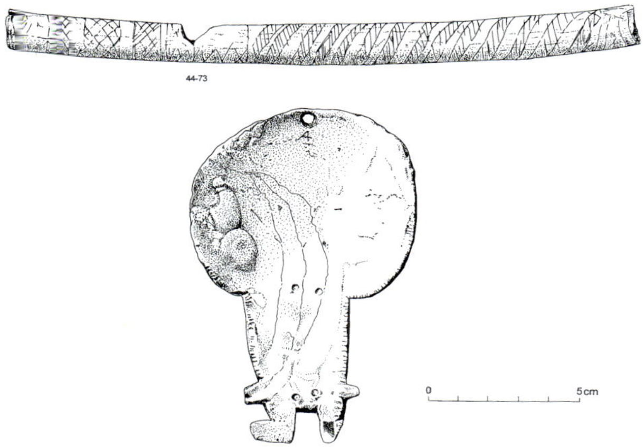

Figure 6.12. Bird bone whistle (44-73) and "banjo" (SCR-44-1) abalone pendant from CA-SCR-44 at Watsonville (redrawn from drawing by Anna L. Runnings in Breschini and Haversat 2000). These artifacts show striking resemblance to finds from the San Francisco Bay area and may distinguish Ohlone presence in the region in contrast with earlier cultures.

Dietz 1991]) and a site near Davenport in Santa Cruz County (CA-SCR-7 [Jones and Hildebrandt 1990]) harbor Early Period materials, but the samples available from them are more limited. The Early Period is not well represented in the central Monterey Bay area owing to the interval between 3000 and 2000 B.C. when Elkhorn Slough was abandoned due to an intrusion of freshwater. The most substantial Early Period occupation was identified at CA-MNT-234, which produced contracting-stemmed and square-stemmed points (fig. 6.11b). Early Period assemblages, marked by stemmed points and bowl mortars, are significantly different from those of the preceding period and are often attributed to the Hunting Culture, which was present throughout central coastal California.

In contrast to the Early Period, the Middle Period (600 B.C.–A.D. 1000) is well represented at Elkhorn Slough, with substantial remains found at CA-MNT-228, -229, -234, and -1570. The Vierra Site (CA-MNT-229), at the mouth of Elkhorn Slough, although also occupied earlier, produced the greatest array of Middle Period implements and therefore was used to define a Vierra Phase in Monterey Bay area prehistory. Dated to approximately 1000 B.C.–A.D. 1000 via an extensive series of radiocarbon assays and obsidian hydration readings (Dietz, Hildebrandt, and Jones 1988; T. Jones and Jones 1992), the assemblage from this component is marked by stemmed projectile points (fig. 6.11c), saucer-shaped Olivella beads, bowl mortars, well-made cylindrical pestles, bone tubes, whistles, awls (fig. 6.11d), and serrated bone tools (fig. 6.11e). T. Jones (1993) ascribes Vierra to the Hunting Culture, but this is a point of contention among Monterey Bay archaeologists.

The Late Period has been difficult to define in the Monterey Bay area because the most common late sites were coastal processing stations that have produced few formal artifacts (Dietz and Jackson 1981; Breschini and Haversat 1991a). Recent findings from two sites at Rancho San Carlos in the upper Carmel Valley (CA-MNT-1485/H and -1486/H) helped to fill this void (Breschini and Haversat 1992b). Both of these residential sites were occupied during the Late Period and into the Protohistoric Period, but the sample size was greatest at CA-MNT-1486/H. Calibrated and corrected radiocarbon assays date occupation of this site to between approximately A.D. 1050 and 1650. Beads, obsidian hydration, and projectile points confirm that it was occupied between the Middle/Late Transition, Late, and early Protohistoric Periods, although hydration results suggest the possibility of a Middle Period component only weakly represented in the excavation sample. The Middle/Late Transition assemblage from CA-MNT-1486/H shows typical co-occurrence of both earlier large-stemmed projectile points with newer, smaller leaf-shaped types associated with the bow and arrow. During the Late Period, the larger types were replaced completely by small arrow points, mostly representing the well-known Desert side-notched type. The Late Period assemblages also included mortars, pestles, handstones, earspools, and a single unperforated plummet-shaped charmstone (Breschini and Haversat 1992b, 84–106). The Late materials from CA-MNT-1486/H represent a proposed Rancho San Carlos Phase in Monterey Bay prehistory. The Late Period is also well represented at a site in the Santa Cruz Mountains (CA-SCR-20 [Roop 1976]). At Elkhorn Slough, Late Period occupation was indicated by radiocarbon dates at CA-MNT-234, but the only artifact associated with the occupation was a Desert side-notched projectile point.

Recent findings from CA-SCR-44, located adjacent to College Lake in Watsonville (Breschini and Haversat 2000), are informative for the Late Period. Unlike most sites in the area, CA-SCR-44 produced radiocarbon evidence for occupation during both the Middle and Late Periods, although the exact span of the site's use is unclear due to the availability of only three radiocarbon dates. Corrected and

Figure 6.13. Summary of settlement history at Elkhorn Slough showing abandonment between 3500 and 2000 B.C., and settlement disruption during the Medieval Climatic Anomaly A.D. 850-1350.

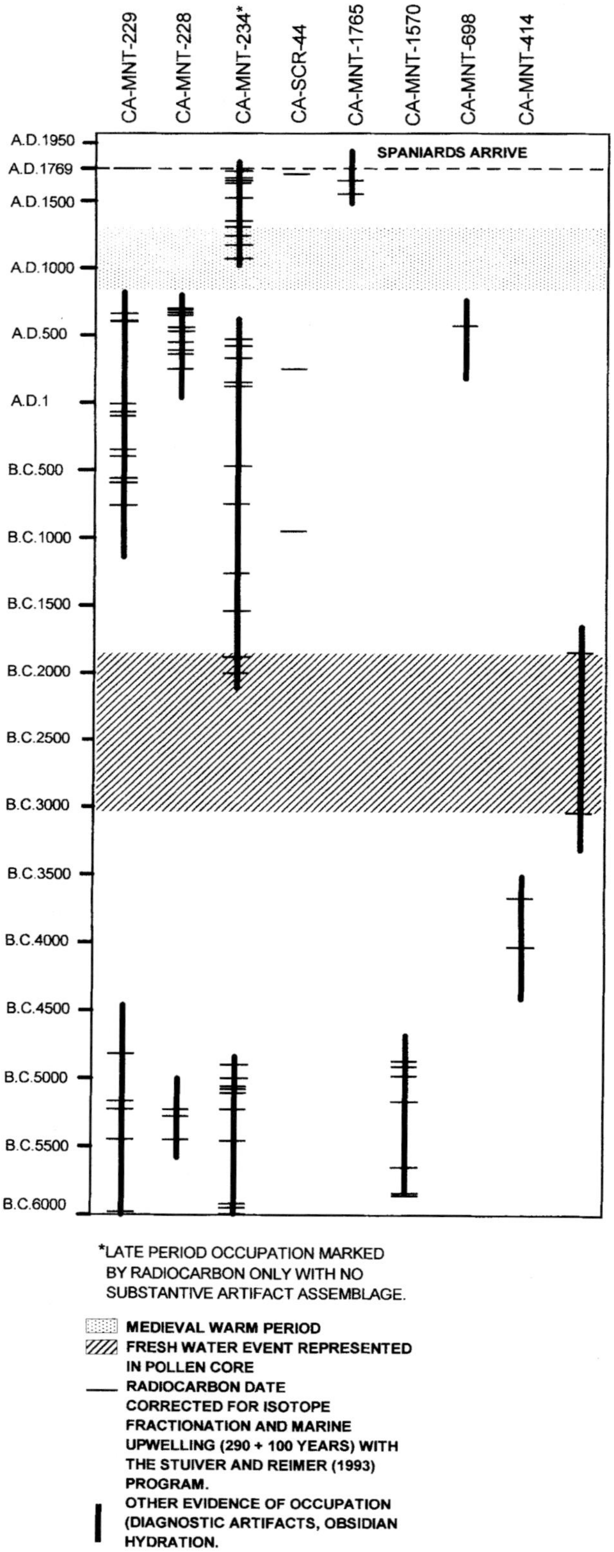

calibrated, these dates indicate site use ca. 920 B.C., A.D 260 and again ca. A.D. 1640. Whether this was an ongoing occupation over nearly 3,000 years (a continuity in single-site use as yet unseen in the area) or an intermittent one with a gap during the Medieval Climatic Anomaly is unclear. Some of the 33 burials uncovered from this location included distinctive bone whistles and abalone shell pendants (fig. 6.12).

Diet and Settlement Patterns

Evidence from archaeological sites has helped researchers develop an idea of Monterey Bay area natives' diets, seasonal movements, and settlement patterns beginning with the Milling-stone Period. The condition of Elkhorn Slough's fishery appears to have played a major role in natives' use of sites near the slough; degradation of this fishery and other marine resources brought on by closure of the slough mouth and other environmental changes may have contributed to site abandonment.

A growing body of data indicates that the summer fishery was Elkhorn Slough's most important prehistoric resource. During the summer spawning migrants enter the system, and it is then that the slough harbors its greatest numbers and variety of fishes (Yoklavich et al. 1991). Based on analysis of prolific fish remains and *otoliths* from sites near Bennett Slough and along the old Salinas River channel (CA-MNT-228 and -234), we know that natives occupied the slough during the seasonal peak in fish populations. A similar occupation pattern was identified at Morro Bay in San Luis Obispo County (T. Jones et al. 1994). The stone tool assemblages from sites at Bennett Slough and the mouth of Elkhorn Slough (CA-MNT-228 and -229) lack obvious fishing paraphernalia (e.g., fishhooks), but natives probably used traps, nets, and baskets, which do not preserve well. Shell hooks were most commonly used to fish from rocky shores (Strudwick 1986), although nine hook fragments recently reported from the site along the old Salinas River channel (CA-MNT-234 [Breschini and Haversat 1995, 5[illegible]]) indicate that they were also used in protected settings.

Site location patterns reflect the importance of fisheries and other wetland resources, beginning with the region's earliest inhabitants. The oldest archaeological sites in the central coast region are concentrated at lakes in Santa Clara and Santa Cruz Counties (CA-SCL-119/SBN-24/H, CA-SCL-178, CA-SCR-177) and at estuaries (CA-MNT-228, -229, -234, -1570; but see Hildebrandt 1997 for an opposing view). Both settings were probably occupied for some part of the year as part of a highly mobile settlement system in which natives harvested the most desirable resources available in each setting. At a minimum, natives occupied Elkhorn Slough in the summer, hunting and harvesting fish, shellfish, some marine mammals, rabbits, and tule elk. At lakes, they exploited important commodities such as cattail pollen, bullrush seeds, freshwater shellfish, and tule elk (found at CA-SCL-178 [Hildebrandt 1983]). Millingstone Period components at sites on or near Elkhorn Slough (CA-MNT-228, -234, and -1570) indicate that fish were the most important dietary items, followed by shellfish.

Food gathering along the central California coast seems to have intensified at mid-Holocene (the onset of the Early Period, 3500 B.C.) with people becoming more sedentary and focusing on smaller, more labor-intensive resources. These developments have been attributed to population growth and circumscription, as access to some resources was limited by the simple presence of greater numbers of people on the landscape (T. Jones 1995, 1996; T. Jones and Waugh 1995, 1997). However, mid-Holocene economic changes on the central coast may also reflect coastward migration of interior peoples during the mid-Holocene Warm Period (Mikkelsen, Hildebrandt, and Jones 1999). Many settlements appear on rocky shores of Monterey and Santa Cruz Counties at this time, fish were exploited more heavily, and the mortar and pestle were introduced, supplementing assemblages that previously included only slabs and handstones for grinding. The mortars and pestles suggest an increased emphasis on foods that required intensive processing. At Elkhorn Slough, site abandonments during much of the Early Period, which coincided with the intrusion of freshwater species represented in the pollen record (fig. 6.13), seem to reflect deterioration of marine habitats and a subsequent decline in the fishery and other resources between 4000 and 2000 B.C., caused by closure of the slough's outlet to the sea. Findings from several sites in the area suggest that both Elkhorn Slough and the Salinas River estuary were temporarily abandoned during this interval, a change that may have led to greater settlement of the open coast.

By the Middle Period, natives were again living at Elkhorn Slough. Fisheries were apparently reestablished, as fish remains feature prominently in excavation findings. Middle Period diets and settlement practices seem best interpreted as outgrowths of a fishing focus that began during the Early Period. Fish are a resource that can yield increased returns as more sophisticated, labor-intensive technologies are applied. Otoliths from sites at Bennett Slough and the old Salinas River channel (CA-MNT-228 and -234) suggest that Elkhorn Slough fisheries were exploited from spring through fall; settlement does not seem to reflect year-round occupation, as the sites may not have been used in winter.

Both marine and terrestrial mammals also made up a significant part of the diet during the Middle Period. Mammal remains from the Middle Period component at Bennett Slough (CA-MNT-228) feature deer, rabbit, sea otter, and harbor seal (table A-6.5) in proportions similar to those found at the Vierra Site near the mouth of Elkhorn Slough (CA-MNT-229 [Dietz, Hildebrandt, and Jones 1988]). Findings from the old Salinas River channel site (CA-MNT-234) show a similar emphasis on rabbits and sea otters, although northern fur seals (*Callorhinus ursinus*) were also important. The dominance of deer, rabbits, and otters is consistent with Middle Period and Middle/Late Transition assemblages from the outer coast of Monterey County, including CA-MNT-63 (table A-6.5).

Many sites around Elkhorn Slough were apparently abandoned toward the end of the Middle Period or during the Middle/Late Transition Period (ca. A.D. 1250), supporting a regional trend of site abandonment about this time that has been attributed to the Medieval Climatic Anomaly (T. Jones et al. 1999). Few sites in the Elkhorn Slough area show signs of use during the Late Period, and none show continual occupation from the late Middle through Late Period. A site on Moro Cojo Slough (CA-MNT-1765) was occupied during the Late Period, but only beginning in approximately A.D. 1530 (Fitzgerald et al. 1995). CA-MNT-234 was occupied from A.D. 1200 to 1730, following a 700-year hiatus between A.D. 470 and 1180. In the southern Santa Clara Valley, to the east, four sites on the western edge of the valley (CA-SCL-308/H, -577/H, -639, and -698) were also abandoned during the Middle/Late Transition Period. Only one site, at San Felipe Lake in Santa Clara County (CA-SCL-119/SBN-24/H), showed continued occupation into the Late Period.

The Middle/Late Transition in the Monterey Bay area has in recent years been repeatedly interpreted as a consequence of continued population growth as people turned to acorns and inland habitats (Hildebrandt and Mikkelsen 1993; Hildebrandt and Jones 1992; Hylkema 1991; T. Jones 1992) to meet food requirements that could not be met with fish and other marine foods. The so-called "collector" economy of the Late Period,

(Above) Archaeologist (left to right) Nina Ilic, Peter Johnson, and Deborah Jones sifting soils during excavation of CA-MNT-229 in 1985. (Right) Archaeologist Randall Milliken shoveling at CA-MNT-229 in 1985. Photo credit: Dian Duchin.

marked by abalone processing stations on the Monterey Peninsula, is interpreted as a period when natives settled inland and returned to the coast only for such specific tasks as harvesting abalone. There is growing evidence, however, that Late Period lifeways reflect a departure from a previous trajectory of intensification, and that the Middle/Late Transition marks a significant disruption of local settlement systems, not simply an adaptive outgrowth from previous food gathering practices. The number of sites abandoned is simply too large, and many Late Period middens do not show evidence for increased numbers of people or greater use of labor-intensive foods.

Inland residential sites indeed seem to have been favored during the Late Period, as supported by evidence from CA-MNT-1765 at Moro Cojo Slough, although the recent findings from CA-MNT-234 show some coastal presence as well—albeit one so minimal that it resulted in very few artifacts. Other Late Period inland middens include a site on the UCSC campus (CA-SCR-160 [Edwards, Smith, and Macko 1991]) and two sites in Big Sur (CA-MNT-759/H and -1236). While terrestrially focused, these deposits are neither larger nor do they suggest more permanent occupation than Middle Period sites. Their appearance may be more related to environmental problems during the Middle/Late Transition, specifically widespread drought during the Medieval Climatic Anomaly (fig. 6.13) (Graumlich 1993; Stine 1994; T. Jones and Kennett 1999; T. Jones et al. 1999) that seems to have precipitated abandonment of many coastal sites and initial settlement of others. Most of the inland sites were probably inhabited during the Little Ice Age when the climate ameliorated. Elkhorn Slough may have again lost its outlet to the sea during this interval, stimulating the abandonment of sites along Bennett Slough and at the mouth of Elkhorn Slough (CA-MNT-228 and -229).

Ethnolinguistic Group Migration

Some researchers have proposed that the Monterey Bay area's original inhabitants spoke a language in the Hokan grouping, a linguistic family that includes the ethnographic Pomo, Salinan, and Esselen. Sometime in the past these Hokan speakers are thought to have been replaced by those speaking Costanoan—the language spoken at the time of historic contact (Breschini 1983; Breschini and Haversat 1980,; Dietz and Jackson 1981; Dietz, Hildebrandt, and Jones 1988; and Moratto 1984) Others, however, argue that the original inhabitants adapted the Costanoan language (Gerow with Force 1968;. Dietz, Hildebrandt, and Jones (1988) suggested that a population replacement took place at the end of the Middle Period, when the Vierra Site at the mouth of Elkhorn Slough (CA-MNT-229) was abandoned. Hildebrandt and Mikkelsen (1993 suggest that none of the changes in tool styles or subsistence at the end of the Middle Period were extreme enough to suggest population replacement. Instead, they proposed a merger between Hokan and Costanoan speakers, consistent with a theory originally proposed by Gerow with Force (1968).

Growing evidence for environmental problems (Graumlich 1993; Stine 1994; T. Jones and Kennett 1999) and significant disruption of local settlement and exchange systems in the Monterey Bay area during the Middle/Late Transition could reflect population movements precipitated by drought, resource scarcity, population decline, and subsequent in-migration of peoples from the north and the interior. In the Gilroy and Elkhorn Slough areas, no fewer than seven sites were abandoned during this period, in apparent response to the recession of Lake San Felipe and environmental change at Elkhorn Slough. Tool assemblages show significant changes, as large stemmed points and bowl mortars were replaced by Desert side-notched points and bedrock mortars. Exchange networks apparently deteriorated, as indicated by the regional *obsidian profile*. Small, isolated middens of the Late Period further seem inconsistent with the predictions of economic intensification, which anticipate larger sites, increased population, and greater archaeological visibility. Breschini and Haversat (1995), however, endorse an earlier date for population replacement in the Monterey Bay area, ca. 1000–600 B.C. at the Early/Middle Transition. Jones (1995) argues against major population movement at this time in light of the continuity in tool assemblages from the Early through Middle Periods, which he classifies as the Hunting Culture.

Recent findings from CA-SCR-44 at Watsonvillle (Breschini and Haversat 2000) are important with respect to alternative dates for Ohlone entrance into the region. The distinctive bone whistles and abalone shell pendants recovered from burials at this site are the first in the Monterey Bay District to show unquestionable stylistic similarities to the San Francisco Bay area and clearly reflect local Ohlone presence. One of the abalone ornaments representing the so-called "banjo" style (see fig. 6.12) produced a calibrated radiocarbon date of A.D. 1640. The mortuary assemblage contrasts with findings from Hunting Culture cemeteries (e.g., CA-MNT-229 and -391) and may reflect Ohlone entrance into the Monterey Bay area, most likely during the Late Period. Breschini and Haversat (2000), however, feel that CA-SCR-44 was continually used during the Middle and Late Periods, suggesting Ohlone entrance into the area earlier, ca. 900 B.C.

Management Issues and Research Recommendations

Despite a significant amount of archaeological research over the last two decades, the prehistory of Elkhorn Slough is still poorly documented, and issues remaining for further research are numerous. The following is a brief discussion of topics for future studies at the slough.

Paleoenvironmental History

The paleoenvironmental history of the central Monterey Bay area and its relationship to the southern Santa Clara Valley and San Francisco Bay remain poorly understood. The present-day landscape of this area is replete with features that suggest alternative hydrographic configurations in the past, including McClusky Slough, the Monterey Submarine Canyon, and Elkhorn Slough itself. In the interior of the southern Santa Clara Valley, the Laguna Seca and Llano de Tequiquista, both named by the Spanish, apparently represent bodies of water that were more extensive in the past. If the Elkhorn Valley was at one time the drainage outlet for the San Joaquin and Sacramento Rivers, the chronology of the shift of that outlet to its current location at the Golden Gate is entirely undocumented. Some geologists are hesitant to attribute the Monterey Submarine Canyon exclusively to the earlier course of a very large drainage (see Starke and Howard 1968), but the absence of a major canyon at the Golden Gate and the relatively shallow bathymetry there suggest such a shift may have occurred, possibly as recently as the late Pleistocene/early Holocene, when human beings entered the area. The relatively shallow prehistoric record of the San Francisco Bay shell mounds could reflect the recentness of this shift and the later maturation of the bay's estuarine habitat relative to Elkhorn Slough. Mudflats suitable for estuarine mollusks were well established at Elkhorn Slough by 8,000 years ago, but archaeological clamshells are no more than 5,000 years old at San Francisco Bay. The remains of fishes endemic to the Sacramento and San Joaquin drainages in the archaeological sites at Elkhorn Slough (see Gobalet 1990, 1993) also suggest a former connection between the Pajaro drainage and San Francisco Bay, as does the following observation by Bayard Taylor, an early traveler to California (recorded between 1848 and 1869): "According to an Indian tradition, of comparatively modern origin, the waters of San Francisco Bay once communicated with the bay of Monterey by the valley of San Jose (and the Rio del Pajaro)" (Taylor 1949, 131).

To secure a regional hydrographic history sufficiently fine-grained to provide a context for the human behavior represented in the archaeological record, more paleoenvironmental research needs to be completed in and around Elkhorn Slough, particularly studies that track potential climatic/ habitat indicators (pollen, diatoms, radiolaria, isotopes, etc.) in the strata of wetland deposits.

Archaeological Inventory and Site Conservation

Only 10–15% of the acreage making up the lower Pajaro River, Salinas River, and Elkhorn Slough watersheds has been inventoried for archaeological sites; doubtless, many sites remain to be discovered. Review of records on file at the Northwest Information Center of the California Historical Resources Information System shows that impact-related archaeological surveys are being conducted on a regular basis, so that the threats of site destruction are not nearly as great as they were prior to the passage and implementation of the California Environmental Quality Act (CEQA). Nonetheless, it is likely that some local sites are suffering from erosion (common in coastal settings) and other impacts not covered by CEQA review. Because contemporary archaeological research recognizes that sites can be damaged as much by unnecessary or poorly thought out archaeological excavation as they can by construction or earth-moving projects, future research should concentrate on deteriorating and/or threatened sites. Among these are partially inundated deposits, occupied when sea level was lower.

It is difficult to evaluate patterns in the local surface archaeology because of incomplete available documentation. Much of the land in the Elkhorn Slough area is privately owned, and access to possible sites has always been a problem. Perhaps as partnerships develop between the private sector and the Elkhorn Slough National Estuarine Research Reserve, opportunities will develop for larger-scale survey.

Chronology and Culture History

While archaeological excavations have been completed at nearly a dozen sites in the vicinity of Elkhorn Slough, it cannot be overemphasized that meaningful data are available from no more than five of these, and the current picture of local prehistory is based on a relatively meager record. Only one phase in the Monterey Bay occupational sequence (Vierra, dating 1000 B.C.–A.D. 1000) is based on a component from Elkhorn Slough, and assemblages marking other periods are virtually undocumented. While radiocarbon dates testify to human habitation as early as 8,000 years ago, initial human colonization probably occurred two to three millennia earlier; however, sites testifying to such habitation remain to be discovered. More basic, preliminary research needs to be completed to flesh out the time/space grid of Elkhorn Slough archaeology and prehistory.

Native American Concerns

Intimately related to archaeology's conservation ethic are the concerns of Native Americans interested in preserving archaeological sites as the physical remnants of their cultural heritage. Local descendants of the native inhabitants of the Monterey Bay area have been active proponents of site preservation. While they commonly collaborate with local archaeologists, they are increasingly opposed to excavation not related to a proposed construction project or other impact. Excavation in areas containing human remains is particularly sensitive, and the Public Resources Code requires consultation with a local descendant (mediated by the county coroner and Native American Heritage Commission) when human remains are encountered.

Acknowledgments

Thanks to Gary Breschini and William Hildebrandt for providing insightful comments on earlier drafts of this paper, and graphic artist Rusty van Rossman for compiling all maps, charts, and figures. Thanks are also due to Martin Brown and Mark Silberstein for commissioning the contribution and shepherding it through to completion.

Appendix 6.1: Mission Records and Tribelet Locations

The first convert from the Castroville area was recorded in February 1782, twelve years after Mission San Carlos began to baptize natives. The delay in enrolling neophytes from more distant areas was due to the availability of converts from villages close to the mission in Carmel Valley. After conversion of those Indians was complete, the missionaries moved on to more far-flung villages, including Calenda Ruc.

Baptism and marriage records reported by Milliken (1988, 67) indicate that Calenda Ruc was located in the vicinity of the mouths of the Pajaro and Salinas Rivers. Baptism #856, recorded by Father Noriega at Mission San Carlos, originated "de la ranchería de Kalenda-Ruc en los esteros de la entrada al Mar del Río de Monterey" (from the village of Calenda Ruc in the sloughs at the mouth of the Monterey River; Milliken 1988, 67). As the river today known as the Salinas was referred to by the Spanish as the Monterey, the locational implications of this record are fairly clear. Another baptismal record (#1060) states that the convert was "de la Ranchería de Kalenda-Ruc en los esteros de la entrada de los ríos Monterey y el Pajaro" (from the village of Kalenda-ruc in the sloughs at the mouth of the Monterey and Pajaro rivers; Milliken 1988, 67). This passage is significant in its suggestion that the mouths of the Salinas and Pajaro Rivers were joined at the time of its writing. That such a confluence existed in the recent past is also indicated on an 1854 map of the central Monterey Bay, which depicts an old bed of the Pajaro River between Elkhorn Slough and the current mouth of the Pajaro (Dietz, Hildebrandt, and Jones 1988, 43). At that time, the Salinas River also emptied to the north of the present outlet of Elkhorn Slough. The map and the baptismal record suggest that sometime between 1782 and 1854 all three drainages emptied through a common outlet near the present mouth of the Pajaro River. Mission records referring to the mouth of the Monterey River, therefore, actually connote the area near what is now the mouth of the Pajaro. Yet another record from Mission San Carlos (Marriage #26) stated the couple's origins as follows: "Calenda Ruc cerca de las salinas como a cinco leguas del real Presidio siguente a la costa" (Calenda Ruc near the salt marshes about 5 leagues [22.5 kilometers, 14 miles] on the road from the presidio following the coast; Milliken 1988, 67). A 14-mile route from the presidio of Monterey along the coast winds up in the vicinity of Castroville; as the crow flies, however, it is nearly 14 miles from the presidio to Elkhorn Slough.

While some of these passages (e.g., baptism #856) suggest that Calenda Ruc was a village, later records clearly indicate that it was also a territory or tribelet, given that individual villages are described as being within Calenda Ruc. Baptism #1424 from Mission San Carlos, for example, describes a neophyte as "de la Ranchería de Lucuyusta en Kalenda-Ruc" (from the village of Locuyusta in Calenda Ruc; Milliken 1988, 67). Other villages attributed to the Calenda Ruc territory are Tiuvta, Mustac, and Culul. Most frequently mentioned were Tiuvta and Locuyusta, but unfortunately, the locations of these settlements are not indicated in any records. Another important notation accompanied baptism record #2174: "de la ranchería de Santa Cruz de Tiuvta" (from the village of Tiuvta of Santa Cruz; Milliken 1988, 68). Association of this place-name with Santa Cruz suggests that the village was located in the vicinity of Watsonville, because the sphere of influence of Mission Santa Cruz reached its southern limit in that area (Milliken 1988, 68). Recent findings from archaeological site CA-SCR-44 at Watsonville (Breschini and Haversat 2000) suggest it could represent the village of Tiuvta.

Overall, the Mission San Carlos records indicate, in a fairly consistent and coherent fashion, the presence of a tribelet community, Calenda Ruc, among the sloughs and salt marshes in the central portion of Monterey Bay, in the vicinity of the mouth(s) of Elkhorn Slough, Pajaro River, and Salinas (Monterey) River. At least seven villages are associated with that area in mission records: Calenda Ruc, Mustac, Culul, Locuyusta, Tusquesta, Chalicta, and Tiuvta (Milliken 1988, 67–68).

Complicating this portrait are references to another place-name, Guachirron, which seems to apply to much of the same general area. C. King (1974) felt this name represented a tribelet, distinct from Calenda Ruc, that occupied the Castroville area. Several historical references indeed link this name with Castroville: Pinart, collecting place-name and linguistic information in the 1870s, was told that the "pueblo of Guacaron" was situated "near the present site of Castroville," while Alexander Taylor reported that the "Watcharanuka lived on the east side of the Salinas River" (Milliken 1988, 68). Only nine baptisms recorded at Mission San Carlos between 1782 and 1806 refer to the Guachirron, and none of these give locational information.

The name Guachirron appears much more frequently in the records at Mission San Juan Bautista, where a distinction was made between *Guachir-rones de la playa* (Guachirrones of the beach), and *Guachirrones de las montañas* (Guachirrones of the mountains). Milliken (1988, 71) believes the latter were synonymous with the Pagsin of the Hollister area. The former, however, were a coastal group, members of which were consistently classified as distinct from residents of Calenda Ruc. Baptisms for both groups were numerous, and the records provided many village names not mentioned at Mission San Carlos. For Calenda Ruc, these include Teharun, Hueneren, Tiguita, and los Corralitos (Milliken 1988, 71). The latter is particularly important as a reference to the historic/contemporary community of Corralitos, 8 kilometers (5 mi) north of Watsonville. Milliken (1988, 72) believes that Tiguita is probably synonymous with Tusquesta, or possibly with Tiuvta. For Guachirron, Milliken mentions the village of Juarto.

A final indication of the presence of two discrete tribelets in the central Monterey Bay area is found in Father Amoroso's response to the "*interrogatorio*" of 1811. He stated that seven nations of Indians were present at Mission San Carlos, among whom were the Guachirron and Calenda Ruc (Kroeber 1908, 20).

Appendix 6.2. History of Archaeological Research in the Elkhorn Slough Region

In general, much of the archaeological research in the Elkhorn Slough region completed before 1980 was limited to the basic location and description of resources. The dating of sites and artifacts needed to address more complex research issues was lacking in most of the early survey-oriented studies and pre-1980 excavation reports. In contrast, research accomplished in the last twenty years includes several projects (e.g., Dietz and Jackson 1981; Dietz, Hildebrandt, and Jones 1988; Breschini and Haversat 1989a) that made meaningful contributions to the understanding of local prehistory. The pace of research increased in the 1990s, as a series of larger, more substantive, longer-term studies were completed by Breschini and Haversat (1995), Cartier (1993a, 1993b), Hylkema (1991), Hildebrandt and Mikkelsen (1993), T. Jones (1993), and T. Jones et al. (1996). More detailed histories of archaeological investigation in the region are available (e.g., Breschini, Haversat, and Hampson 1983; Dietz 1987; Dietz and Jackson 1981; Dietz, Hildebrandt, and Jones 1988; Hildebrandt and Mikkelsen 1993; T. Jones et al. 1989; Moratto 1984; Patch and Jones 1984; Welch 1992) for those interested in specific aspects of local research.

Evaluating radiocarbon dates from the slough area has been an ongoing problem in interpreting local prehistory. Until the 1990s, the California archaeological community questioned the reliability of radiocarbon dates obtained from shell samples (see Erlandson 1994). This suspicion eventually proved unfounded, but it gave way to uncertainty over sample selection criteria and procedures for calibrating shell-derived dates. Dates obtained from multiple shell samples remain troublesome, if not fully untrustworthy, because of the partially mixed character of local sites and the likelihood that multiple shells come from different time periods (see Glassow 1996); their combination in single dates can bring about a deceptive averaging effect. All single-specimen radiocarbon dates obtained from sites in the Elkhorn Slough area are listed in table A-6.3. All dates referred to in the following discussion are calibrated calendric ages A.D./B.C.

For the purposes of integrating the Monterey Bay area into the broader fabric of California prehistory and archaeology, T. Jones (1993) established the Monterey Bay District as a discrete cultural historical province within the central California coastal region. The district encompasses the greater shoreline of Monterey Bay, from roughly Point Sur to Point Año Nuevo. District boundaries are not firm, nor do they correspond precisely to prehistoric sociopolitical boundaries. Instead, they generally correspond to the territory held by southern Costanoan speakers, and further represent a cohesive environment within which cultural patterns can be evaluated.

Elkhorn Slough and the other former wetlands associated with the lower courses of the Pajaro and Salinas Rivers are recognized as a discrete locality within the Monterey Bay District. Early surveys were completed in this area by E. Gifford (1913), Golomshtok (1922), Hill (1929), Wood (1930), and Fischer (1935). Arnold Pilling conducted extensive surveys in the Elkhorn Slough environs in the 1940s, assigning many sites their permanent trinomials (Lönnberg and Morris 1981, 132). Over the last twenty years, a series of small-scale archaeological surveys have been carried out in the Elkhorn Slough locality in compliance with the historic preservation mandates of the California Environmental Quality Act (CEQA) and section 106 of the Historic Preservation Act. A review of records on file at the Northwest Information Center of the California Historical Resources Information System at Sonoma State University shows that as of 1996, approximately 10–15% of the acreage in the lower Pajaro River, Salinas River, and Elkhorn Slough watersheds had been systematically examined by archaeologists. Based on this survey, and the earlier work done by University of California archaeologists, we can determine that 99 prehistoric archaeological sites have been identified. Larger surveys include those completed for the Elkhorn Slough National Estuarine Research Reserve, where John King (1981) recorded 12 sites on 980 acres, and the Rubis Ranch, where Dietz and Jackson recorded 12 sites on 600 acres in 1977. Five prehistoric sites were identified during a survey for a proposed realignment of Highway 1 in 1990 (Snethkamp and York 1990). Overall, with so few acres systematically examined, there can be little doubt that many sites remain to be discovered.

The first subsurface data from this area were shellfish column samples reported by Greengo (1951) from CA-MNT-229. Further substantive excavation work did not take place until 1974, when volunteers from the University of California and the Santa Cruz Archaeological Society conducted test excavations at CA-MNT-414 and -415 (reported in Gifford 1977 and Patch 1979). Later, Patch and Jones (1984) reported radiocarbon dates from CA-MNT-414 and -698, along with the results of shellfish analyses from CA-MNT-414, -415, and -698.

In 1976, Dietz and Jackson conducted a limited test program at CA-SCR-101, near the mouth of the Pajaro River. Although no chronometric data are available, the site yielded a diverse assemblage indicating use as a residential base. Artifacts included projectile points, a pestle, a hammerstone, flake tools, cores, Olivella beads, and a cache of seven notched stone net weights. The invertebrate fauna, while limited to residues retained in 1/4-inch (6-mm) mesh, was dominated by estuarine clam and cockle shells, indicating exploitation of a now-defunct Pajaro River estuary/embayment.

In 1979, Peak conducted limited testing in the vicinity of CA-MNT-228, and a modest test excavation was subsequently completed there by Spanne (1979), who later submitted shell samples for radiocarbon analysis, the results of which were reported by Breschini, Haversat, and Erlandson (1992).

CA-MNT-229, situated near the current mouth of Elkhorn Slough, was investigated by Dondero et al. (1984) and Dietz, Hildebrandt, and Jones (1986, 1988). These excavations are the most substantive yet completed in the slough area, and CA-MNT-229, the Vierra Site, is still the only deposit from which meaningful, well-dated artifact assemblages have been defined. In the original data recovery report, the site was thought to represent a single-component Middle Period occupation marking the Vierra Phase (see table A-6.4 for dating of cultural periods in the Monterey Bay area), but a later assessment of the site's chronometric data identified an early Holocene component at the base of the deposit dating to 6000–4000 B.C. (T. Jones and Jones 1992). Unfortunately, there is inter-component mixing at this site, and 12 of the original 28 radiocarbon dates were obtained from multiple shell samples.

Milliken et al. (1999) recently reported important findings from CA-MNT-234, the Moss Landing Hill Site, situated near the present-day Moss Landing Marine Laboratories. Their research followed earlier testing by Breschini and Haversat (1995). The two studies revealed a sequence of human occupations similar to other sites in the slough area, although an extensive radiocarbon dating program allowed for refinement of chronological patterns. Although some of the radiocarbon dates (e.g., those from multiple shell and bulk soil samples) were unreliable, 30 dates show three discrete intervals of occupation: 6010–5000 B.C., 200 B.C.–A.D. 470, and A.D. 1100–1700. A single glass bead indicates some ephemeral site use during the historical era. Fish dominated the faunal assemblage from all time periods. This is the first site to reveal substantial evidence for occupation during the Early Period at Elkhorn Slough, but it still shows the distinctive early Holocene gap in occupation that has been found at all other sites so far investigated at the slough. Like other sites, CA-MNT-234 was abandoned after 5000 B.C., and was not reoccupied until after the freshwater event in the slough's pollen profile. Other sites in the area were not reoccupied until ca. 1000 B.C., but the extensive research at CA-MNT-234 shows reoccupation 1,000 years earlier (fig. 6.13). The site shows another interval of abandonment between A.D. 470 and 1100. Unlike other sites so far investigated, CA-MNT-234 shows evidence of occupation during the Late Period, as the investigators obtained a series of dates between 1100 and 1700. This more recent use of the site, however, was decidedly different from that of previous occupations, in that it resulted in almost no formal artifacts. Use of the site seems to have been very infrequent and nonintensive after 1100, which is consistent with the widespread evidence of settlement change on the central coast at this time.

An interesting sidelight from CA-MNT-234 was the recovery of a large number of bones from northern fur seals (*Callorhinus ursinus*) (Burton 2000). Prior to the establishment of a small herd on San Miguel Island during the 1960s, this animal was known to breed only on islands along the coast of Alaska. During the non-breeding season (largely fall and winter), immature males and females migrate south, many traveling as far as central California, where they remain offshore unless they become sick or injured. In addition to an unusually high frequency of northern fur seal bones, Burton's (2000) analysis of the CA-MNT-234 materials also shows a significant number of pups among the seal remains. Jaw length measurements from these specimens, when compared to reference samples with known age of death, showed that some of the pups were less than three months old when they died, suggesting exploitation of a rookery somewhere close to the archaeological site (because pups don't swim effectively at this age, a migration from another rookery can be ruled out). Furthermore, analysis of stable isotope profiles showed that 90% of the adult individuals were consuming fish exclusively from the central California coast (i.e., no evidence for the consumption of Alaskan fauna was found), and analysis of the pup remains showed that they were feeding one trophic level higher than the adults, indicating that they were nursing at the time of death (Burton 2000). The rookery was most likely located somewhere on the sand spit that is separated from the mainland by

Table A-6.1. Archaeological shellfish findings from Elkhorn Slough.

	% of shell by weight							
	Millingstone Period (6500–3500 B.C.)				Middle Period (1000 B.C.– A.D. 1200)			Late Period (post-1200)
Shellfish	**MNT-228**	**MNT-229***	**MNT-1570***	**MNT-234***	**MNT-228**	**MNT-229***	**MNT-234***	**MNT-176[illegible][1]**
Protected habitat clams	65.6	74.9	68.2	42.3	16.8	53.4	14.4	41.9
Pismo clams	0.0	0.1	0.0	0.0	0.0	0.1	0.0	0.0
Oyster	3.0	1.7	3.7	3.8	0.0	1.1	0.3	0.1
Mussel	28.6	20.5	15.0	50.8	81.7	42.2	84.5	31.6
Barnacle	0.3	0.6	0.4	0.9	1.1	0.6	0.4	0.1
Moonsnail	1.7	0.1	T	0.0	0.1	0.0	0.0	T
Other/unidentified	0.8	2.1	12.7	2.2	0.3	2.6	0.4	26.3
Totals	100.0	100.0	100.0	100.0	100.0	100.0	100.0	100.0

*Partially mixed component.

T = trace. [1]Remains from 1/4-inch (6-mm) mesh only. All other findings from 1/8-inch (3-mm) mesh. Sources: T. Jones et al. (1996), Breschini and Haversat (1995), Fitzgerald et al. (1995).

the old Salinas River channel (Jones et al. 1996, 190), a site that provided the animals with some protection from nonhuman terrestrial predators. The abundance of northern fur seal bones at CA-MNT-234, as well as at several other sites in the Monterey Bay area, suggests that prehistoric northern fur seals had a greater presence on the mainland than they do today, and it is has been proposed that native overhunting may have contributed to the disappearance of this species from mainland settings (Burton 2000; Hildebrandt and Jones 1992; T. Jones and Hildebrandt 1995).

Fitzgerald et al. (1995) reported test excavation results from CA-MNT-1765, located 8 kilometers inland near Moro Cojo Slough (see fig. 6.1). While the testing program was small, findings are nonetheless significant because the deposit appears to harbor a single Late Period component. Until this discovery, the Late Period had been largely unrepresented in the Elkhorn Slough locality. The vertebrate fauna, though limited to residues from 1/4-inch (6-mm) screening, was dominated by fish and birds, with some deer and rabbit. Invertebrates were dominated by bay mussel (*Mytilus trossulus*, 31.6%) and littleneck clam (*Protothaca staminea*) (23.5%). Combined representation of clams and cockles (*Protothaca, Clinocardium, Tresus*, and *Macoma* spp.) was 41.8%, while oysters represented only 0.2% of the assemblage. Recovery was highlighted by a hearth feature, described as a dense concentration of burned shell, charcoal, and floral remains in the floor of Unit 1. Flotation analysis showed an abundance of live-oak-wood charcoal, suggesting the hearth was used briefly to process mussels or small game, while charred seeds suggest it was used in the late spring or summer. Radiocarbon results from the feature indicate it was used between ca. A.D.1530 and 1650 (Fitzgerald et al. 1995, 35).

Most recently, Breschini and Haversat (2000) reported results of archaeological salvage excavations at CA-SCR-44, located adjacent to College Lake in Watsonville (see fig. 6.5); the 2000 study supplements an earlier testing project (Breschini and Haversat 1989b). While project results were somewhat limited due to constraints on the area that could be examined, findings were highlighted by recovery of 33 human burials, several of which produced artifacts unusual for the Monterey Bay area. Unlike many sites in the area, CA-SCR-44 produced evidence for occupation during both the Middle and Late Periods, although the exact span of site use could not be determined owing to a limited radiocarbon record of only three dates.

Table A-6.2. Costanoan botanical resources, based on ethnohistoric and archaeological data.

Common name	Taxon	Part used	Purpose
horsetail	*Equisetum* spp.	roots	basketry
bracken fern	*Pteridium aquilinum*	young fronds	food
redwood	*Sequoia sempervirens*	burl sprouts	basketry
California hazel	*Corylus californica*	nuts	food
		wood	basketry
tanbark oak	*Lithocarpus densiflora*	nuts	food
oaks	*Quercus* spp.	nuts	food,
		wood	fuel, utensils
cottonwood	*Populus* sp.	inner bark	food
arroyo willow	*Salix lasiolepsis*	shoots	basketry
valley willow	*S. hindsiana*	shoots	basketry
		twigs	fuel
willow	*Salix* spp.	bark	rope
		branches	house poles
California bay	*Umbellularia californica*	nuts	food
sycamore	*Platanus racemosa*	inner bark	food
chamise	*Adenostoma fasciculatum*	wood	arrow shafts
strawberry	*Fragaria* spp.	fruit	food
Islay	*Prunus ilicifolia*	pits	food
		wood	bows
blackberry	*Rubus* spp.	fruit	food
suncup	*Oenothera ovata*	foliage	food
Durango root	*Datisca glomerata*	root	dye
currant	*Ribes* spp.	fruit	food
California broom	*Lotus scoparius*	foliage, branches	house thatching
lupine	*Lupinus* spp.	seeds*	food
clover	*Trifolium* spp.	foliage	food
stone crop	*Sedum* sp.	leaves, stems	food
poison oak	*Toxicodendron diversilobum*	shoots	basketry
red maids	*Calandrinia ciliata*	foliage, seeds	food
miner's lettuce	*Montia perfoliata*	foliage	food
Pacific oenanthe	*Oenanthe sarmetosa*	stems	food
western red dogwood	*Cornus californica*	stems	basketry
California buckeye	*Aesculus californica*	nuts	food
madrone	*Arbutus menziesii*	fruits	food
manzanita	*Arctostaphylos* spp.	fruits	food
milkweed	*Asclepias* sp.	stem	cordage
yerba santa	*Eriodictyon californicum*	leaves	woven clothing
chia	*Salvia columbariae*	seeds	food
elderberry	*Sambucus caerulea*	fruits	food
sunflower	*Helianthus annuus*	seeds	food
tarweed	*Madia* sp.	seeds**	food
coast tarweed	*Hemizonia corymbosa*	seeds	food
mule ears	*Wyethia angustifolia*	seeds	food
western rye grass	*Elymus glaucus*	seeds	food
native barley	*Hordeum* sp.	seeds**	food
reedgrass	*Calamagrostis* sp.	seeds**	food?
fescue	*Festuca* spp.	seeds	food
sedge	*Carex* spp.	roots	basketry
chufa	*Cyperus esulentus*	tubers	food
tule	*Scirpus* spp.	roots	food, basketry
		stems	thatch, watercraft
cattails	*Typha latiflolia*		roots, shoots, pollen food
rush	*Juncus* spp.	stems, leaves	textiles, cordage
		leaves	basketry
wild onion	*Allium* sp.	bulbs	food
brodiaea	*Brodiaea* spp.	bulbs	food
soap root	*Chlorogalum pomeridianum*	bulb	fish poison
		young leaves	food

*Recorded etnohistorically and represented archaeologically at CA-MNT-1765 (Fitzgerald et al. 1995 29).
**Not recorded ethnohistorically but represented archaeologically at CA-MNT-1765 (Fitzgerald et al. 1995 29).
Source: Bocek 1984.

Table A-6.3. Calibrated radiocarbon dates from Elkhorn Slough archaeological sites and sediment cores.

Site	Laboratory Number	Measured Radiocarbon Age	C-13 Adjusted Age (Conventional Date)	Reservoir Effect Corrected Date (Delta R=325+35)	2 Sigma Probability Range
CA-MNT-228	Beta 48652	6850+90	7250+90	5453 B.C.	5600–5290 B.C.
CA-MNT-228	Beta 48653	6570+90	6980+90	5215 B.C..	5360–5010 B.C.
CA-MNT-228	Beta 51906	6650+90	7060+90	5291 B.C.	5450–5120 B.C.
CA-MNT-228	Beta 51908	1890+70	2270+70	A.D. 453	A.D. 270–630
CA-MNT-228	Beta 48654	1960+90	2360+90	A.D. 367	A.D. 130–580
CA-MNT-228	Beta 48655	1770+50	2180+50	A.D. 571	A.D. 440–670
CA-MNT-228	Beta 48656	1970+100	2350+100	A.D. 381	A.D. 120–610
CA-MNT-228	Beta 51911	1740+60	2120+60	A.D. 634	A.D. 470–720
CA-MNT-228	Beta 48657	1690+70	2080+70	A.D. 662	A.D. 520–790
CA-MNT-228	Beta 49773	1900+120	2280+120	A.D. 445	A.D. 150–690
CA-MNT-228	Beta 49774	1650+90	2040+90	A.D. 685	A.D. 520–690
CA-MNT-228	Beta 48660	1790+70	2200+70	A.D. 553	A.D. 380–680
CA-MNT-228	Beta 51912	1650+60	2040+60	A.D. 685	A.D. 580–810
CA-MNT-229	WSU-3297	1920+130	2330+150	A.D. 410	A.D. 50–690
CA-MNT-229	WSU-3298	7700+90	8110+120	6222 B.C.	6480–6010 B.C.
CA-MNT-229	WSU-3299	1700+70	2110+100	A.D. 640	A.D. 420–820
CA-MNT-229	WSU-3300	6510+80	6920+110	5190 B.C.	5340–4900 B.C.
CA-MNT-229	WSU-3301	6820+100	7230+120	5460 B.C.	5640–5230 B.C.
CA-MNT-229	WSU-3302	6240+150	6650+160	4830 B.C.	5210–4460 B.C.
CA-MNT-229	WSU-3303	1760+80	2170+110	A.D. 590	A.D. 340–770
CA-MNT-229	WSU-3304	2070+90	2070+120	100 B.C.	390 B.C.–A.D. 130
CA-MNT-229	WSU-3308	2780+200	3190+210	740 B.C.	1190–120 B.C.
CA-MNT-229	WSU-3310	6580+80	6990+110	5220 B.C.	5410–4980 B.C.
CA-MNT-229	WSU-3311	1380+100	1460+140	A.D. 600	A.D. 260–870
CA-MNT-229	WSU-3312	1980+75	2060+100	95 B.C.	370 B.C.–A.D. 120
CA-MNT-229	WSU-3313	1980+70	2060+100	95 B.C.	370 B.C.–A.D. 120
CA-MNT-229	WSU-3314	7020+170	7100+180	5985 B.C.	6380–6310 B.C.
CA-MNT-229	WSU-3320	2720+140	3130+160	580 B.C.	930–180 B.C.
CA-MNT-229	WSU-3321	2270+135	2680+150	20 B.C.	380 B.C.–A.D. 350
CA-MNT-234	Beta-46928	1900+70	2310+70	A.D. 425	A.D. 240–610
CA-MNT-234	Beta-46929	6880+50	7290+50	5450 B.C.	5570–5360 B.C.
CA-MNT-234	Beta-46930	6490+50	6900+50	5110 B.C.	5240–4950 B.C.
CA-MNT-234	Beta-46932	6310+70	6720+70	4900 B.C.	5070–4730 B.C.
CA-MNT-234	Beta-47606	2130+70	2540+70	A.D. 140	A.D. 30–350
CA-MNT-234	Beta-47608	1850+70	2260+70	A.D. 470	A.D. 280–650
CA-MNT-234	Beta-47611	3480+703	890+701	490 B.C.	1680–1320 B.C.
CA-MNT-234	Beta-50269	6460+60	6870+605	5070 B.C.	5230–4910 B.C.
CA-MNT-234	Beta-50270	3270+90	3680+90	1260 B.C.	1470–980 B.C.
CA-MNT-234	Beta-50271	6200+80	6610+804	780 B.C.	4970–4570 B.C.
CA-MNT-234	Beta-82757	2160+60	2570+60	A.D. 120	A.D. 40–270
CA-MNT-234	Beta-82758	3520+60	3930+70	1520 B.C.	1730–1375 B.C.
CA-MNT-234	Beta-116532		1340+50	A.D. 1370	A.D. 1310–1420
CA-MNT-234	Beta-116533		4300+130	2000 B.C.	2190–1850 B.C.
CA-MNT-234	Beta-116534		1090+60	A.D. 1530	A.D. 1440–1680
CA-MNT-234	Beta-116535		930+70	A.D. 1690	A.D. 1510–1870
CA-MNT-234	Beta-116536		1010+60	A.D. 1650	A.D. 1490–1800
CA-MNT-234	Beta-116537		1000+50	A.D. 1660	A.D. 1570–1690
CA-MNT-234	Beta-116538		1580+70	A.D. 1180	A.D. 1060–1250
CA-MNT-234	Beta-116539		880+40	A.D. 1730	A.D. 1640–1870
CA-MNT-234	Beta-116540		1410+70	A.D. 1310	A.D. 1190–1430
CA-MNT-234	Beta-120173		6890+80	5090 B.C.	5270–4900 B.C.
CA-MNT-234	Beta-120175		7780+80	5930 B.C.	6090–5730 B.C.

Table A-6.3. (cont'd)

Site	Laboratory Number	Measured Radiocarbon Age	C-13 Adjusted Age (Conventional Date)	Reservoir Effect Corrected Date (Delta R=325+35)	2 Sigma Probability Range
CA-MNT-234	Beta-120176		7800+190	5940 B.C.	6350–5570 B.C.
CA-MNT-234	Beta-120177		7900+120	6010 B.C.	6310–5780 B.C.
CA-MNT-234	Beta-120178		4200+110	1880 B.C.	2190–1590 B.C.
CA-MNT-234	Beta-124006		6810+90	5000 B.C.	5230–4780 B.C.
CA-MNT-234	Beta-124007		7000+90	5220 B.C.	5400–5000 B.C.
CA-MNT-234	Beta-130521		1480+50	A.D. 1270	A.D. 1150–1340
CA-MNT-234	Beta-130522		1640+40	A.D. 1080	A.D. 1000–1220
CA-MNT-414	UCR-0797	5540+160	5950+160	4040 B.C.	4410–3690 B.C.
CA-MNT-414	UCR-1075	5200+100	5610+100	3680 B.C.	3940–3490 B.C.
CA-MNT-698	UCR-0796	1760+110	2170+110	A.D. 560	A.D. 320–790
CA-MNT-1570	Beta-51904	6350+90	6750+90	4930 B.C.	5190–4720 B.C.
CA-MNT-1570	Beta-51905	7320+90	7730+90	5870 B.C.	6050–5670 B.C.
CA-MNT-1570	Beta-51907	7300+70	7710+70	5850 B.C.	5990–5680 B.C.
CA-MNT-1570	Beta-51909	6550+90	6950+90	5190 B.C.	5340–4930 B.C.
CA-MNT-1570	Beta-51910	6280+130	6690+130	4870 B.C.	5200–4550 B.C.
CA-MNT-1570	Beta-54391	6340+90	6760+90	4940 B.C.	5200–4740 B.C.
CA-MNT-1570	Beta-54392	7110+90	7530+90	5670 B.C.	5870–5500 B.C.
CA-MNT-1765	Beta-76313	690+60	1090+60	A.D. 1530	A.D. 1440–1680
CA-MNT-1765	Beta-76314	250+70	250+70	A.D. 1650	A.D. 1480–1954
CA-SCR-44	Beta-30291	2725+55	2805+65	920 B.C. 960 B.C. 970 B.C.	1210–810 B.C.
CA-SCR-44	Beta-141194	620+40	1030+40	A.D. 1640	A.D. 1500–1690
CA-SCR-44	Beta-141195	1760+40	1760+40	A.D. 260 A.D. 280 A.D. 290 A.D. 300 A.D. 320	A.D. 140–410
Pollen core (313 cm)	WSU-3358	3550+70	1830 B.C. 1840 B.C. 1880 B.C.	2110–1662 B.C.	
Pollen core (344 cm)	Beta-63515	4410+60	2940 B.C. 2950 B.C. 3040 B.C. 3050 B.C. 3070 B.C.	3340–2890 B.C.	
Pollen core (683 cm)	Beta-63514	5540+60	4350 B.C. 4430 B.C. 4440 B.C.	4490–4250 B.C.	

Note: Data for archaeological sites represent single-shell, bone, and charcoal samples only.

Sources: Breschini and Haversat 1989b, 1995, 2000; Fitzgerald et al. 1995; T. Jones et al. 1996; T. Jones and Waugh 1997, 117; Milliken et al. 1999; Patch and Jones 1984; West 1988.

All dates were corrected for isotopic fractionation (either on a sample-specific basis as completed by Beta Analytic or through addition of 410 years as suggested by Stuiver and Polach [1977]), and for marine/atmospheric ^{14}C imbalance using the Stuiver and Reimer (1993) computer program and a local upwelling correction factor of 325+35 years. This value, proposed by T. Jones and Jones (1992), differs only slightly from another value of 290+35 recently developed by Ingram and Southon (1996).

Table A-6.4 Cultural periods of the central California coast.

Period	Dating	Fredrickson (1974) Equivalent North Coast Ranges	King (1990) Equivalent Santa Barbara Channel	Bennyhoff and Hughes (1987) Equivalence Sacramento Valley	Monterey Bay District Phase
PaleoIndian	9000–6500 B.C.	PaleoIndian			
Millingstone	6500–3500 B.C.	Lower Archaic	Ex, Eya		
Early	3500–1000 B.C.	Middle Archaic	Eyb	Early	Saunders
Early/Middle Transition	1000–600 B.C.		Ez		
Middle	600 B.C.– A.D. 1000	Upper Archaic	M1-M5a	E/M Transition, Middle, Middle/Late, Transition	Vierra
Middle/Late Transition	A.D. 1000–1200	M5b, M5c			Rancho San Carlos
Late	A.D. 1200–1500	Emergent, Phase 1	L1	L1	Rancho San Carlos
Protohistoric	A.D. 1500–1769	Emergent, Phase 2	L2	L2	
Historic	post- A.D.1769		L3		

Source: T. Jones 1993, 1995.
E=Early; M= Middle; L= Late

Table A-6.5 Summary of mammal findings from Middle Period components on the central California coast.

Component	Deer NISP	%	Rank	Otters NISP	%	Rank	Rabbits NISP	%	Rank	Harbor seals NISP	%	Rank
Elkhorn Slough												
CA-MNT-228	85	34.7	1	35	14.3	3	37	15.1	2	20	8.2	4
CA-MNT-229*	403	50.4	1	89	11.1	4	114	14.3	2	17	2.1	6
CA-MNT-234**	2	1.7	10	23	19.3	3	31	26.1	1	2	1.7	10
Open Coast												
CA-MNT-63	38	24.4	2	14	8.9	4	52	33.3	1	10	6.4	5
CA-MNT-1233[1]	203	78.4	1	2	0.7	4	21	8.1	2	4	1.5	3
CA-SCR-9**	499	80.7	1	2	0.3	9	55	8.9	2	11	1.8	4
Total	1229			142			286			62		

*Tule elk ranked third.
**Northern fur seal ranked second.
[1]Middle/Late Transition
NISP= Number of identified specimens.

Note: Rank = Relative importance in the assemblage based on proportion.

References

Adam, D. P., R. Byrne, and E. Luther. 1981. A late Pleistocene and Holocene pollen record from Laguna de las Trancas, northern coastal Santa Cruz County, California. *Madroño* 28: 255–277

Antevs, E. 1948. Climatic changes and pre-white man. *University of Utah Bulletin* 38: 168–191.

———. 1952. Climatic history and the antiquity of man in California. *University of California Archaeological Survey Reports* 16: 23–31.

Arnold, J. E. 1992a. Complex hunter-gatherer-fishers of prehistoric California: Chiefs, specialists, and maritime adaptations of the Channel Islands. *American Antiquity* 57: 60–84.

———. 1992b. Cultural disruption and the political economy in Channel Islands prehistory. In T. Jones, ed., *Essays on the Prehistory of Maritime California*, pp. 129–144. Center for Archaeological Research, Publication No. 10, University of California, Davis.

Atwater, B. F., E. J. Helley, and C. W. Hedel. 1977. Late Quaternary depositional history, Holocene sea level changes, and vertical crustal movement, southern San Francisco Bay, California. *United States Geological Survey Professional Papers* 1014: 1–15.

Axelrod, D. I. 1981. Holocene climatic changes in relation to vegetation disjunction and speciation. *American Naturalist* 117: 847–870.

Beaton, J. M. 1985. Evidence for coastal occupation time-lag at Princess Charlotte Bay (North Queensland) and implications for coastal colonization and population growth theories for aboriginal Australia. *Archaeology in Oceania* 20: 1-20.

Bennyhoff, J. A., and R. E. Hughes. 1987. Shell bead and ornament exchange networks between California and the western Great Basin. *American Museum of Natural History Anthropological Papers* 64 (2).

Bickel, P. M. 1978. Changing sea levels along the California coast: Anthropological implications. *Journal of California Anthropology* 5: 6–20.

Binford, L. 1980. Willow smoke and dog's tails: Hunter-gatherer settlement systems and archaeological site formation. *American Antiquity* 45: 4–20.

Bocek, B. R. 1984. Ethnobotany of Costanoan Indians, California, based on collections of John P. Harrington. *Economic Botany* 38: 240–255.

Breschini, G., and T. Haversat. 1980. *Preliminary Archaeological Report and Archaeological Management Recommendations for CA-MNT-170, on Pescadero Point, Monterey County, California.* Archaeological Consulting. Report on file, Northwest Information Center, Sonoma State University, Rohnert Park.

———. 1989a. *Archaeological Investigations at CA-MNT-108 at Fisherman's Wharf, Monterey, Monterey County, California.* Coyote Press Archives of California Prehistory No. 29. Salinas, CA: Coyote Press.

———. 1989b. *Archaeological Investigations at CA-SCR-44, Northeast of Watsonville, Santa Cruz County, California.* Archaeological Consulting. Report on file, Northwest Information Center, Sonoma State University, Rohnert Park.

———. 1990. *Archaeological and Osteological Analysis of a Prehistoric Burial Recovered from Site CA-MNT-228, Moss Landing, Monterey County, California.* Report to California Department of Transportation, District 5, San Luis Obispo.

———. 1991a. Archaeological investigations at three late period coastal abalone processing sites on the Monterey Peninsula, California. *Coyote Press Archives of California Prehistory* 33: 31–62.

———. 1991b. Early Holocene occupation of the central California coast. In J. M. Erlandson and R. H. Colten, eds., *Hunter-Gatherers of Early Holocene Coastal California*, pp. 125–132. Institute of Archaeology, University of California, Los Angeles.

———. 1992a. Archaeological excavations at CA-MNT-108, Fisherman's Wharf, Monterey County, California. In H. Dallas Jr. and G. S. Breschini, eds., *Archaeological Investigations of Some Significant Sites on the Central Coast of California*, pp. 39–47. Coyote Press Archives of California Prehistory No. 37. Salinas, CA: Coyote Press.

———. 1992b. Baseline archaeological studies at Rancho San Carlos, Carmel Valley, Monterey County, California. Ms. on file, Archaeological Consulting, Salinas.

———. 1995. Archaeological evaluation of CA-MNT-234, at the site of the proposed Moss Landing Marine Laboratory, Moss Landing, Monterey County, California. Ms. on file, Archaeological Consulting, Salinas.

———. 2000. *Archaeological Data Recovery at CA-SCR-44 at the Site of the Lakeview Middle School, Watsonville, Santa Cruz County, California.* Coyote Press Archives of California Prehistory No. 49. Salinas, CA: Coyote Press.

Breschini, G., T. Haversat, and J. Erlandson. 1992. *California Radiocarbon Dates.* 7th ed. Salinas, CA: Coyote Press.

Breschini, G., T. Haversat, and R.P. Hampson. 1983. *A Cultural Resource Overview of the Coast and Coast Valley Study Areas.* Salinas, CA: Coyote Press.

Broadbent, S. M. 1972. The Rumsen of Monterey, an ethnography from historical sources. *Contributions of the University of California Archaeological Research Facility* 14: 45–94.

Burton, R. 2000. Ecology of the northern fur seal (*Callorhinus ursinus*) from the Middle and Late Holocene of California. Ph.D. diss., University of California, Santa Cruz.

California Division of Mines. 1956. Peat. *Mineral Information Service* 9: 1–5

Cartier, R.. 1989. Scotts Valley chronology and temporal stratigraphy. *Proceedings of the Society for California Archaeology* 2: 81–112

———. 1992. The Santas Village site CA-SCR-239: Data recovery report of the Borland property on Santas Village Road in the City of Scotts Valley, County of Santa Cruz. Ms. on file, Northwest Information Center of the California Historical Resources Information System, Sonoma State University, Rohnert Park.

———. 1993a. *The Saunders Site: MNT-391, a Littoral Site of the Early Period.* Scotts Valley Historical Society Monograph No. 1.

———. 1993b. *The Scotts Valley Site: CA-SCR-177.* Santa Cruz Archaeological Society Monograph No. 1.

Costansó, M. 1911. The Portolá expedition of 1769–1770: Diary of Miguel Costansó. Edited by F. J. Teggart. *Publications of the Academy of Pacific Coast History* 2: 161–327.

Crespí, J. 1927. *Fray Juan Crespí—Missionary Explorer on the Pacific Coast, 1769–1774.* Edited by Herbert E. Bolton. New York: AMS Press.

Currey, D. R., and S. R. James. 1982. Paleoenvironments of the northeastern Great Basin and northeastern basin rim region: A review of geological and biological evidence. In D. B. Madsen and J. F. O'Connell, eds., *Man and Environment in the Great Basin*, pp. 27-52. Society for American Archaeology Papers 2. Washington, D.C.

Curry, R. R. 1969. Holocene climatic and glacial history of the central Sierra Nevada, California. *Geological Society of America Special Papers* 123: 1–47.

Dietz, S. A. 1987. Archaeological test excavations at CA-MNT-101, CA-MNT-298, CA-MNT-929, and El Castillo at the Presidio and City of Monterey, California. Ms. on file, Northwest Information Center of the California Historical Resources Information System, Sonoma State University, Rohnert Park.

———. 1991. Final report of archaeological investigations at Pescadero Point, data recovery excavations and monitoring of CA-MNT-170. Ms. on file, Northwest Information Center of the California Historical Resources Information System, Sonoma State University, Rohnert Park.

Dietz, S. A., W. R. Hildebrandt, and T. Jones. 1986. Final report of archaeological data recoverey program at CA-MNT-229. Ms. on file, Northwest Information Center of the California Historical Resources Information System, Sonoma State University, Rohnert Park.

———.1988. *Archaeological Investigations at Elkhorn Slough: CA-MNT-229, a Middle Period Site on the Central California Coast.* Papers in Northern California Anthropology No. 3. Northern California Anthropological Research Group, Berkeley.

Dietz, S. A., and T. L. Jackson. 1976. Archaeological test excavations and impact evaluations Pajaro Headlands, Santa Cruz County, California. Ms. on file, Northwest Information Center of the California Historical Resources Information System, Sonoma State University, Rohnert Park.

———. 1977. Archaeological Resource Inventory of the Rubis Ranch, Moss Landing, California. Ms on file, Northwest Information Center of the California Historical Resources Information System, Sonoma State University, Rohnert Park.

———. 1981. Report of Archaeological Excavations at Nineteen Archaeological Sites for the Stage 1 Pacific Grove-Monterey Consolidation Project of the Regional Sewerage System. Archaeological Consulting and Research Services. Submitted to Engineering-Science, Berkeley. Report on file, Northwest Information Center of the California Historical Resources Information System, Sonoma State University, Rohnert Park.

Dondero, S. B., D. Noble, J. Offerman, and T. Schuster. 1984. A preliminary report on archaeological testing CA-MNT-229. Ms. on file, Northwest Information Center of the California Historical Resources Information System, Sonoma State University, Rohnert Park.

Edwards, R., C. S. Smith, and M. Macko. 1991. Archaeological excavation at CA-SCR-160, University of California, Santa Cruz, California. Ms. on file, Northwest Information Center of the California Historical Resources Information System, Son-oma State University, Rohnert Park.

Elsasser, A. B. 1978. Development of regional prehistoric cultures. In R. F. Heizer, ed., *California: Handbook of North American Indians,* vol. 8, pp. 37–57. Washington, D.C.: Smithsonian Institution.

Erlandson, J. M. 1994. *Early Hunter-Gatherers of the California Coast.* New York: Plenum Press.

Fagan, B. 2000. *The Little Ice Age.* New York: Basic Books.

Fages, P. 1911. Expedition to San Francisco Bay in 1770: Diary of Pedro Fages. *Publications of the Academy of Pacific Coast History 2:* 141–159.

———. 1937. *A Historical, Political, and Natural Description of California by Pedro Fages: Written for the Viceroy in 1775.* Translated by Herbert Ingram Priestley. Berkeley: University of California Press.

Fenenga, G. L. 1993. The evolution of technology and proposed cultural chronology. In R. Cartier, ed., *Scotts Valley Site, CA-SCr-177,* pp. 243–265. Santa Cruz Archaeological Society Monograph 1.

Fisher, E. M. 1935. Shell deposits of the Monterey Peninsula: Notes on their distribution and fauna. Ms. on file, Archaeological Research Facility, University of California, Berkeley.

Fitzgerald, R. T. 1998. Archaeological data recovery at CA-SLO-1797, the Cross Creek Site, San Luis Obispo County, California, Coastal Branch Phase II Project. Ms. on file, California Department of Water Resources, Sacramento.

Fitzgerald, R. T., J. L. Edwards, J. M. Farquar, and K. Loeffler. 1995. Archaeological test excavation at CA-MNT-1765, for the Moro Cojo Standard Subdivision Project (SH 93001), Monterey County, California. Report prepared by Biosystems Analysis and submitted to Community Housing Improvement Systems Planning Association, Salinas. Ms. on file, Northwest Information Center of the California Historical Resources Information System, Sonoma State University, Rohnert Park.

Fitzgerald, R. T., and T. L. Jones. 1999 The Milling Stone horizon revisited: New perspectives from northern and central California. *Journal of California and Great Basin Anthropology* 21: 65–93

Fredrickson, D. A. 1974. Cultural diversity in early central California: A view from the North Coast Ranges. *Journal of California Anthropology* 1: 41–54.

Gallegos, D. 1987. A review and synthesis of environmental and cultural material for the Batiquitos Lagoon region. In Dennis Gallegos, ed., *San Dieguito–La Jolla: Chronology and Controversy*, pp. 23–34. San Diego: San Diego Archaeological Society.

———. 1992. Patterns and implications of coastal settlement in San Diego County: 9,000 to 1,300 years ago. In T. L. Jones, ed., *Essays on the Prehistory of Maritime California*, pp. 205–216. Center for Archaeological Research, Publication Number 10, University of California, Davis.

Gerow, B. A., with R. W. Force. 1968. *An Analysis of the University Village Complex with a Reappraisal of Central California Archaeology.* Stanford: Stanford University Press.

Gifford, D. G. 1977. Report on the archaeological assemblage for CA-Mnt-414 (Elkhorn Slough). Ms. on file, Northwest Information Center of the California Historical Resources Information System, Sonoma State University, Rohnert Park.

Gifford, E. W. 1913. Monterey County shell mounds. University of California Archaeological Survey Manuscript 22, Berkeley.

Glassow, M. A. 1996. *Purisemeno Chumash Prehistory: Maritime Adaptations along the Southern California Coast.* Orlando, Fla.: Harcourt Brace College Publishers.

Gobalet, K. W. 1990. Prehistoric status of freshwater fishes of the Pajaro-Salinas River system of California. *Copeia* 1990: 680–685.

———. 1993. Additional archaeological evidence for endemic fishes of California's Central Valley in the coastal Pajaro-Salinas Basin. *Southwestern Naturalist* 38: 218–223.

Gobalet, K., and T. L. Jones. 1995. Prehistoric Native American fisheries of the central California coast. *Transactions of the American Fisheries Society* 124: 813–823.

Golmshtok, E. 1922. Monterey, Pacific Grove, and Salinas mounds. Manuscript No. 374, University of California Archaeological Research Facility, Berkeley.

Gordon, B. L. 1979. *Monterey Bay Area: Natural History and Cultural Imprints.* Pacific Grove, Calif.: Boxwood Press.

Graumlich, L. J. 1993. A 1,000-year record of temperature and precipitation in the Sierra Nevada. *Quaternary Research* 39: 249–255.

Greengo, R. E. 1951. Molluscan subspecies in California shell middens. *University of California Archaeologi-cal Survey Reports* 13: 1–29.

Heizer, R. F., ed. 1974. *The Coastanoan Indians: The Indian Culture from the Mouth of the Sacramento River, South to Monterey, and Inland Past the Salinas River.* Cupertino, CA: California History Center, De Anza College.

Heusser, L. 1978. Pollen in Santa Barbara, California: A 12,000-year record. *Geological Society of America Bulletin* 89: 673–678.

Hildebrandt, W. R. 1983. Archaeological research of the Southern Santa Clara Valley Project. Report submitted by Daniel, Mann, Johnson, and Mendenhall to the California Department of Transportation, San Francisco. Contract No. 42921-E.P. On file, Caltrans Environmental Branch, Oakland.

Hildebrandt, W. R. 1997. The relative importance of lacustrine and estuarine resources to prehistoric hunter-gatherer populations: A view from southern Santa Clara Valley, California. *Journal of California and Great Basin Anthropology* 19: 197-225.

Hildebrandt, W. R., and T. L. Jones. 1992. Evolution of marine mammal hunting: A view from the California and Oregon coasts. *Journal of Anthropological Archaeology* 11: 360–401.

Hildebrandt, W. R., and P. Mikkelsen. 1993. Archaeological test excavations of fourteen sites along Highways 101 and 152, Santa Clara and San Benito Counties, California. Vol. 1: Prehistory. Report prepared by Far Western Anthropological Group and submitted to California Department of Transportation, District 4. Contract No. 04E633-E.P. On file, Caltrans Environmental Branch, Oakland.

Hill, W. W. 1929. Monterey County sites, general. University of California Archaeological Survey Manuscript 38, Berkeley.

Hylkema, M. 1991. Prehistoric Native American adaptations along the central California coast of San Mateo and Santa Cruz Counties. Master's thesis, San Jose State University.

Ingram, B. L., and J. R. Southon. 1996. Reservoir ages in eastern Pacific coastal and estuarine waters. *Radiocarbon* 38: 573–582.

Inman, D. L. 1983. Application of coastal dynamics to the reconstruction of paleocoastlines in the vicinity of La Jolla, California. In P. M. Masters and N. C. Flemming, eds., *Quaternary Coastlines and marine Archaeology*, pp. 1–49. New York: Academic Press.

Jenkins, O. P. 1973. Pleistocene Lake San Benito. *Journal of California Geology* 26: 151–163.

Johnson, D. L. 1977. The late Quate-rnary climate of coastal California: Evidence for an ice age refugium. *Quaternary Research* 8: 154–178.

Jones, D. A. 1992. The forager-collector model and Monterey Bay prehistory. In T. L. Jones, ed., *Essays on the Prehistory of Maritime California*, pp. 105–113. Center for Archaeo-logical Research, Publication Number 10, University of California, Davis.

Jones, D. A., and W. R. Hildebrandt. 1990. Archaeological excavations at Sand Hill Bluff: Portions of prehistoric site CA-SCR-7, Santa Cruz County, California. Report prepared by Far Western Anthropological Research Group, Inc., and submitted to Pacific Mariculture Inc., Santa Cruz, California.

Jones, T. L. 1991. Marine-resource value and the priority of coastal settlement: A California perspective. *American Antiquity* 56: 419–443.

———. 1992. Settlement trends along the California coast. In T. L. Jones, eds., *Essays on the Prehistory of Maritime California*, pp. 1–37. Center for Archaeological Research, Publication Number 10, University of California, Davis.

———. 1993. Big Sur: A keystone in central California culture history. *Pacific Coast Archaeological Society Quarterly* 29 (1): 1–78

———. 1995. Transitions in prehistoric diet, mobility, exchange, and social organization along California's Big Sur coast. Ph.D. diss., University of California, Davis.

———. 1996. Mortars, pestles, and division of labor in prehistoric California: A view from Big Sur. *American Antiquity* 61: 243–264.

Jones, T. L., S. Anderson, M. Brown, A. Garsia, K. Hildebrand, and A. York. 1989. *Surface Archaeology at Landels-Hill Big Creek Reserve and the Gamboa Point Properties.* University of California Santa Cruz Environmental Field Program No. 18. Santa Cruz, CA: Environmental Field Program.

Jones, T. L., G. Brown, L. M. Raab, J. McVicar, G. Spaulding, D. Kennett, A. York, and P. L. Walker. 1999. Environmental imperatives reconsidered: Demographic crises in western North America during the Medieval Climatic Anomaly. *Current Anthropology* 40: 137–156.

Jones, T. L., K. Davis, G. Farris, S. D. Grantham, T. W. Fung, and B. Rivers. 1994. Toward a prehistory of Morro Bay: Phase II archaeological investigations for the Highway 41 widening project, San Luis Obispo County, California. Report prepared for the California Department of Transportation, Sacramento. Ms. on file at Central Coast Information Center, University of California, Santa Barbara.

Jones, T. L., and W. R. Hildebrandt. 1995. Reasserting a prehistoric tragedy of the commons: Reply to Lyman. *Journal of Anthropological Archaeology* 14: 78–98.

Jones, T. L., and M. Hylkema. 1988. Two proposed projectile point types for the Monterey Bay Area: The Año Nuevo long-stemmed and the Rossi square-stemmed. *Journal of California and Great Basin Anthropology* 10: 163–185.

Jones, T. L., and D. A. Jones. 1992. Elkhorn Slough revisited: Reassessing the chronology of CA-MNT-229. *Journal of California and Great Basin Anthropology* 10: 153–186.

Jones, T. L., and D. J. Kennett. 1999. Late Holocene sea temperatures along the central California coast. *Quaternary Research* 51: 74–82.

Jones, T. L., T. Van Bueren, S. Grantham, J. Huddleson, and T. Fung. 1996. Archaeological test excavations for the Castroville bypass project, Monterey County, California. Report prepared for the California Department of Transportation, Environmental Division, Sacramento. Draft Ms. on file, Caltrans Environmental Program, Sacramento.

Jones, T. L., and G. Waugh. 1995. *Central California Coastal Prehistory: A View from Little Pico Creek.* Los Angeles: Institute of Archaeology, University of California, Los Angeles.

———. 1997. Climatic consequences or population pragmatism? A mid-Holocene prehistory of the central California coast. In J. M. Erlandson and M. A. Glassow, eds., *The Archaeology of the California Coast during the Middle Holocene,* pp. 111–128. Los Angeles : Institute of Archaeology, University of California, Los Angeles.

King, C. D. 1974. Calendaruc ethnohistory. In R. L. Edwards and M. E. Farley, eds., An Assessment of the Lower Pajaro River Basin, California. Report submitted to the U.S. Army Corps of Engineers. Ms. on file, Northwest Information Center, California Archaeological Site Inventory, Sonoma State University, Rohnert Park, California.

——— 1990. *The Evolution of Chumash Society: A Comparative Study of Artifacts Used in Social System Maintenance in the Santa Barbara Channel Region before* A.D. *1804.* New York: Garland Publishing.

King, J. M. 1981. Cultural resources of the Elkhorn Slough Estuarine Sanctuary. Ms. on file, Northwest Information Center of the California Historical Resources Information System, Sonoma State University, Rohnert Park.

Koerper, H. C., J. S. Killingley, and R. E. Taylor. 1985 The Little Ice Age and coastal southern California human economy. *Journal of California and Great Basin Anthropology* 7: 99–103.

Kroeber, A. L. 1908. A mission record of the California Indians. *University of California Publications in American Archaeology and Ethnology* 8: 1–27.

———. 1925. *Handbook of the Indians of California.* Bureau of American Ethnology Bulletin 78. Washington, D.C.

———. 1955. Nature of the land-holding group. *Ethnohistory* 2: 303–314.

———. 1962.The nature of land-holding groups in aboriginal California. *University of California Archaeological Survey Reports* 56: 19-58.

Latham, R. C. 1856 On the languages of northern, western, and central America. *Transactions of the Philological Society of London* for 1856: 57–115.

LaMarche, V. C., Jr. 1974. Paleoclimatic inferences from long tree-ring records. *Science* 183: 1043–1048.

Levy, R. S. 1978. Costanoan. In R. F. Heizer, ed., *California: Handbook of North American Indians*, vol. 8, pp. 485–495. Washington, D.C.: Smithsonian Institution.

Lönnberg, A., and J. Morris. 1981. Archaeological background. In S. A. Dietz and T. L. Jackson, eds., Report of archaeological excavations at nineteen archaeological sites for the Stage 1 Pacific Grove–Monterey Consolidation Project of the Regional Sewerage System. Report prepared by Archaeological Consulting and Research Services and submitted to Engineering-Science, Berkeley. Ms. on file, Northwest Information Center of the California Historical Resources Information System, Sonoma State University, Rohnert Park.

Menzies, A. 1924. Menzies's California journal. Annotated by Alice Eastwood. *California Historical Society Quarterly* 2: 265–340.

Merriam, C. H. 1967. *Ethnographic Notes on California Indian Tribes.* Edited by Robert F. Heizer. University of California Archaeological Reports 68. Berkeley.

Mikkelsen, P., W. R. Hildebrandt, and D. A. Jones. 1999. Prehistoric adaptations on the shores of Morro Bay Estuary: A report on excavations at CA-SLO-165. Report on file Caltrans, Sacramento.

Milliken, R. T. 1981. Ethnohistory of the Rumsen: The Mission Period. In Report of archaeological excavations at nineteen archaeological sites for the Stage 1 Pacific Grove–Monterey Consolidation Project of the Regional Sewerage System. Report prepared by Archaeological Consulting and Research Services and submitted to Engineering-Science, Berkeley. Ms on file, Northwest Information Center of the California Historical Resources Information System, Sonoma State University, Rohnert Park.

———. 1988. Ethnographic context. In S. A Dietz, W. Hildebrandt, and T. Jones, *Archaeological Investigations at Elkhorn Slough: CA-MNT-229, a Middle Period Site on the Central California Coast.* Papers in Northern California Anthropology No. 3. Northern Calfiornia Anthropological Research Group, Berkeley.

———. 1991. An ethnohistory of the Indian people of the San Francisco Bay area from 1770 to 1810. Ph.D diss., University of California, Berkeley.

———. 1993. Early history of native peoples. In S. Waechter, ed., Archaeological test excavations at fourteen sites along Highways 101 and 152, Santa Clara and San Benito Counties, California, vol. 2: History, ethnohistory, and historic archaeology, pp. 61–93. Report prepared by the Far Western Anthropological Research Group. On file, Caltrans, Sacramento.

Milliken, R. T., J. Nelson, W. R. Hildebrandt, and P. Mikkelsen 1999. The Moss Landing Hill site: A technical report on archaeological studies at CA-MNT-234. Ms. on file, Northwest Information Center of the California Historical Resources Information System, Sonoma State University, Rohnert Park.

Moratto, M. J. 1984. *California Archaeology*. San Francisco: Academic Press.

Osborn, A. 1977. Strandloopers, mermaids, and other fairy tales: Ecologi-cal determinants of marine resource utilization—the Peruvian case. In L. R. Binford, ed., *For Theory Building in Archaeology*, pp. 157-206. San Francisco: Academic Press.

Patch, D. 1979. The paleoecology of Elkhorn Slough: Implications for human subsistence. Bachelor's thesis, University of California, Santa Cruz.

Patch, D., and T. Jones. 1984. Paleoenvironmental change at Elkhorn Slough: Implications for human adaptive strategies. *Journal of California and Great Basin Anthropology* 6: 19–43.

Peak, A. S. 1979. Archaeological test excavations of CA-Mnt-228 and CA-Mnt-229, Moss Landing, Monterey County, California. Ms. on file, Northwest Information Center of the California Historical Resources Information System, Sonoma State University, Rohnert Park.

Pisias, N. G. 1978. Paleoceanography of the Santa Barbara Basin during the last 8,000 years. *Quaternary Research* 10: 366–384.

Pohorecky, Z. S. 1976. Archaeology of the South Coast Ranges of California. *Contributions of the University of California Archaeological Research Facility* 34: 1–235.

Prichard, W. E. 1968. Preliminary archaeological investigations at El Castillo, Presidio of Monterey, Monterey, California. Report submitted to National Park Service and Army Corps of Engineers, San Francisco.

Rivers, B. N.d. Asilomar State Beach and Conference Grounds general plan. Ms. on file, Resource Protection Division, California Department of Parks and Recreation, Sacramento.

Roop, W. C. 1976. Adaptation on Ben Lomond Mountain: Excavations at CA-SCr-20. Master's thesis, San Francisco State University.

Schwaderer, R. 1992. Archaeological test excavation at Duncans Point Cave, CA-SON-348/H. In T. L. Jones, ed., *Essays on the Prehistory of Maritime California*, pp. 55–71. Center for Archaeological Research, Publication Number 10, University of California, Davis.

Schwartz, D. L. 1983. Geologic history of Elkhorn Slough, Monterey County, California. Master's thesis, San Jose State University.

Schwartz, D. L., H. T. Mullins, and D. F. Belknap. 1986. Holocene geologic history of a transform margin estuary: Elkhorn Slough, central California. *Estuarine, Coastal, and Shelf Science* 22: 285–302.

Shackleton, N. J., and N. D. Opdyke. 1973. Temperatures and ice volumes on a 10^5- and 10^6-year scale. *Quaternary Research* 3: 39–55.

Snethkamp, P. E., and A. York. 1990. Archaeological survey report for the widening of Highway 1 from Castroville, California, to the Santa Cruz County Line. Report prepared for Environmental Planning Branch, California Department of Transportation, San Luis Obispo.

Spanne, L. W. 1979. An extended archaeological survey of portions of MNT-229 for the Struve Road channelization project, Monterey County, California. Report prepared for the State of California, Department of Transportation, San Luis Obispo.

Stanger, F. N., and A. K. Brown. 1969. *Who Discovered the Golden Gate?* San Mateo, CA: San Mateo County Historical Association.

Starke, G. W., and A. D. Howard. 1968. Polygenetic origin of the Monterey submarine canyon. *Geological Society of America Bulletin* 79: 813–826.

Stine, S. 1990. Late holocene fluctuations of Mono Lake, eastern California. *Palaeogeography, Palaeoclimatology, Palaeoecology* 78: 333-381.

—— 1994. Extreme and persistent drought in California and Patagonia during Mediaeval time. *Nature* 369: 546–549.

Strudwick, I. 1986. Temporal and areal considerations regarding the prehistoric circular shell fishhook. Master's thesis, California State University Long Beach.

Stuiver, M., G. W. Pearson, and T. Braziunas. 1986. Radiocarbon age calibration of marine samples back to 9000 cal. yr. B.P. *Radiocarbon* 28: 980–1021.

Stuiver, M., and H. A. Polach. 1977. Reporting of 14C data. *Radiocarbon* 19: 355–363.

Taylor, Bayard. 1949. *Eldorado, or Adventures in the Path of Empire.* Lincoln: University of Nebraska Press.

Weisner, P. 1982. Style and social information in Kalahari San projectile points. *American Antiquity* 48: 253–276.

Welch, P. 1992. Cultural resources overview for the Pajaro Valley Water Augmentation Study. Report prepared by the Bureau of Reclamation Mid-Pacific Region, Sacramento, California. Ms. on file, Northwest Information Center of the California Historical Resources Information System, Sonoma State University, Rohnert Park.

West, G. J. 1988. Exploratory pollen analysis of sediments from Elkhorn Slough. In S. A Dietz, W. Hildebrandt, and T. Jones, eds., *Archaeological Investigations at Elkhorn Slough: CA-MNT-229, A Middle Period Site on the Central Califirnia Coast,* pp. 25–56. Papers in Northern California Anthropology No. 3. Northern California Anthropological Research Group, Berkeley.

Wood, A. 1930. Monterey Bay mounds. Manuscript No. 380, University of California Archaeological Research Facility, Berkeley.

Yesner, D. 1980. Maritime hunter-gatherers: Ecology and prehistory. *Current Anthropology* 21: 727–750.

Yoklavich, M. M., G. M. Cailliet, J. P. Barry, D. A. Ambrose, and B. S. Antrim. 1991. Temporal and spatial patterns in abundance and diversity of fish assemblages in Elkhorn Slough, California. *Estuaries* 14: 465–480.

7

CHAPTER SEVEN

History of Land Use

Mark Silberstein, Chela Zabin, Louise Newberry, Daniel Mountjoy, Les Strnad, and Jane Caffrey

From the pre-Columbian era to the present, human activities have shaped the landscape of Elkhorn Slough and its surrounding watershed. Use of fire by the slough's native inhabitants modified vegetation in ways beneficial to humans. The Spanish and Mexican settlers brought livestock and plants that supplanted native species and permanently changed the region's flora and fauna. Later settlers logged the watershed and diked and drained the slough to create farming and grazing land, eliminating wetlands and altering the slough's hydrography. Populations of native wildlife dwindled and in some cases disappeared under hunting and fishing pressure.

These activities, combined with industrial and residential development, have produced major changes in Elkhorn Slough's natural resources. In recent decades, new attitudes toward the natural landscape have brought a focus on resource conservation and restoration, a change that will in part shape future land use in the Elkhorn Slough watershed.

B. L. Gordon, in his landmark book *Monterey Bay Area: Natural History and Cultural Imprints*, identifies several waves of people who have moved through the Monterey Bay area, each helping shape the landscape we see today. "Historically the Monterey Bay area has been occupied successively by three major cultural groups: American Indians of the Costanoan group, Spanish-Mexicans, and Americans. The last-named is plainly a very general grouping; it includes diverse elements, each of which has made contributions to the area's development: Americans from the eastern seaboard, Chinese, Italians, Japanese, Portuguese, Yugoslavians, and others. The existing cultural landscape is the cumulative effect of occupancy of the area by these several groups" (Gordon 1996).

In this chapter we chronicle the ways that the succession of Monterey Bay area settlers used Elkhorn Slough's land and marine resources, how these uses influenced the landscape, and how research activities and preservation of the slough's ecosystems have become a priority in recent years. This chapter does not offer a cultural analysis of the human activities that have taken place at the slough; rather, it provides a timeline of those activities. From these observations we may be better able to understand current conditions and the options available for land use in the future. We close with suggestions for research to increase our understanding of the slough's human history and its impacts on natural resources.

Early Period of Human Use: Prehistory to 1848

For thousands of years before the arrival of Spanish explorers and missionaries, Costanoan-speaking residents (or Ohlone) of the Monterey Bay area harvested plants and wildlife and used fire to improve hunting and foraging. Although their activities had some impact on the region's resources, it was the plants, animals, and land use practices introduced by the Spanish and later by Mexican cattle ranchers that most dramatically altered the Monterey Bay landscape, including the area around Elkhorn Slough.

Ohlone Land Use

Chapter 6, "Archaeology and Prehistory," presented the complex archaeological picture of the hunter-gatherers who occupied the Elkhorn Slough area beginning 8,000 years ago. The shell middens that have been excavated reveal a people who used a wide variety of plants and animals for their food, clothing, tools, and shelter. Over this long period, changing sea levels, shifting river courses, and other geological and climatic changes provoked drastic and repeated alterations in the slough's environment. By itself, the archaeological record does not necessarily show the real causes of, for example, a marked decline of oyster shells in shell mounds. Does it indicate human impact or a natural change in the environment of the slough?

The historical record does, however, provide a picture of one way the hunter-gatherers of the Monterey Bay area influenced its landscape. Early explorers and settlers reported extensive, intentional fires close to the bay, carried out especially during the fall. Gordon (1996) writes, "when the Spanish first saw the Monterey Bay area, its potential natural vegetation was not everywhere in existence and manmade fires had long been an important ecological factor." The Ohlone tribes of the area regularly burned the land to keep space open for the grasses and wildflowers whose seeds they collected. The practice also flushed game for hunting and created habitat for small game such as rabbits that flourished in meadow environments.

While the Ohlone certainly affected the abundance of some sea life and much game, the fact that they ate a wide variety of foods meant fewer severe impacts on local biodiversity. Gordon (1996) explains: "It seems likely that the Costanoans were a principal control of animal population sizes in the littoral zone, particularly of mollusks and pinnipeds, and that this control was maintained for centuries. On the other hand, the Indians' subsistence demands were spread over virtually the entire biotic spectrum, in contrast to those of later inhabitants, whose special and limited preferences (e.g., for the red abalone and sea otter) may have generated imbalances by making heavy demands on only a few species."

Mission Period

Spanish explorers were the first Europeans to make contact with the native peoples of the Monterey Bay area. Juan Rodriguez Cabrillo sailed north along the coast in 1542, and in 1602 Sebastián Vizcaíno described and mapped the shores of Monterey Bay. By 1770 Spanish missionaries had established a mission in Carmel and begun the conversion of the hunter-gatherers of the region to Christianity. Seventeen years later the mission of San Juan Bautista was established; it would become the focus of efforts to settle the Costanoan-speaking people of the Elkhorn Slough area.

The missionaries sought to convert the Ohlone not only to their religion, but to a completely new way of life: a seminomadic people were to adopt the sedentary life style of the new settlers. This change led to new ways of using the land. According to the padres' original plans, natives would be brought into the missions for ten-year periods, during which they would learn how to raise crops, care for livestock, and build masonry homes. Then the Ohlone would be given land, and villages of their productive, small, tidy farms would spring up around the missions (Margolin 1978). The padres gathered the natives around the missions and put them to work cultivating food in the mission vegetable gardens and caring for the livestock that grazed on the huge plots of mission-owned lands. Plants that were to become familiar features of the California landscape, including grapes, olives, figs, walnuts, and almonds, were grown in the mission's gardens and orchards.

The Spanish authorities disapproved of many Ohlone customs and habits. They regarded their lack of agriculture and of permanent settlements as evidence of shiftlessness and lack of industry, and interpreted the native peoples' seasonal cycles of feasting and fasting as signs of gluttony and indolence. But the Europeans reserved particular scorn for the practice of intentionally setting fires. In 1793 Governor Arrillaga in Santa Barbara sent a message to the Father President of Missions: "Because of…the serious damage that results from the fires that are set each year in the pastures by Christian and Gentile Indians…[the mission fathers should warn] the Christian Indians, and particularly the old women…threatening them with the rigors of the law" (Clar 1957, in Gordon 1996). Even so, the practice was slow to die among the Ohlone, and the Spanish cattlemen who were already beginning to settle in the region during the mission period adopted the technique to clear pastures for their cattle.

A history of Santa Cruz County published in 1879 gives an interesting statistical picture of the mission at San Juan Bautista

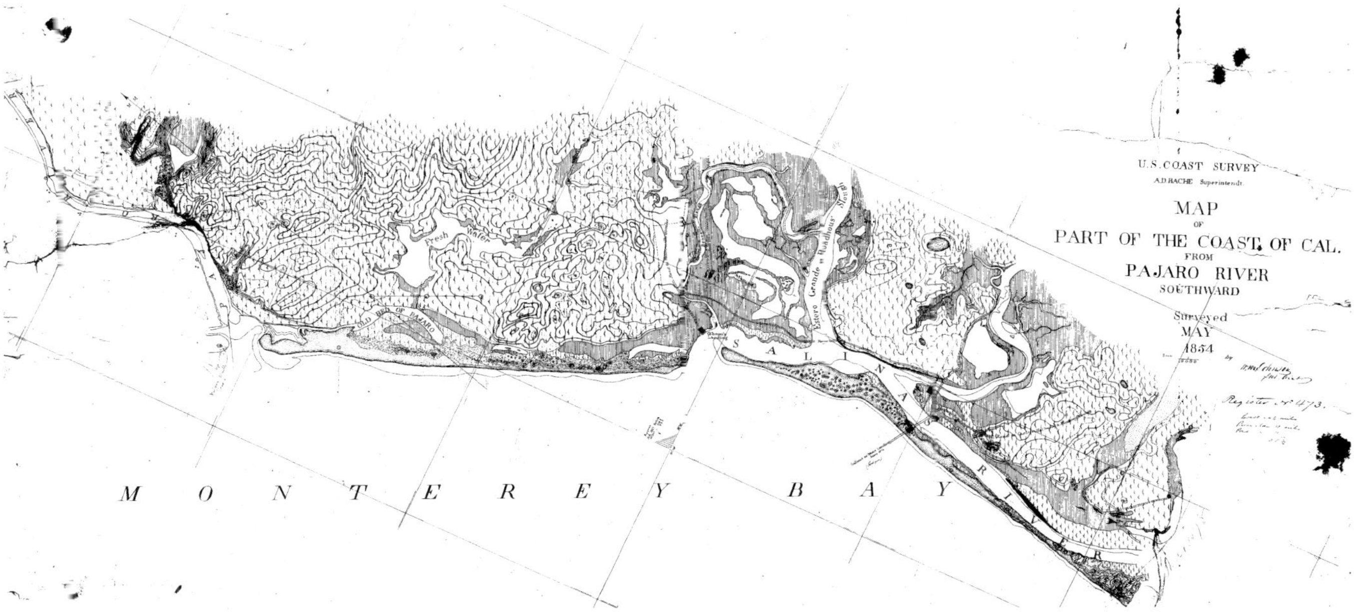

Figure 7.1 1854 Coast and Geodetic Survey Map. This earliest map of the slough shows the Salinas River and Elkhorn Slough emptying into Monterey Bay through a common mouth. The first recorded names of the slough were Estero Grande or Roadhouse Slough, named for the Roadhouse family who were early settlers in the region.

in 1832, near the end of the mission period. Among the missions of the central coast, it ranked first in population, with 987 inhabitants. Roughly seven thousand head of cattle and another seven thousand sheep grazed there, and the harvest of grain—wheat, corn, and barley—was substantial that year. Zenas Leonard, an American traveler who visited the mission a year later, reported a population of six to seven hundred people, nearly all of them Indian. Leonard observed that "some of the natives live well, as they cultivate pumpkins, beans, and some of them Indian corn" (Leonard 1934, in Gordon 1996).

Most of the Ohlone who went to the missions did not return to freedom at the end of ten years. Deprived of their usual foods and devastated by lethal European diseases, their populations quickly dwindled. With the secularization of the missions in 1834, the remaining Ohlone found their lands occupied; unable to return to their traditional lifestyle, they worked as laborers and servants for the European settlers. American occupancy was even more devastating to the Ohlone population, and by the early twentieth century only a few survived. Today, descendants of the region's Costanoan-speaking populations are working to resurrect and transmit their language and traditions.

Mexican Cattle Ranching

When Mexico won its independence from Spain in 1821, it took over the territory of Alta California. As had happened in the mission period, the change in authority brought a change in land management. The Mexican government secularized the mission lands and divided them into large land grants, which it gave to settlers from Mexico and Spain to be used almost exclusively for grazing cows, sheep, and horses. The land grants of Bolsa Nueva y Moro Cojo and Bolsa de San Cayetano included the land surrounding Elkhorn Slough, which was then called Estero Grande or Roadhouse Slough (fig. 7.1). Presumably, grazing prevailed wherever the marshy conditions allowed.

The Spanish-speaking settlers built adobe homes in Aromas, Corralitos, and San Juan Bautista. The Vallejo family alone built several adobes between 1820 and 1830 on the Bolsa de San Cayetano—one near the Pajaro River, two on Werner's Hill, and three on San Juan Road (King 1982). These cattle ranchers did not fence their land; indeed, many travelers remarked on the absence of fences between Monterey and San Francisco. Yet the land grant boundaries have persisted so thoroughly that one can still trace their edges on modern aerial photographs. Writes Gordon (1979), "the old grant boundaries often stand out strongly because field strips, furrow, and plant rows abut them at different angles on opposite sides. In hilly

and potatoes was raised, which commanded enormous prices—barley and wheat about ten cents and potatoes sixteen cents." The following fall, "large numbers of settlers" followed and "took possession of lands on the various ranches in regular squatter style". After 1852 the market for potatoes collapsed and wheat became the dominant—almost the only—crop.

Unsuitable for early row crop cultivation, lands adjacent to Elkhorn Slough were used for dairy herds. Elkhorn Dairy in foreground, with Elkhorn Slough in the distance, circa 1940s. Bob Bowen collection.

Over the next three decades farmers introduced tobacco, hops, and sugar beets on a commercial scale and began to cultivate for trade the mustard that had grown wild since its introduction by the mission fathers (Lydon 1985). By 1879, crops of apples, apricots, pears, currants, and blackberries grew in the valleys, and late in the century today's dominant crop, strawberries, became a commercial crop. Strawberry acreage expanded rapidly with the introduction of irrigation a few years later (see page 104, for a detailed history of the strawberry industry). The less fertile, sandy soils bordering Elkhorn Slough itself, ill suited to early cultivation methods, were used primarily for dairying by later Portuguese and Italian immigrants.

In the late 1800s production of fresh vegetables and fruits in the lower valleys expanded in response to new technologies that facilitated large-scale irrigation, processing, and transportation. Wetland areas were diked and drained to create cropland, and streams were diverted for irrigation. Plowing and tilling furthered the spread of nonnative species such as hemlock, sweet fennel, Bermuda grass, and common groundsel. Tillage in the hills also led to soil erosion in the watershed.

Another nonnative species was introduced in this period, as both the state and federal governments promoted planting of eucalyptus; the trees were thought to purify the air of malaria-causing agents and were also considered a potential source of hardwood. While the eucalyptus failed on both these counts, they did provide valuable firewood and made successful windbreaks. By 1874 approximately a million of these Australian trees had been planted in California (Gordon 1996). Many of them and their descendants still stand along the edges of Elkhorn Slough. While eucalyptus provide habitat for some birds and butterflies, they shelter a much sparser biota than did the stands of oaks they replaced.

Boats and Railroads

The prosperity of commercial agriculture after the 1850s encouraged central coast growers to seek new ways to ship their products to San Francisco and other markets. Elkhorn Slough served a new use in this trade by providing a transport route between inland landings and the coast. In the early 1860s, James Brennan and Captain Robert Sudden built two landings in the area, Gibson's Landing, also called Salinas Landing, and Watsonville Landing (the deed transferring title to Goodall Nelson and Perkins Steamship Co. on January 26, 1875, can be found in Mohlo 2000). Watsonville Landing, a major shipping point for Pajaro Valley farmers, later became known as Hudson's Landing because of the local agent, Mark Hudson. The *Salinas*, a 157-ton steamship, made regular runs between Watsonville Landing, Santa Cruz, and San Francisco during the 1860s and 1870s (*Pajaro Times,* August 1, 1863, in Mohlo 2000).

In 1866 Charles Moss, a local sea captain who was farming near Corralitos, leased land south of the mouth of the Salinas River. Moss and his construction engineer, Cato Vierra, built a number of warehouses and a wharf from which to transfer produce to coastal schooners bound for San Francisco (*Santa Cruz County Times* and *Pajaro Times*, November 5, 1866, in Mohlo 2000). Vierra also built a bridge across the Salinas River to the warehouses and wharf. Moss and his partner, Donald Beadle, bought the steamer *Santa Cruz*, which made regularly scheduled runs between San Francisco, Santa Cruz, and Moss Landing (*Santa Cruz County Times*, April 3, 1869, in Mohlo 2000). During this period, the Moss Landing wharf was expanded, as was Castroville Landing, with numerous steamers stopping at Moss, Gibson's, and Watsonville Landings.

The Pacific Coast Steamship Company, organized in 1876 by Robert Sudden and three partners, also provided freight and passenger service along the coast and had partial interests in Watsonville, Pajaro, and Gibson's Landings (Best 1964). The *Vaquero*, one of the best-known ships of the company's fleet, was a 100-ton steam-powered sternwheeler. The *Vaquero* plied the slough, transferring grain between Watsonville Landing and Salinas and Moss Landings until 1882 when it sank on a bar at the slough's mouth after a boiler explosion (Mann 1972). When Charles Moss sold his holdings to the Pacific Steamship Company in 1881, they acquired a virtual trade monopoly. All that remains of Moss's empire is the town that still bears his name.

By the early 1880s the erosion that accompanied intensive farming and logging had already led to enough silting to make lighter barges a more practical choice for slough shipping. The Vierra family added a shipping business to their other enterprises, running a barge operation along the length of the slough.

The Southern Pacific Railroad became a major shipping competitor in the area when it built tracks for part of its Oakland–Los Angeles line adjacent to Elkhorn Slough in 1872. Within a few decades trains replaced barges and schooners as the primary means of moving crops and produce. New towns rose along the railroad's route as well. Although the towns themselves were located away from the slough's edges, setting a roadbed in the marshy slough had a huge impact on the land. Southern Pacific dumped thousands of carloads of gravel along the slough to build the roadbed. Despite these efforts, the roadbed continued to subside and even today, periodically elevating the road bed is an ongoing maintenance task for the railroad.

Another rail line, the narrow-gauge Pajaro Valley Consolidated Railroad, paralleled the shore behind the dunes near the mouth of the Elkhorn Slough. Owned and operated by C. Spreckels Sugar Co., the trains made daily round trips to transport sugar beets from Pajaro to the processing plant in the town of Spreckels, just south of Salinas. In the same area, the Vierras built a bridge across the mouth of the slough, replacing the ferry that had operated there (near the present-day Highway 1 bridge) (Gordon 1996; King 1982).

Looking inland from a schooner tied up at Moss Landing pier to load produce from surrounding farms (late 1800s). Photo credit: Carlos Vierra.

Gun Clubs

Elkhorn Slough in the early 1900s harbored large numbers of waterfowl and fish. Its recreational potential captured the imaginations of San Francisco sportsmen, and in 1902 a group of businessmen from that city purchased land around the slough, set up several duck blinds, and established the Empire Gun Club. The club consisted of a lodge and caretaker's house and was a Southern Pacific Railroad flagstop.

The club maintained an extensive wildlife management program, which included "a full-time resident game keeper, dike and pond construction, water-level controls, and baiting practices" (King 1982). The ponds were baited twice a week with grain—amounting to a train carload and a half a year. The club also strictly regulated hunting practices. At the end of twelve years a report stated that "there has been no marked diminution in the total annual bag. There has, however, been a change in the makeup of the bag in that certain larger ducks are now taken in smaller numbers" (King 1982).

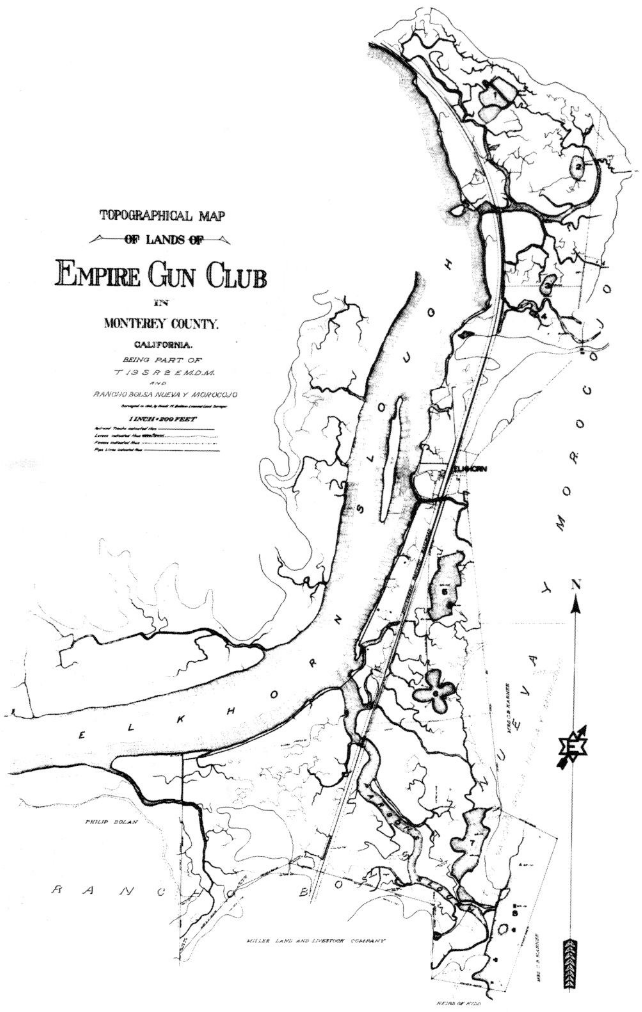

Figure 7.2. 1913 map of the southeast corner of Elkhorn Slough, now part of the Elkhorn Slough National Estuarine Research Reserve.

Dairies

In 1906 two Empire Gun Club members—J. Henry Meyer, a San Francisco banker, and his son-in-law and business partner, Frank Buck—bought 194 hectares (480 acres) at the site of the present-day Elkhorn Slough National Estuarine Research Reserve (ESNERR) and built their own hunting lodge. Nine years later, Meyer expanded his already substantial business interests by buying a herd of Ayrshire cows for the property. He later added about 202 hectares (500 acres) to form the Elkhorn Dairy, which produced, processed, and marketed milk from the 1920s to the 1950s. At its peak the dairy supplied 75 percent of Watsonville's retail dairy sales and for a time was the sole supplier for Stanford University, of which Meyer and Buck were

Elkhorn Dairy, circa 1940s. Bob Bowen Collection.

prominent benefactors. Calcagno's Moon Glow Dairy and Silva Dairy were also established in the early 1900s on lands adjacent to the slough.

Slough Modifications

During the early twentieth century, human activities and natural events combined to radically alter the nature of Elkhorn Slough. Throughout the slough, wetlands were lost to diking and draining as the land was converted to farmland and pasture. The present ESNERR property provides a good example. As a 1913 map of the Empire Gun Club shows (fig. 7.2), marshland accounted for 310 hectares (757 acres), with only about 40 hectares (100 acres) in highland. After Meyer and Buck expanded their dairy operations, much of the marsh was diked off and converted to pasture.

In 1886 S. N. Laughlin of the Pacific Coast Steamship Company deeded the right of way over Moro Cojo Slough to the County of Monterey so that a dam and floodgate could be built next to the old bridge to prevent the free tidal exchange of water. Although it is unclear exactly when the dam was built, A. T. Vierra walked over it on his way to school as a youngster in 1900. He remembered that the slough became stagnant after the dam was built and that "fishing degenerated to schools of large brown carp in the upper reaches to a few ancient striped bass in the lower slough" (letter of May 30, 1972, to Garth V. Lacey). A dramatic change in slough hydrology occurred following a series of winter storms in 1908, when the Salinas River, which had previously joined Elkhorn Slough before emptying into the bay north of Moss Landing, broke

Aerial view of the central Monterey Bay, circa 1930s, prior to the dredging of the Moss Landing Harbor channel. Old Salinas River on left, Moro Cojo on the lower right, with Elkhorn Slough in the upper right. (no photo credit available)

through the dunes approximately five miles farther south. Its new mouth, which persists today, is just south of both Mulligan Hill and the head of the Monterey submarine canyon (see fig. 2.5 in chapter 2, "Geology"). The slough, meanwhile, continued to drain into the bay via the old Salinas River channel, and to receive saltwater from the bay on returning tides, though its main supply of freshwater was now cut off.

The Salinas River has shifted its course repeatedly over recent geological time, emptying both north and south of the submarine canyon, but until 1909 it had held steady in the northern position for most of the historical period. Records from the time indicate that it was moving ever northward toward the Pajaro River, but that it sometimes breached the dunes south of Mulligan Hill during floods. With agriculture increasing along the river and on its floodplain, it had become common practice by the turn of the century, as it still is today, to prevent the Salinas River from flooding in wet winters by opening a channel to the bay at the site of the present river mouth (Gordon 1996).

The change in location of the Salinas River mouth had an enormous impact on Elkhorn Slough. The Salinas River had contributed large amounts of freshwater and sediments to the slough ecosystem, particularly during wet winters. Long a seasonally freshwater or brackish estuary, Elkhorn Slough was now transformed into a saltwater system. As the naturalist G. E. MacGinitie noted in 1935, "The connection between the Salinas River and the Slough is disestablished to all intents and purposes: Elkhorn Slough at the present time may be considered strictly a saltwater estuary."

Oyster Culture and Shellfish Use

The change from a river mouth to a saltwater environment opened new possibilities for economic use of the slough. By the early 1900s the once-thriving oyster industry of San Francisco Bay had fallen on hard times, probably because of increasing pollution. Raising and marketing oysters is, for a variety of reasons, a precarious business. The animals require suspended nutrients in their habitats but are vulnerable to being silted over. They are also temperature-sensitive when it comes to reproducing—hence to sustaining a population beyond one generation. The commercially profitable species of oysters, grown from "seed" or "spat" (very young, newly settled juveniles) imported from the eastern United States or Japan, cannot reproduce in cold California waters, even though they grow well here. The delicious native West Coast oyster (*Ostrea lurida*) could be the commercial species of local choice, but it is small and "all shell." Oyster growers face not just high initial investments, but also the continuing cost of bringing in healthy new seed to keep the crop going, along with the risky business of raising, shipping, and marketing the animals.

Unpublished Monterey County records show oyster leases in the lower slough and Salinas River mouth as early as 1904 (Mark Silberstein, pers. comm). In spite of the early establishment of leases, the history of oyster farming in Elkhorn Slough during the 1920s and 1930s is a complicated picture of mostly failed experiments. In 1923 Consolidated Oyster Company experimented with raising oysters, probably from the East Coast. Evidently they did not find suitable beds, for the oysters were "easily silted over" (Barrett 1963). In 1929 another experiment, this time with oysters from the west coast of Mexico (Barrett 1963), also ended in failure. A year later West Coast Oyster Farms tried the American method of mudflat or bottom culture, using Pacific oyster seed from Japan. In 1932 they adopted the Japanese method of hanging seed oysters from floats suspended over the deepest parts of the slough. This innovation was so successful that after only one year in the slough's relatively warm water the oysters had reached a marketable size and could be harvested. During the 1930s the slough brought in a significant portion of the state's harvest, peaking at about 16 percent in the first half of the decade (Barrett 1963).

Aerial view of Elkhorn Slough, circa 1930s. Salt evaporating ponds are in the upper left, and oyster racks are visible in the lower center of the main channel. (no photo credit available)

In spite of this success, the company (now called San Francisco International Fish Co.) made its last planting in 1936. Barrett (1963) suggests that the involvement of Japanese personnel in the operation may have made the oysters difficult to market in those prewar years. Lydon (1997) likewise notes that the anti-Japanese movement targeted all niches developed by the Japanese, but particularly the fishing industry.

According to Gordon (1996), Pacific oysters could still be found in the slough as late as 1970. The most significant long-term effect of these oyster farming efforts, however, was the accidental introduction into the slough of plants and animals imported with the oyster seed. Among them are the Atlantic oyster drill snail, which preys on mollusks; the soft-shell clam; the Japanese cockle; a Japanese mussel and horn snail; and perhaps also the bay mussel and shipworm. (See chapter [illegible], "Invertebrates," for more information on introduced invertebrate species.)

Other shellfish were also harvested from slough waters. Native clams such as the Washington and gaper as well as cockles were plentiful in intertidal mudflats. In the 1920s and 1930s shellfish harvesting became a significant enterprise; in 1931 and 1933, for example, Elkhorn Slough supplied 45 percent and 31 percent, respectively, of all the shellfish harvested in the state (Foster et al. 1967).

Salt Ponds

In addition to oyster farming, diversion of the Salinas River mouth created other commercial possibilities. Although the Vierras had had a small salt pond near the mouth of Elkhorn

Harvested salt from evaporating ponds at Elkhorn Slough, late 1800s or early 1900s. Photo credit: Carlos Vierra.

Slough in the late 1800s, the slough's newly reliable salinity made salt production viable on a larger scale. The Monterey Bay Salt Company began production in 1916, using approximately 65 hectares (162 acres) near the slough mouth (site of the present-day ponds on the Moss Landing Wildlife Management Area) for evaporating and crystallizing ponds (Gordon 1996; Hart 1966); it continued production until 1973. Some of this salt was used by the local fish canneries (Browning 1972).

Industrialization and Changing Perspectives: 1942-2000

America's entry into World War II had a direct impact on Elkhorn Slough. The Kaiser Refractory, which extracted magnesium from seawater for making war materials, was the first large-scale industrial enterprise developed at the slough. More significantly, wartime efforts to improve Moss Landing Harbor by opening a direct channel to Monterey Bay permanently altered the slough environment. Following the war, commercial enterprises and housing expanded at Elkhorn Slough and throughout the watershed. Pacific Gas & Electric Company (PG&E) built a major power plant at the slough's mouth, and developers proposed other large-scale commercial and industrial plans for the area, including a major oil refinery and a nuclear power plant.

As pressure for development increased, many began to appreciate the slough's intrinsic value as an important ecological system that was in danger of being lost. It was already clear that land use such as farming and dairy operations were affecting slough habitats. Beginning in the 1970s, legislation mandating conservation of coastal resources and private efforts to protect the slough and its watershed combined to change the direction of land use at Elkhorn Slough, leading to the mix of commercial and conservation uses we see today.

Industrial Development

The 1940s and 1950s marked the start of industrial development at Elkhorn Slough. Kaiser Refractories built a plant in 1942 to extract magnesium from seawater; the magnesium was used to make incendiary bombs used in World War II (T. K. McCarthy, pers. comm.). After the war the magnesium plant was converted to making refractory materials and bricks used in steel mills. In the late 1980s the plant became part of National Refractories and Minerals. The plant closed in 2000.

In 1947 PG&E began construction of the Moss Landing power plant in a broccoli field bordered by Highway 1 and Dolan Road, just inland of Moss Landing Harbor. The plant began operations in 1950, with three generating units at the mouth of Elkhorn Slough producing 330 megawatts of power. Two more units (producing 240 megawatts) were built in 1952, and units 6 and 7 (producing 1,500 megawatts) were added in 1968. Today, the plant burns entirely natural gas, which is brought in by a pipeline constructed in 1950. Oil delivered by tankers and stored in tanks on the property was

Moss Landing power plant, 1987, with its two 500-foot-tall stacks and multiple smaller stacks. Note stratified wind shear, with onshore flow below 200 feet and offshore flow at higher elevations. Photo credit: Michael Kenna.

used as a backup fuel when natural gas was in short supply. The power plant removed water from the harbor for cooling turbines, and released about 25 percent of the heated water (about 5°C [41°F] above ambient temperatures) back into the slough, with the remainder released 1 kilometer (0.62 mi) offshore in Monterey Bay.

During the 1970s and 1980s the power plant could produce up to 2,000 megawatts for use in surrounding areas, depending on demand. By the early 1990s, units 1 to 5 from the original plant became increasingly unreliable and were much less efficient than units 6 and 7. In 1995, several operational changes were made: pumping units 1–5 were taken offline, and oil burning capacities were eliminated. Since units 1–5 were the sole dischargers into Elkhorn Slough, cooling water discharges into the slough ended at this time. Duke Power purchased the plant in 1998 and built a new plant (see below). With its two active 500-foot-tall stacks, the power plant continues to be the most visually dominant industry in the area.

Moss Landing Harbor

In 1943, the secretary of war initiated a project to improve Moss Landing Harbor "in the interest of the wartime need for increase in the production of fishery commodities" (Senate Doc. No. 50, 1945, quoted in Browning 1972). The 1945 River and Harbor Act authorized the harbor improvement, on which the Army Corps of Engineers eventually spent $33[illegible],215 (Browning 1972). In the summer and fall of 1946 the Corps

Strawberry Hills

History of the Strawberry Industry

Strawberries were first grown in the Elkhorn Slough region at the turn of the nineteenth century by Americans of northern European descent (Wells 1991). Labor shortfalls in the expanding agricultural sector led to the importation first of Chinese and then Japanese laborers. The Japanese who came to work in the area brought techniques of intensive farming and a cultural value system that placed high esteem on farming (Iwata 1962). Traditional forms of Japanese labor organization allowed the workers to negotiate for higher earnings and thereby accumulate sufficient savings to begin farming as tenants. Before long they began to acquire land of their own on which they produced strawberries and other labor-intensive crops.

The local Anglo farming community saw the Japanese success as a threat and, with the passing of the Alien Land Law in 1913, soon prevented Japanese from owning land. The Japanese responded by directing their efforts to sharecropping and tenant farming on small acreages using intensive production methods. Continuing to draw upon their established social relations to effectively compete in financial and sales markets, they prospered. A few Japanese formed a strawberry marketing cooperative with Anglo landowners in 1917 in order to strengthen their combined position within the industry (Wells, 1996). By the beginning of World War II, the Japanese were producing 90 percent of the strawberries in California (Saloutos 1976).

With the war came the internment of all Japanese farmers and a stagnation of the strawberry industry. Upon their eventual release, many resumed sharecropping or tenant farming, but in 1952, with the repeal of the Alien Land Laws, the strawberry industry took on a new form. The Japanese moved to acquire small acreages, primarily in the lower-priced coastal bluffs and hill lands surrounding Elkhorn Slough. Many former rangelands were converted, new wells were bored, and intensive hillside strawberry production was begun. This new form of production required careful management but yielded high returns to the hardworking Japanese entrepreneurs. Over the next two decades, the number of Japanese strawberry growers expanded to three-quarters of all central coast strawberry producers (Wells 1991). The other 25 percent of producers were Anglo farmers, consisting of a few well-established strawberry families and a new infusion of large-scale corporate producers.

Beginning in the 1950s, new strawberry varieties and production methods resulted in higher yields but demanded more labor than the farm family could provide. Seasonal Mexican migration, which had begun after the Mexican Revolution in 1910 (Saloutos 1976), now became the primary source of California agricultural labor under the federally managed Bracero program. These sanctioned seasonal laborers, and the later undocumented migrants, dredged a new entrance to Moss Landing Harbor, cutting through the sandbar between the harbor and the slough, thus opening the estuary directly to Monterey Bay.

The original plan was for the Corps to construct a minimum harbor with wooden jetties that could be put up rapidly at a reduced cost. After the war, the Corps would use local matching funds to construct a more permanent harbor with rock jetties, as well as tide gates at Bennett, Moro Cojo, and Elkhorn Sloughs that would minimize tidal incursion. The total estimated cost for the larger harbor project was $965,000, with half that amount to be contributed by the Moss Landing Port District (precursor to the present-day Moss Landing Harbor District). In a memo dated June 9, 1945, Henry S. Pond, the Corps district engineer, wrote to his superiors that,

> Representatives of the Moss Landing Port ... are frank in acknowledging that the district is attempting to obtain the construction of the minimum harbor and to delay the authorization of the larger harbor in order to evade the burden of making the required contribution toward the cost of the latter. If they succeed, they will place the responsibility for later replacing the wooden jetties with permanent jetties and also of installing permanent drainage structures on Elkhorn, Moro Cojo and Bennett Sloughs onto the United States.

---

began to learn the process of strawberry production from their Anglo and Japanese employers. Some became sharecroppers in the 1960s, and in the 1970s publicly sponsored production co-ops provided other Mexican laborers the opportunity to became independent producers (Rochin 1986). The rise of Mexican laborers into independent strawberry production occurred primarily in the Elkhorn Slough watershed on small hilly farms. Some took over land given up by Japanese who had moved to flatter land, while others initiated new cultivation of steeper undeveloped property.

An increased demand for fresh strawberries in the mid-1970s led to the founding of new strawberry shipping companies that were willing to finance farmers' production costs in order to capture a share of the berry market. The formation of these new shipping companies fortuitously corresponded with a decline in sharecropping and the failure, owing to inadequate experience, capital, and business relationships, of many of the Mexican cooperatives. By offering production loans to these farmers, the shippers made independent farming possible for the Mexicans, who otherwise lacked the requisite social ties or collateral to obtain commercial loans. As a result of these new markets and production financing, Mexicans became the most numerous strawberry growers along the central coast by the late 1970s (Wells 1991). Today in the Elkhorn Slough watershed, 80 percent of the strawberry farmers are of Mexican ancestry, and they farm 87 percent of the strawberry acreage. This stands in sharp contrast to state and county proportions of Hispanic independent producers in other agricultural sectors (table 7.1; Mountjoy 1996a).

— *Daniel Mountjoy*

Table 7.1. Ethnic distribution of farmers in California, Monterey County, and the Elkhorn Slough watershed.

	California	Monterey County	Elkhorn Slough Watershed	
Ethnicity	% of all Farm Operators	% of Strawberry Farmers	% of Strawberry Farmers	% of Strawberry Acreage
White (Anglo)	89.0	20.7	5.6	4.4
Asian (Japanese)	4.1	31.7	14.0	8.5
Hispanic (Mexican)	4.2	47.6	80.4	87.1
Others	2.7	0	0	0

Note: Ethnic categories vary for each of the different data sources. The first headings reflect the inclusive categories used by the U.S. Bureau of the Census (1989). Parenthetic categories are the ethnic terms applied by Wells (1988) and Mountjoy (1996b). The "Others" category includes the census categories of Black, American Indian, and Other, none of which are encountered in the local strawberry industry.

Sources: for all farm operators in California, Bureau of the Census 1989, table 17: Selected Characteristics of Farms Operated by Females, Persons of Spanish Origin, and Specified Racial Groups, 1987; for Monterey County strawberry farmers, 1985 Monterey County Agricultural Commissioners' pesticide enforcement records, as reported in Wells 1991, 745; for Elkhorn Slough strawberry farmers, 1991 Monterey County Agricultural Commissioners' pesticide use permits. Empirical acreage totals (of lands within the watershed only) from survey respondents were projected to watershed totals from sample population as proportion of total population (Mountjoy 1996a).

The original timber jetties were damaged by the winter storms of 1946 and replaced by rock jetties in 1947 and 1948. Thus, the temporary minimum became a permanent harbor without the local contribution of funds or the construction of tide gates. In fact, the permanent project was never authorized. In a memo from August 25, 1989, G. H. Yanagihara of the U.S. Army Corps of Engineers, San Francisco District, acknowledged that the loss of marshland and erosion of the slough was a direct result of the opening of the harbor and that "the Federal Government constructed the project without taking measures to safeguard Elkhorn Slough, which the Corps knew would be adversely affected by the construction of the project."

Opening the slough to Monterey Bay expanded the volume of water exchanged with each tide and increased the speed of tidal currents, thus subjecting the slough's lower reaches to significant scouring and erosion (see chapter 4, "Hydrography"), which has affected many plant and animal species (see, for example, chapter 10, "Fishes"). In 1989, construction of a rock sill across the mouth of Elkhorn Slough was proposed to reduce tidal incursion at an estimated cost of $1 million (Yanagihara memo, August 25, 1989); later estimates projected a cost of over $3.6 million, which did not include feasibility studies, detailed modeling, or environmental review (Phillip Williams & Associates, 1992). The sill has never been built.

Commercial Fishing

Commercial fishing out of Moss Landing became significant after 1935, and by 1952 the fishery supported eight canneries and reduction plants at Moss Landing (Browning 1972). Fishing fleets based in Moss Landing and Monterey Harbors pursued the bay's seemingly limitless schools of sardines. Some thirty to forty boats—primarily purse seiners, but also trawlers and small trolling boats—were docked in Moss Landing Harbor during the height of the fishery (Browning 1972).

Most of the canneries went out of business following the decline of the sardine fishery that began in 1947 (Gordon 1996). By 1967 only the Santa Cruz Canning Company remained open (it closed in 1991), although three other plants continued to process fresh fish for market (Foster et al. 1967). Nevertheless, commercial fishing continued to play an important role in Moss Landing's economy, with 15 million pounds of fish worth $770,000 landed in 1970 (Heimann and Carlisle 1970). That year more than two hundred commercial fishing boats used Moss Landing Harbor as their home port.

Part of the commercial fishing fleet at Moss Landing Harbor. Photo credit: Mark Silberstein

In recent years, salmon, albacore, anchovies, squid, rockfish, jack mackerel, herring, sole, sanddabs, and sablefish have been the principal species taken by commercial fisheries in offshore waters (U.S. Department of Commerce 1979). Landings at Moss Landing fluctuated dramatically in the 1990s, ranging from a low of 6.9 million pounds in 1993, valued at $[illegible].1 million, to a high of 44.6 million pounds in 1997, valued at $9.4 million (National Marine Fisheries Service web site). In 1999 the harbor supported 115 vessels involved in charter and commercial fishing (Linda G. Horning, pers. comm.).

Salt Pond Operations

The Monterey Bay Salt Company continued production at its ponds until 1973, after which the ponds were used for several years to produce brine shrimp. The ponds affected the surrounding areas by creating regions of high salinity and became habitats for aquatic invertebrates such as brine shrimp and for the many species of shorebirds that feed on them (Gordon 1996). The California Department of Fish and Game acquired this property and some of the salt marshes east of the salt ponds in 1984 to create the Moss Landing Wildlife Management Area. Because some of the dikes began to fail, a new levee and water control structures were built in the 1990s, splitting the salt ponds in half. Water levels in the inner half of the salt ponds can be controlled; the outer half next to the slough, however, is exposed to tidal inundation since the old levee system failed.

Growth in Agriculture

Agriculture in the Elkhorn Slough watershed expanded and diversified in the postwar years. By the 1960s, greenhouses for flowers, nursery plants, and mushrooms became a common sight while improved irrigation technology made even more extensive strawberry farming possible. Artichokes and brussels sprouts and other row crops were concentrated on the heavier soils of the flatter, low-lying areas, while the watershed's steeper soils supported orchards and annual or multiple-year crops of strawberries, raspberries, and flowers.

The most extensively grown crop is strawberries (table 7.2), which tend to be concentrated on the steep, erosion-prone hillsides surrounding the slough. The moderate climate, well-drained soils, and southern orientation of these hills provide an ideal environment for berry production. The acreage indicated in table 7.2 covers all the land devoted to strawberry production, although only a portion is in production at any one time due to fallowing and crop rotations. The area in production in 1992 was estimated to be 1,113 hectares (2,750 acres). This represents 40 percent of the total Monterey County strawberry-producing acreage and 11.7 percent of total California strawberry acreage (Monterey County 1994). Strawberry acreage in Monterey County more than doubled between 1930 and 1990 in response to consumer demand. Between 1990 and 2000, strawberry acreage increased by only 20% due to declining profit margins and lack of new land. Expansion in the steeper hills around Elkhorn Slough has been curtailed by county zoning laws, which require permits for new cultivation on slopes exceeding 10 percent. Some growers have voluntarily stopped producing on steeper lands.

An abandoned brine shrimp harvesting boat, Elkhorn Slough salt ponds, 1987. Photo credit: Michael Kenna.

Dairy and Cattle Operations

Cattle operations begun in the early 1900s continued to produce both meat and milk during and following the war, and

Table 7.2. Estimated land use and crop distribution in Elkhorn Slough watershed.

Land Use	Acres*	Percent of Total Watershed	Acres Cropped	Percent of Crop Acreage
Native vegetation	26,000	57.9		
Pasture	400	0.9		
Water bodies	1,000	2.2		
Urban, suburban, rural residential, highway, idle	7,200	16.0		
Total cropland	10,300	22.9		
Field crops			300	2.9
Bushberries			400	3.9
Orchards			500	4.9
Flowers, nurseries			900	8.7
Artichokes			1,300	12.6
Strawberries (including rotation)			3,600	35.0
Other crops			3,300	32.0
TOTAL	**44,900**	**100.0**	**10,300**	**100.00**

Source: USDA 1994. *There may be minor discrepancies between these figures and current GIS-based data (Van Dyke and Contreras 2001).

are still active in the region. A 1993 survey found 3,000 – 4,000 cows in the Elkhorn Slough watershed (Mark Silberstein, pers. comm.). In 1998 an estimated 1,600 cows were being grazed along the shores of the slough, with 1,000 of these at Moon Glow Dairy (the former Calcagno Dairy), another 500 on property off Dolan Road, and smaller herds on other slough-side properties (Andrea Woolfolk, pers. comm.).

The use of slough-side land for cattle grazing began to decline in the 1980s. Elkhorn Dairy, which had a herd of some 650–700 cows in the 1960s (Foster et al. 1967), ceased operations in 1972 and the pastures were leased out for cattle grazing until 1982. The present-day ESNERR was established on these lands in 1979 (see below and chapter 1, "Introduction," for more on the ESNERR's establishment). The Estrada Marsh was grazed until the early 1990s, when it became part of the ESNERR. A small herd of cattle grazed at the Packard Ranch but were removed in the late 1990s. Cattle that had grazed on the southwestern side of Moro Cojo were removed when the land was purchased by the Elkhorn Slough Foundation in 1998. An active sustainable grazing program was established in the early 1990s on the Porter Ranch, owned by the Foundation.

Residential Development

The changes in the overall population of Monterey County are perhaps representative of the changes within the Elkhorn Slough watershed. Monterey County has grown on average about 3 percent per year since 1900, with the greatest increases occurring in the periods 1920–1930 and 1940–1960 (fig. 7.3). The population nearly doubled between the 1920 and 1930 censuses, and it nearly tripled between 1940 and 1960. In 1990, the U.S. Census Bureau estimated that 37,314 people lived around the slough, of whom 32,179 are in Monterey County and 5,135 in San Benito County. Castroville (population 5,272) is the largest town in the watershed.

Proposed Developments

During the 1960s and 1970s many plans to develop Elkhorn Slough were initiated (fig. 7.4). In 1965, Humble Oil (now Exxon) proposed building a 50,000-barrel-a-day refinery just south of Moss Landing. In looking back on the proposal 20 years later, *Monterey Peninsula Herald* reporter Ken Peterson wrote, "[The Humble Oil refinery would have been] the cornerstone of what some hoped in 1965 would become a

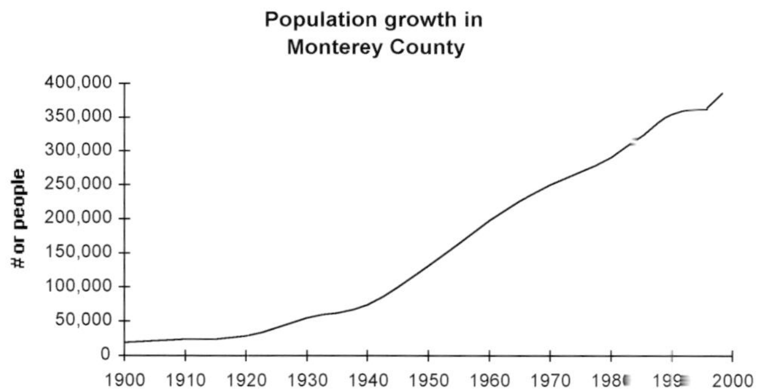

Figure 7.3. Population growth in Monterey County since 1900

3,800-acre industrial zone around Moss Landing, perhaps even the start of a Moss Landing–Salinas industrial corridor," (*Monterey Peninsula Herald*, October 20, 1985).

The refinery proposal galvanized environmental groups, which gathered 15,000 signatures in six months opposing the project, while proponents who wanted to see economic development in the region gathered 8,000 signatures (*Monterey Peninsula Herald*, October 20, 1985). Monterey County planning commissioners rejected the project 5 to 4, a vote that was later overturned by the board of supervisors. Because of the intense opposition to the refinery, Humble Oil moved the project to Benicia, on San Francisco Bay. Had the refinery been built at Moss Landing, a deepwater port for supertankers to offload crude oil would probably have been developed.

During this same era, PG&E considered building a nuclear power plant next to the existing power plant. There were also proposals by real estate developers to build a marina in the salt ponds and hundreds of condominiums on the site of the present-day Packard Ranch (*North County News*, October 6, 1976). Caltrans proposed rerouting Highway 1 across the middle of the slough, through what are now the Packard Ranch and Elkhorn Native Plant Nursery. This proposal was finally dropped in 1996 thanks in part to the efforts of Wil Smith of the Elkhorn Slough Foundation Board, who petitioned the Monterey County Transportation Agency to remove this proposed realignment as an option (Mark Silberstein, pers. comm.).

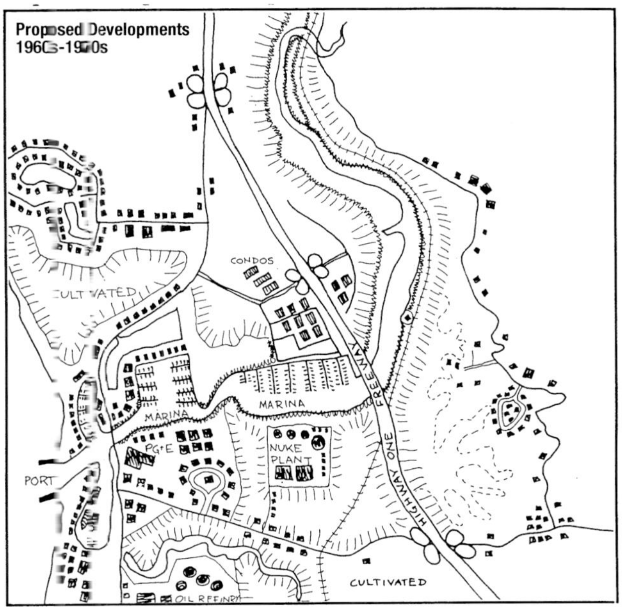

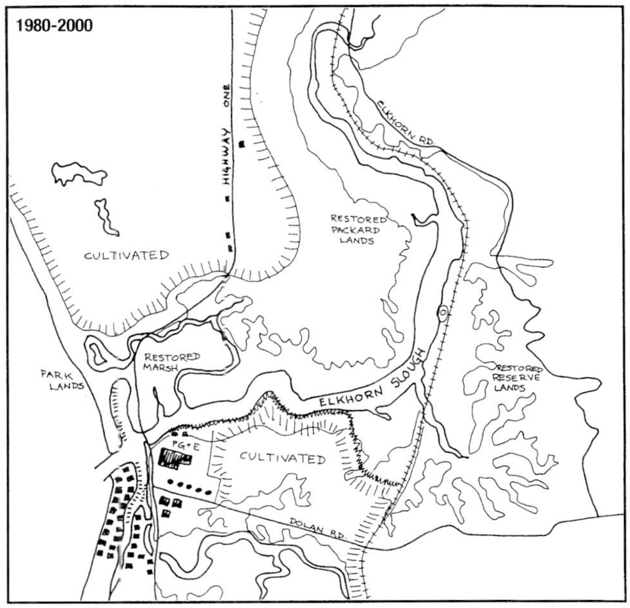

Figure 7.4 This map shows what the slough might have been like had developers and industrialists prevailed. During the 1960s and 1970s, several development schemes were proposed for the slough. PG&E considered building a nuclear power plant. The Humble Oil Company (now Exxon, Inc.) drew up plans for a major oil refinery that would adjoin a deep-water harbor at Moss Landing where giant oil tankers would pump oil. Real-estate developers proposed building hundreds of condominiums on the site of the current Packard Ranch, complete with a marina. Government agencies also had plans for the slough. Cal Trans suggested rerouting Highway 1 by building an overpass that would bring hundreds of cars over the slough each day.

Fortunately, biologists, planners, conservationists, business people and public citizens had a different vision for the slough. In 1978, the slough became one of 20 sites in the United States to be nominated as a national estuarine research reserve. Today, over 4,000 acres of the slough is managed by public agencies, private land-holders or conservation groups dedicated to preserving the land. Instead of a nuclear power plant and housing tracts, a wildlife area, state park and an undeveloped ranch are the slough's nearest neighbors. Thanks to the efforts of the current landowners, parts of the slough are returning to their natural state.

Designation of the Elkhorn Slough National Estuarine Research Reserve

The development pressures outlined above catalyzed the early environmental movement in the Elkhorn Slough area. On a larger scale, the history of the slough's designation as an estuarine research reserve is intertwined with the beginnings of environmental awareness, both locally and globally. In 1968, the World Health Organization (WHO) released a demographic study predicting that by the beginning of the twenty-first century 80 percent of the world's population would reside within 20 kilometers (12.4 miles)of the coastline, and urged member nations to examine issues of sustainability connected with such a population shift. In the United States, response to this document took the form of increasing infrastructure. By the early 1970s, however, writers such as Rachel Carson had begun to influence public thinking about the nation's natural resources and the need to protect them.

Between 1970 and 1972, this growing awareness led to the passage of three major pieces of federal legislation: the National Environment Policy Act, the Clean Water Act, and the Marine Mammal Protection Act. Additionally, a voluntary program, the Coastal Zone Management Act, was passed in 1972, providing financial incentives to states for establishing estuarine sanctuaries. A parallel program established marine sanctuaries. The act established the Federal Marine Sanctuaries Program to develop a nationwide network of estuarine reserves and preserve them for research and education. The program made available up to $2 million to coastal states for the planning, purchase, and startup operation of these preserves; matching state funds were required.

At the state level, the California Coastal Plan, written in response to the Coastal Conservation Initiative (Proposition 20, passed in 1972), included the recommendation that a management plan be prepared for the entire Elkhorn Slough

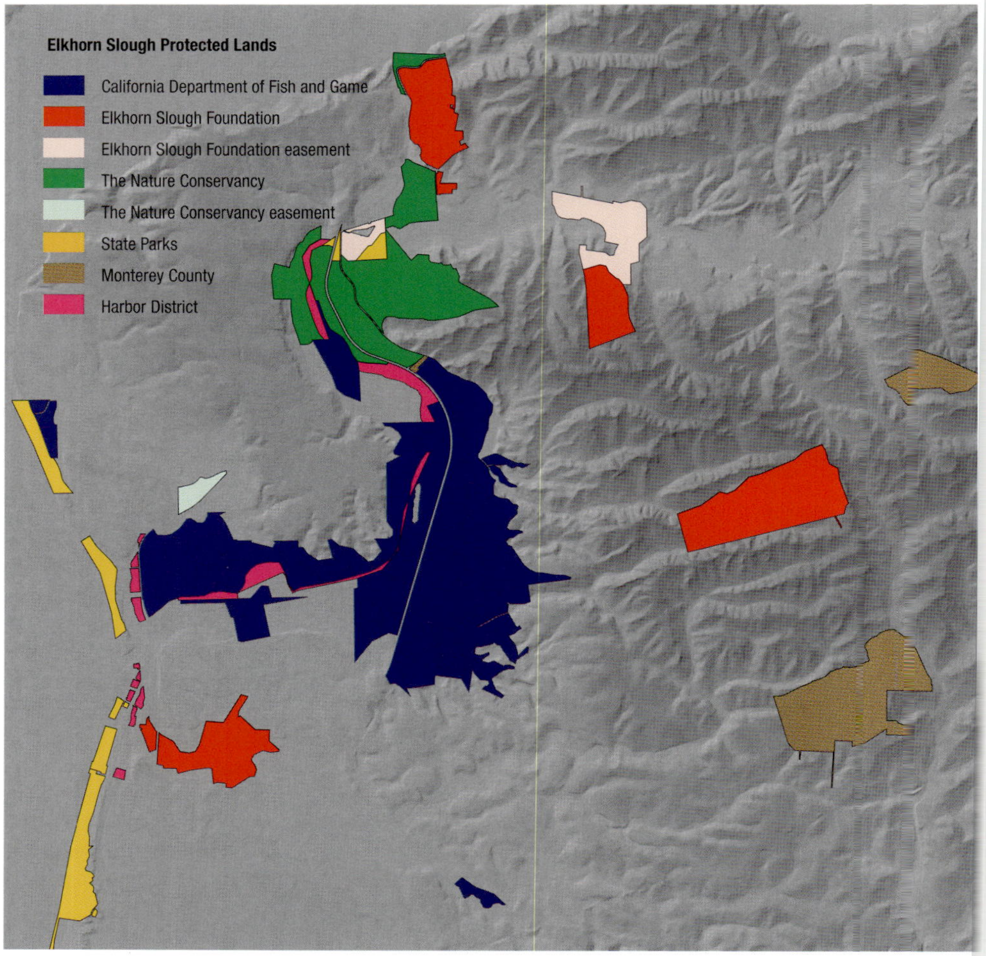

Figure 7.5 Lands protected by public agencies or conservation organizations as of 2001.

watershed, and that Elkhorn Slough and the Monterey Bay be considered for designation as national sanctuaries.

The federal Bureau of Sport Fishery, which would later become the Fish and Wildlife Service, was interested in obtaining Elkhorn Slough and surrounding lands for a preserve. The bureau intended to use eminent domain where necessary to acquire the lands. These plans met with opposition from many property owners and developers and from the Moss Landing Harbor District.

The Nature Conservancy had also begun conservation work in the slough. Having identified the area as an important ecosystem in need of protection, the organization quietly began to purchase land parcels and easements around the slough, including 57 hectares (140 acres) in the upper slough in 1971 (*Monterey Peninsula Herald*, May 5, 1971).

Meanwhile, a number of environmental groups and local regulatory agencies also began to push for the designation of Elkhorn Slough as a national estuarine sanctuary. Their plan was to purchase some of the lands surrounding the slough and promote stewardship by private landowners in the watershed. Wetland management would be carried out as a joint effort of UC Berkeley, California State University through its Moss Landing Marine Laboratories facility, and the Monterey County Planning Department.

In 1974 the California Coastal Commission, acting on recommendations from the public and local agencies, nominated Elkhorn Slough, along with Upper Newport Bay and Tomales Bay, for designation as a federal estuarine sanctuary. Among the arguments made for Elkhorn Slough's designation was the need for applied wetland and estuarine research in the state and the opportunity that the slough, with its wide variety of microhabitats, provided for such research as an important "living laboratory." In 1975 the Fish and Wildlife Service pulled out of the process, leaving local organizations and the state Department of Fish and Game to pursue sanctuary status. In 1980, 405 hectares (1,000 acres) of Elkhorn Slough lands located on the old Elkhorn Dairy were purchased by the Department of Fish and Game and named the Elkhorn Slough Ecological Reserve. Shortly thereafter, the slough received federal estuarine sanctuary status and was renamed Elkhorn Slough National Estuarine Sanctuary, which was later changed to the Elkhorn Slough National Estuarine Research Reserve. The reserve has since grown to 567 hectares (1,400 acres) in the intervening years.

The original goal of the federal sanctuary program was to provide natural, outdoor laboratories in estuarine environments for research (both scientific and policy-oriented) and educational use (Madrone Associates 1979). To carry out this objective, maintenance and restoration of the slough ecosystem, as well as long-term preservation were also identified as critical goals. At the time of Elkhorn Slough's federal designation, the slough was actively used for recreational and commercial purposes. It was agreed that these activities would be allowed to continue as long as they did not interfere with the sanctuary's primary goal of long-term protection for education and research.

The Elkhorn Slough Foundation was established in 1982, shortly after the designation of the Elkhorn Slough Reserve, and has collaborated with the reserve and the Department of Fish and Game in program development. The Foundation is dedicated to the conservation and restoration of Elkhorn Slough and to promoting the wise use of coastal resources through research, education, restoration, and habitat protection. It has been a critical partner with the reserve, helping to protect the land and waters of Elkhorn Slough. In 1997 the Foundation assumed responsibility for acquiring and managing conservation lands in the slough, and currently owns and/or manages over 800 hectares (2,000 acres) of land and easements.

Incremental land acquisition efforts begun in the 1980s have allowed for protection of natural cover, groundwater conservation and recharge, and restoration and enhancement of natural communities both within the ESNERR and on surrounding lands in the watershed. These projects have included marsh and riparian restoration, as well as erosion prevention, research, and education efforts with local landowners and farmers (see chapter 14, "Management Issues," for details).

Protection of other sensitive habitats within the region (fig. 7.5) has been ongoing since the 1960s, when California State Parks established the Salinas River State Beach (1960), Zmudowski State Beach (1962), and Moss Landing State Beach (1972). These parks protect the fragile dunes fronting Monterey Bay. The Moss Landing Wildlife Management Area was established in 1984 on 259 hectares (640 acres). In the mid-1980s David Packard purchased over 480 hectares (1,185 acres) of prime agricultural lands on the Springfield Terrace. Some of the land remains in agricultural production, some has become a native plant nursery, and the remainder has been restored. The Blohm and Azevedo Ranches were purchased by The Nature Conservancy in the early 1990s. More recent acquisitions by the Elkhorn Slough Foundation include 103 hectares (254 acres) of the Porter Ranch plus more than 80 hectares (200+ acres) in conservation easements, approximately 86 hectares (212 acres) along Moro Cojo Slough, and the 172-hectare (425-acre) Long Valley property, which includes extensive fragile maritime chaparral habitat. The Elkhorn Slough Foundation also protected the old 3M Ranch, totalling 134 hectares (329 acres).

Development of Research Institutions

Although the history of scientific research at Elkhorn Slough dates to the 1920s, when George MacGinitie conducted his studies of tide flat communities, on-site laboratory facilities weren't developed for another four decades. The Beaudette Foundation for Biological Research was the first research laboratory established at the slough. Palmer Beaudette, a philanthropist who had traveled throughout South America, hoped to use the ocean's productivity to address world hunger. Located initially in southern California, Beaudette's foundation moved to Moss Landing in 1963, where an abandoned sardine cannery was converted into a laboratory complete with library, seminar room, and both wet and dry laboratory space. The foundation's mission was to study marine plants and animals of the northeastern Pacific. Several scientists were hired to conduct marine research and to publish the journal *Pacific Naturalist*. Between 1958 and 1963 the foundation mounted six major field expeditions from California to Peru using their research vessel *Neptunus Rex*.

By the early 1960s, San Francisco State University (SFSU) had begun an effort to acquire a marine field station. The SFSU faculty heard that Beaudette was interested in selling the laboratory at Moss Landing, and in combination with faculty at San Jose State University they submitted a grant proposal to the National Science Foundation to purchase the facility and set up a teaching consortium with the state colleges at Fresno, Hayward, and Stanislaus. The Beaudette Foundation laboratory was purchased in 1966 and renamed the Moss Landing Marine Laboratories (MLML).

Initially, faculty from the consortium campuses came once a week to teach classes at the lab. After several years, however, MLML received new faculty positions so that professors could be in residence, a change that greatly strengthened the program (Margaret Bradbury, pers. comm.). Research by MLML's faculty and students has added immensely to our understanding of Elkhorn Slough (see chapter 1 for highlights of major studies).

In 1989, the Loma Prieta earthquake destroyed MLML's buildings and the laboratories were forced to relocate to trailers in Salinas and Moss Landing. Groundbreaking for the new MLML buildings in Moss Landing occurred in 1997, and the new lab opened in the spring of 2000. Currently, MLML serves as a coastal field station for a consortium of California state universities, providing both undergraduate- and graduate-level courses toward a master's degree.

In 1995, the Monterey Bay Aquarium Research Institute (MBARI) opened its Moss Landing site, establishing Elkhorn Slough's second major research facility. Founded in 1987 by David Packard as a nonprofit oceanographic research center, MBARI is funded by the David and Lucile Packard Foundation. Through a partnership of scientists, engineers, and operations staff, the institute develops instruments, methods, and systems to conduct deepwater research.

Researchers at MBARI have pioneered underwater exploration using remotely operated vehicles (ROVs) to study the geology, chemistry, and biology of cold seeps and the midwaters of the Monterey Bay Canyon, hydrothermal vents along the midocean ridges, and other areas of the deep ocean. In addition, they have developed new technologies in a variety of fields, from in situ monitoring instruments to track chemical changes in the ocean to DNA probes to identify different phytoplankton species.

Moss Landing Harbor is currently the home port for three major oceanographic research vessels: R/V *Point Sur*, operated by Moss Landing Marine Laboratories, and R/V *Point Lobos* and R/V *Western Flyer*, both owned by the Monterey Bay Aquarium Research Institute. Up to a dozen smaller research vessels are based at the harbor.

Elkhorn Slough Today

Today, Elkhorn Slough, Moss Landing, and the surrounding watershed are a multiuse region of industry, agriculture, residential development, and commercial enterprises existing side by side with lands protected for research, wildlife habitat, and recreation.

Residential and Commercial Activities

There are a number of businesses in Moss Landing, most of them marine- or visitor-related, including marine supply shops, restaurants, and gift and antiques stores. Although most of Moss Landing is residential, and the nearby towns of Prunedale, Pajaro, and Las Lomas are expanding, the area directly adjacent to the slough remains primarily in agriculture. Housing within the immediate slough area is currently low density, but homes are being built on or are planned for a growing number of sloughside parcels. The growth of Silicon Valley and the demand for rural homes may continue to put increasing development pressure on land within the watershed.

Duke Energy, the largest industrial entity at the slough, is building a new generating station to replace units 1–5 (built in the 1950s and retired from use in 1995) with two high-efficiency combined-cycle units. The new units should add 1,080 megawatts of power while drawing less cooling water than the old units (250,000 gallons per minute versus 380,000 gallons per minute); they will also discharge into Monterey Bay instead of into Elkhorn Slough. Additionally, the steam boilers currently in operation, units 6 and 7, will be upgraded. Concern over the impacts of the water intakes on slough organisms is discussed in chapter 9. National Refractories and Minerals, the other major industrial manufacturer in the watershed, closed in 2000. The land is now for sale and there are ongoing discussions about its future use.

Agriculture continues to dominate commercial use of watershed lands, occupying almost 4,400 hectares (11,000 acres), including fallow land and greenhouses (Van Dyke and Contreras 2001). Although grazing adjacent to the slough has declined, three large dairies and several smaller operations continue in the watershed (Henry Gonzales, pers. comm.). On the Porter Ranch, a program of sustainable grazing has been practiced for the past ten years. Thanks to the program, this property has the densest and healthiest stands of native bunchgrasses in the watershed (Mark Silberstein, pers. comm.). Other commercial activities include an active charter and commercial fishing industry based in Moss Landing Harbor.

Public Access and Recreation

As the central coast population continues to grow, outdoor recreation in remaining undeveloped areas has become increasingly popular. Elkhorn Slough offers a variety of recreational opportunities, with fishers, boaters, and nature enthusiasts using various access points to the slough. Moss Landing State Beach includes Moss Landing Harbor's north jetty, and the harbor itself is open to the public for fishing and boating. Several kayak rental businesses operate near the harbor and feature tours of the slough. On an average summer day, kayak concessions rent to 50–150 people. Hiking, birding, and seasonal waterfowl hunting are available in the Moss Landing Wildlife Area near the Bennett Slough branch of Elkhorn Slough. Kirby Park, operated by the Moss Landing Harbor District, includes a pier and a public boat launch and is used for kayaking, canoeing, and birding. The Elkhorn Slough Foundation built and maintains a wheelchair-accessible path along the slough's shore at Kirby Park. An estimated 60,000 visitors use the ESNERR's interpretation facilities and shore access trails each year, including 10,000 students on school trips. Beach visitation at the slough mouth is estimated at over 300,000 per year (Mark Silberstein, pers. comm.).

The bay, harbor, and slough are also used for recreational fishing. Small boats can be taken up into the slough from the harbor or from the boat launch at Kirby Park. Fish commonly caught in the slough include rubberlip surf perch, pile perch, black perch, jacksmelt, sand sole, staghorn sculpin, starry flounder, walleye perch, cabezon, bat ray, leopard shark, and round stingray. Local rod and gun clubs first sponsored annual shark derbies in the 1940s, which became catch-and-release tournaments between 1988 and 1996, the last year of the shark derby. A description of species abundance and diversity from these derbies and from creel surveys is included in chapter 10.

A number of benthic invertebrate species are harvested from mudflats for human consumption, despite a continuous advisory by the Monterey County Health Department since 1969 warning of shellfish contamination by fecal coliform. In recent years, contamination of shellfish by DDT and other pesticides has become a concern as well (see chapter 13 "Land Use and Contaminants"). Among the species taken are gaper, Washington, littleneck, and soft-shelled clams, oysters, and piddocks. Ghost shrimp, which are common inhabitants of

A lone kayaker explores tidal creeks in the mid slough. Photo credit: Paul Zaretsky.

intertidal mudflats, are collected for bait and in recent years have been severely depleted in some areas of the slough (Gardner and Kvitek 1998).

Current Conservation, Restoration, and Land Acquisition Efforts

Protection of Elkhorn Slough has centered on the acquisition of land and conservation easements surrounding the slough, and development of best management practices on agricultural lands within the watershed. Management issues, particularly conservation and restoration efforts, are discussed in detail in chapter 14 and will not be repeated here. The latest land acquisition efforts have been guided by the Elkhorn Slough Watershed Conservation Plan (Scharffenberger 1999). The goal of this plan is –

> … to preserve an intact and interconnected network of natural communities, including over

Elkhorn Slough's diverse migratory and resident bird populations draw birdwatchers year round. Photo credit: Paul Zaretsky.

> 1620 hectares (4,000 acres) of coastal marsh within Elkhorn and Moro Cojo Sloughs, the freshwater wetlands of McClusky Slough, a restored riparian forest in the lower Carneros Creek floodplain and a series of upland ridges with unfragmented maritime chaparral in the Elkhorn Highlands. The Plan envisions these natural communities surrounded by productive, habitat-compatible farmland, scenic vistas, and residences. As a whole, this landscape comprises 9106 hectares (22,500 acres), or approximately one half of the total watershed.

Resource protection, particularly on the few remaining large blocks of connected natural habitat lands, is a key element of the conservation plan. Recommended fee and easement acquisitions include marsh and buffer portions of properties in western Moro Cojo Slough, protecting intact maritime chaparral in the Elkhorn Highlands, and bluff portions of properties north and west of Elkhorn Slough. The plan encourages protection of McClusky Slough, as well as surrounding highly productive farmlands in Springfield Terrace. Priority restoration projects include Moro Cojo Slough marshlands, Porter Marsh, and critical linkages in the Elkhorn Highlands that were once maritime chaparral.

This ambitious undertaking does not rely solely on land acquisition, but supports existing programs and increased outreach to the community to promote compatible management practices by landowners.

Management Issues and Research Recommendations

This chapter has highlighted some of the ways that humans have used Elkhorn Slough's land and natural resources over the last five centuries, including how perceptions and values have changed. Much more research, particularly thorough historical analyses, needs to be done. We hope that this overview of the slough's land use history will further work in a variety of areas. Here we suggest research topics to expand our knowledge of Elkhorn Slough's past and provide information for future restoration efforts.

History of Native American and Early "American Period" Residents

Both here and in the previous chapter we have described our current understanding of Native Americans from the pre-Columbian era to the mid-1850s. A more detailed examination of the cultural history of Native Americans in the region post 1850 is needed.

This chapter includes but a few stories of the early residents of the Elkhorn Slough watershed. Descendents of original Ohlone inhabitants and of many settlers from the "American period" still live in the area and have a wealth of information about their ancestors. Valuable opportunities exist to gather oral histories from people whose heritage goes back to the original inhabitants, and those who are one or two generations removed from the first Europeans and Asians to settle in this area.

Dairies and Agriculture

We have described the early history of agriculture and the rise of strawberry farming in the Elkhorn Slough watershed. However, the role of Chinese and other ethnic groups needs to be incorporated. In addition, the dairies were an important part of the agricultural history, and their role in transforming the landscape deserves greater study. Understanding the dynamics of the current grazing and dairy operations in the slough would also be a valuable research goal.

The mosquito abatement district has had a significant effect on ecology of the area, particularly the alteration of wetlands by ditching and draining. A more detailed description of this agency's activities and their impacts would be useful.

Commercial Activities and Industrial Developments

Other exciting stories waiting to be told in more detail include: the history of commercial fishing in the region, including the whaling industry; development of the railroads, particularly competition between the coastal steamships and railroads; a proposal for an early undersea cable originating at Moss

Landing (S. Lydon, pers. comm.); the origin of the National Refractory plant; and Humble Oil's bid to make Moss Landing an industrial center with a major refinery.

Survey of Human, Cultural, and Ecological History

A scholarship or internship program should be developed to encourage dynamic research into Elkhorn Slough's human, cultural, and ecological history. In San Francisco Bay, this type of research has used old survey maps, sketches, paintings, photographs, and a variety of records from the eighteenth and nineteenth centuries to construct a picture of early bay habitats. Historical reconstructions of Elkhorn Slough's wetland habitats could be turned into Geographic Information Systems (GIS) data layers to guide restoration efforts, increase community awareness of the disappearance of critical habitats, and promote scientific understanding of community change through time.

In addition, a historical archive for the Elkhorn Slough area should be established, which could include photographs, other historical materials and artifacts, and perhaps the Elkhorn Slough Foundation's collection of aerial photographs dating back to 1931. Some of this material could be digitized and posted on the web. Development of this historical archive should be done in collaboration with existing groups, such as the Pajaro Valley Historical Society.

An old dairy barn on the Elkhorn Slough National Estuarine Research Reserve preserves part of the slough's cultural heritage. Photo credit: Paul Zaretsky.

References

Barrett, E. M. 1963. The California Oyster Industry. California Department of Fish and Game *Fish Bulletin* 123.

Browning, B. M. 1972. *The Natural Resources of Elkhorn Slough: Their Present and Future Use.* Coastal Wetland Series #4. Sacramento: California Department of Fish and Game.

Canright, A. 1999. Elkhorn Slough sea otters. *California Coast & Ocean,* Spring issue, pp. 32-37.

Clar, C.R. 1957. *Forest Use in Spanish-Mexican California.* Sacramento: Division of Forestry, Department of Natural Resources.

Dana, R.H. 1911. *Two Years before the Mast.* Boston: Houghton Mifflin Co.

Elliott, W. 1879. *Santa Cruz County, California.* San Francisco: Wallace W. Elliott & Co.

Foster, H. B., H. J. Ongerth, M. E. Stewart, and R. J. Hee. 1967. Sanitary engineering investigation of quality of water and shellfish in the estuary system at Moss Landing, Monterey County, California. California Department of Public Health, Bureau of Sanitary Engineering.

Gardner, M. and R. Kvitek 1998. Tide flat resource restoration and management: Implementation of a novel recovery program in the Elkhorn Slough. NOAA final report. Grant # NA57OR0347

Gordon, B.L. 1996. *Monterey Bay Area: Natural History and Cultural Imprints.* Pacific Grove, CA: Boxwood Press.

Hart, E. W. 1966. *Mines and Mineral Resources of Monterey County, California. County Report 5.* Sacramento: California Department of Mines and Geology.

Heimann, R. F. G., and J. G. Carlisle. 1970. The California marine fish catch for 1968 and historical review 1916-68. California Department Fish and Game. *Fish Bulletin* 149. 70 p.

Iwata, M. 1962. The Japanese immigrants in California agriculture. *Agricultural History* 36(1): 25-37.

King, J. M. 1976. Bio-archaeological reconnaissance of the Elkhorn Slough District. Moss Landing: ERA.

King, J. M. 1982. Elkhorn Slough Estuarine Sanctuary: Cultural resource management. California Department of Fish and Game.

Le Boeuf, B. 1981a. Mammals in *The Natural History of Año Nuevo*, B. Le Boeuf and S. Kaza, eds., p. 289. Pacific Grove, CA: Boxwood Press.

_____. 1981b. History in *The Natural History of Año Nuevo*, B. Le Boeuf and S. Kaza, eds., p. 17. Pacific Grove, CA: Boxwood Press.

Leonard, A. 1934. *Narrative of the Adventures of Zenas Leonard, Written by Himself*, M. M. Quaife, ed.. Chicago: The Lakeside Press.

Lydon, S. 1985. *Chinese Gold: The Chinese in the Monterey Bay Region*. Capitola, CA: Capitola Book Co.

Lydon, S. 1997. *The Japanese in the Monterey Bay Region: A Brief History*. Capitola, CA: Capitola Book Co.

MacGinitie, G. 1935. Ecological aspects of a California marine estuary. *American Midland Naturalist* 16: 629-765.

Madrone Associates. 1979. Pre-acquisition planning study for Elkhorn Slough Estuarine Sanctuary. Prepared for Cali-fornia Department of Fish and Game. Review draft. Novato, CA: Madrone Associates and Jones & Stokes.

Margolin, M. 1978. *The Ohlone Way: Indian Life in San Francisco Bay and Monterey Bay Area*. Berkeley, CA: Heyday Books.

Mohlo, A. 2000. Elkhorn Slough: Waterway to the world: Commerce 1855–1906. Unpublished report. Cabrillo College.

Mountjoy, D. C. 1996a. Culture, capital and contours: Ethnic diversity and the adoption of soil conservation in the Strawberry Hills of Monterey, California. Ph.D. dissertation, University of California, Davis.

Mountjoy, D. C. 1996b. Ethnic diversity and the patterned adoption of soil conservation in the Strawberry Hills of Monterey, California. *Society and Natural Resources* 9: 339–357.

National Marine Fisheries Service. 2001. Summary of commercial fisheries landings. http://www.st.nmfs.gov/st1/commercial/index.html

Philip Williams and Assoc. 1992. Elkhorn Slough tidal hydraulics erosion study. San Francisco, Calif., U.S. Army Corps of Engineers.

Rochin, R. I. 1986. The conversion of Chicano farm workers into owner-operators of cooperative farms, 1970-1985. *Rural Sociology* 51(1): 97–115.

Scharffenberger, T. 1999. Elkhorn Slough conservation plan. Elkhorn Slough Foundation and The Nature Conservancy. July 1999.

Saloutos, T. 1975. The immigrant in Pacific Coast agriculture. 1800–1900. *Agricultural History* 49(1): 182-201.

U.S. Bureau of the Census. 1989. *1987 Census of Agriculture: Volume 1, Geographic Area Series; Part 5, California State and County Data.* Washington, D.C.: U.S. Department of Commerce, Bureau of the Census.

U.S. Department of Agriculture. 1994. Draft watershed plan and environmental assessment, Elkhorn Slough Watershed Project, Monterey and San Benito Counties, California. On file at the USDA Natural Resources Conservation Service office, Salinas, CA

U.S. Department of Commerce. 1979. Final environmental impact statement. Elkhorn Slough National Estuarine Sanctuary. Elkhorn Slough Collection ESTU-3.

Van Dyke, E., and K. Contreras. 2001. Elkhorn Slough Watershed critical resource maps. Unpublished report, on file at Elkhorn Slough National Estuarine Research Reserve.

Villegas, Ygnacio. 1983. Boyhood days: Ygnacio Villegas' reminiscences of California in the 1850s. San Francisco: California Historical Society.

Wells, M. 1988. Commodity systems and family farms. In Gladwin, C. and K. Truman, eds. *Food and Farm: Current Debates and Policy.* Boulder, CO: Westview Press.

Wells, M. 1991. Ethnic groups and knowledge systems in agriculture. *Economic Development and Cultural Change* 39(4): 739-771.

Wells, M. 1996. *Strawberry Fields: Politics, Class and Work in California Agriculture.* Ithaca, NY: Cornell University Press.

8

CHAPTER EIGHT

Primary Producers

Richard C. Zimmerman, Jane M. Caffrey

The Elkhorn Slough watershed and estuary support a rich assemblage of primary producers, those organisms that grow via photosynthesis using sunlight and carbon dioxide to make organic matter. More than 500 species of flowering plants occur in upland (terrestrial), marsh, and subtidal communities, and the estuary supports the seagrass *Zostera marina* L. (eelgrass) and more than 100 species of macrophytic algae and phytoplankton (DeVogelaere et al. 1998).

Primary producers form the base of the food chain and support the rich diversity of animal life found in Elkhorn Slough and its watershed. Grazers, whether they are zooplankton, clams, voles, or deer, depend on primary producers as their source of food, while grazers, in turn, are a food source for higher trophic levels. The annual cycle of primary production sets the timing of reproduction for many species, so that food is available for their offspring.

As sources of food, all primary producers are not alike. Plant and animal species have co-evolved, sometimes to extreme cases where one grazer is totally dependent on a very small number of plant species. In addition, plants provide habitat and refuge for animal species. The replacement of one primary producer with another can greatly affect the composition of the animal community. Thus, conversions such as salt marsh to mudflats, oak forests to eucalyptus forests, or grasslands to agricultural fields have a greater impact than just the replacement of one primary producer species with another.

In all the communities of Elkhorn Slough, primary production is regulated mostly by seasonal variation in light availability, temperature, and moisture (see chapter 3 for a discussion of climate). Terrestrial (nonagricultural) production is determined largely by the duration and extent of the rainy season (November to April). Production peaks in spring (February to June) for many of these communities. Except among some woody perennials, primary productivity is almost nonexistent during the dry summer months. Summer drought also causes early (mid-June) maturation of dominant annual grasses and early dormancy of drought-deciduous perennials. Persistent fog moderates the effects of summer drought in coastal habitats. Winter frost is rare except in isolated canyon basins where cold air can pool at night. In contrast, marine productivity is highest from late spring through summer, when prevailing northwest winds promote strong upwelling of cold, nutrient-rich water near the coast.

Human activities have caused dramatic changes in the watershed's vegetation communities. Introduced nonnative plants now dominate many habitats and have altered primary production patterns. In addition, intensive livestock grazing, row crop cultivation, and higher housing density all contribute to increased nutrient runoff and higher nitrogen levels in slough waters. Although effects of terrestrial nutrient inputs into Elkhorn Slough are poorly understood, *eutrophication* can lead to extensive blooms of nuisance algae and phytoplankton. Decomposition of these algae and phytoplankton reduces oxygen concentrations in the water, ultimately killing fish and invertebrates when concentrations get low enough (see chapter 12, "Biogeochemical Cycling"). Such an occurrence at Elkhorn Slough could affect primary productivity in slough waters and severely restrict fish and invertebrate populations. Restoration efforts to reverse habitat loss and degradation and limit nutrient runoff have taken place at the slough and throughout the watershed, and more efforts are currently underway (see chapter 14, "Management Issues").

In this chapter we discuss the primary producers in nine distinct communities or habitats (table 8.1, fig. 8.1). For each, we identify the extent of the habitat in the watershed (based on Van Dyke and Contreras 2001) and its dominant species, consider historical changes in diversity and abundance, describe the seasonal patterns of primary production, and delineate the factors controlling production. Because little research has been done on the terrestrial habitats of the Elkhorn Slough watershed, much of the descriptive information draws from studies of similar vegetation communities in other parts of California. We close with suggestions for additional research, especially to address a lack of basic knowledge of the watershed's upland areas, and review some of the major management issues that must be addressed to assess and prevent further disturbance to the slough's primary producers.

Upland Communities

The Elkhorn Slough watershed features a terrestrial vegetation assemblage that is unique to the westernmost hills of the Coast Ranges. This is primarily the result of cool, persistent coastal fog, which moderates the effects of summer drought relative to inland habitats. The watershed's steep terrain creates abrupt boundaries between the marsh and upland habitats such as grasslands, woodlands, chaparral, and agricultural fields. Upland habitats also include riparian woodlands and sand dune/beach complexes. The distribution and structure of upland communities is determined primarily by water availability and soil type. In much of the watershed, native plants have been displaced by nonindigenous species from Europe and Asia, altering the original production patterns. The introduction and spread of exotic species continues today.

Due to the prolonged summer drought, fire can be an important structuring force in terrestrial communities. Frequent burns recycle inorganic minerals sequestered in ground surface litter and reduce the overall fuel load. With low fuel, fires burn quickly and produce less heat; thus, while annual species are usually consumed, many woody perennials survive. Fire frequency in the Elkhorn Slough area increased when Ohlone peoples began to occupy the region (see chapter 6, "Archaeology and Prehistory"). Modern fire abatement practices, however, have reduced fire frequency in the past decades. Reduced frequency of burning results in the dangerous accumulation of woody fuels that can produce much hotter and more devastating fires.

Grassland

Grassland occupies 6,174 hectares of the watershed (see table 8.1), dominating the drier, south-facing slopes. Early successional disturbed areas where shrub cover is less than 50% are included in this category (Van Dyke and Contreras 2001). Most grassland areas are characterized by *Vulpia* and *Bromus* spp., mustard (*Brassica nigra*), and coyote brush (*Baccharis pilularis*). Some grassland habitat is dotted by coast live oak (*Quercus agrifolia*) and thus is more accurately categorized as savanna.

More than any other, this community has been severely altered by a century of livestock grazing and row crop agriculture and

Native bunchgrasses are being replanted at restoration sites around Elkhorn Slough. Photo credit: Paul Zaretsky.

the subsequent establishment of exotic plant species (Baker 1989). Native bunchgrasses such as California brome (*Bromus carinatus*) hair grasses (*Deschampsia cespitosa*), and needle grasses (*Nasella* spp.), which once dominated the region, have been replaced by introduced perennial grasses such as harding grass (*Phalaris aquatica*), and other weeds such as poison hemlock (*Conium maculatum*), fennel (*Foeniculum vulgare*), and pampas grass (*Cortaderia* spp.). Grasslands cover 30.8% (266,533 ha) of the terrain throughout Monterey county, but only 170 hectares are still dominated by native California grassland species (Huenneke 1989). Within the Elkhorn Slough watershed, there remain only a few small, fragmented patches of native grassland, characterized by the presence of *Danthonia californica* and other bunchgrasses.

In grassland habitats, native and introduced species are dormant during summer, seeds germinate during fall, and flowering occurs in late spring at the end of the wet season (Chiariello 1989). The growing season mirrors the rainy season, which usually starts in November and continues until June. In winter, light levels and air and soil temperatures are too low, thus peak growth occurs during fall and spring when soil moisture, light, and temperature are optimal (Evans and Young 1989). Warm air or soil temperature and high soil moisture help to maximize seed germination, and herbaceous growth is highest in years with heavy fall or spring rains (Evans and Young 1989).

Woodlands

Woodlands occur in locations that ameliorate the desiccating effects of the summer sun, such as on higher, north-facing slopes, in canyon bottoms, and along river courses. Dry, native oak woodlands characterized by coast live oak (*Quercus agrifolia*), poison oak (*Toxicodendron diversilobum*), and toyon (*Heteromeles arbutifolia*) cover 3,688 hectares within the Elkhorn Slough watershed. Other drought-moderated areas (611 hectares) feature distinct, monotypic stands of introduced blue gums (*Eucalyptus* spp.) (Grove et al. 1997). Like many introduced species, eucalyptus can dominate the landscape but may not provide resources (food, shelter) for native birds or other wildlife (Gordon 1996). Conifers, including Monterey pine (*Pinus radiata*) and Monterey cypress (*Cupressus macrocarpa*), cover a small portion (13 hectares) of the watershed. Riparian woodlands (491 hectares) characterized by perennial species such as willows (*Salix* spp.), buckeye (*Aesculus californica*), and sycamore (*Platanus racemosa*), are usually restricted to areas where water is present on or near the surface year round.

Table 8.1 Habitat types within the Elkhorn Slough watershed.

Habitat	Hectares	(acres)*	% of Total Watershed
Upland (terrestrial)			
Grassland	6,174	(15,257)	25
Woodland			
Oak	3,688	(9,115)	20
Eucalyptus	611	(1,510)	3.3
Conifer	13	(33)	0.07
Chaparral			
Maritime chaparral	689	(1,704)	3.8
Sage scrub	102	(254)	0.6
Dune scrub/beach complex	58	(143)	0.3
Agriculture			
Cultivated	3,900	(9,635)	21.3
Fallow	478	(1,181)	2.6
Greenhouse	61	(150)	0.3
Turf	76	(188)	0.4
Marsh and riparian habitats			
Freshwater riparian	491	(1,214)	2.7
Ponds	83	(206)	0.4
Dry ponds	6	(15)	0.03
Freshwater marsh	108	(267)	0.6
Salt marsh/Tidal wetlands	1,147	(2,833)	6.3
Marine habitats			
Subtidal	113	(279)	0.6
Intertidal	120	(296)	0.7
Development	203	(501)	2.0

*Based on Van Dyke and Contreras 2001

Annual productivity and plant nitrogen accumulation are generally comparable in grassland and oak woodland communities. The communities differ, however, in rates of seedling establishment, seasonal phenology (which is more delayed in understory annuals), dry weight, and nitrogen allocation patterns, with a greater dependence of seed production on currently assimilated rather than stored resources in understory annuals.

Woodland productivity is much less restricted by the annual drought cycle than is grassland production. Many of these

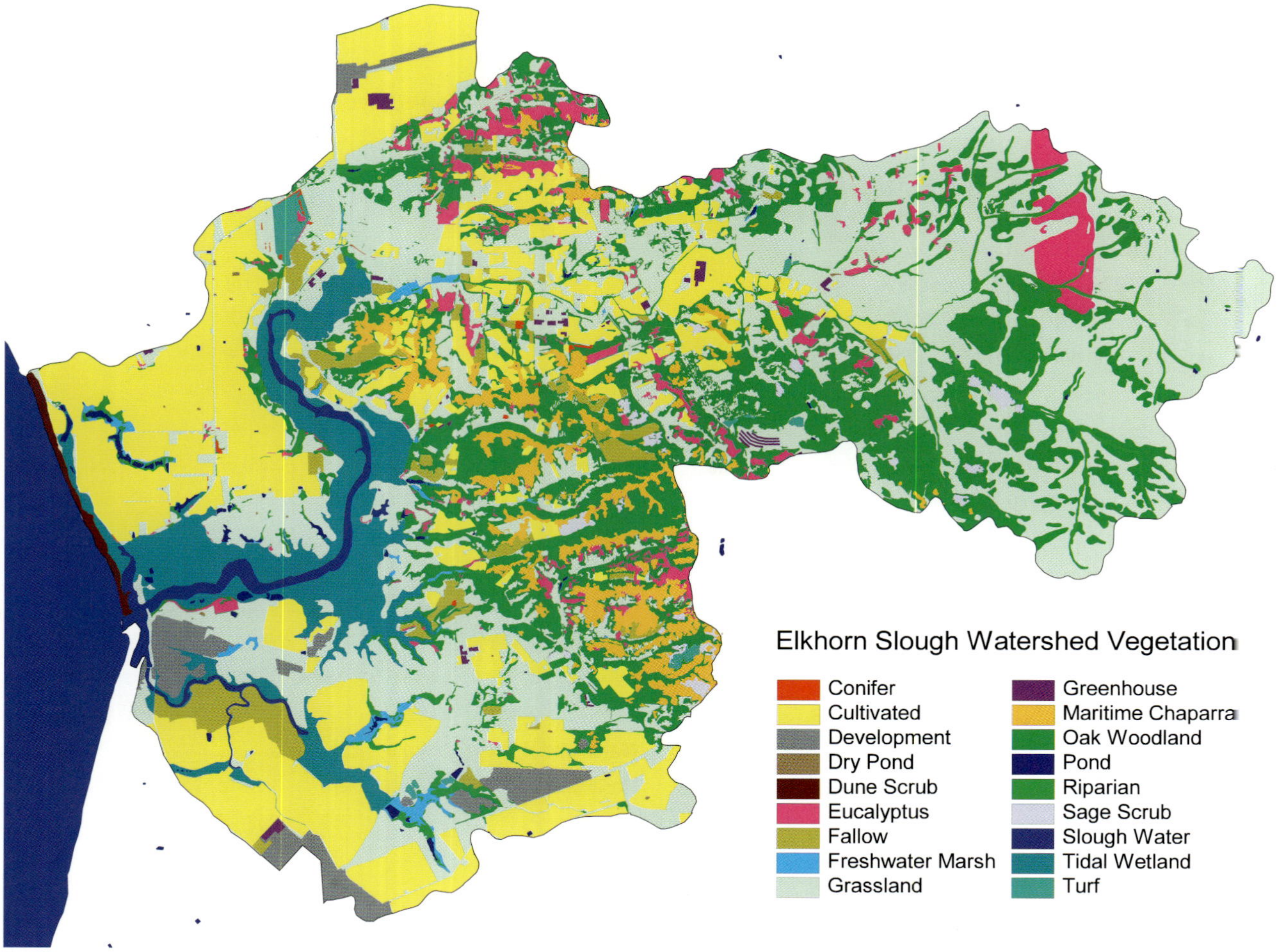

Figure 8.1. Vegetation patterns in the Elkhorn Slough watershed.

water-conserving species are evergreen, with thick, waxy, desiccation-resistant leaves. Vegetative growth occurs throughout the year, but is highest during summer. Leaf and wood production enter the food chain primarily as detritus via litterfall. Acorn and berry production by woodland perennials in summer and autumn provides a major food source for organisms at higher trophic levels. Fallen acorns, berries, and leaf litter are the primary sources of nutrients and thus determine the seasonal patterns of nutrient deposition in the soil. The first autumn rains and acorn fall produce a peak in nutrient deposition, and *throughfall* during early spring leaf emergence generates a second peak in potassium, magnesium, and phosphorus. Late summer leaf fall in response to drought, variable precipitation, and variation in deposition of nonleaf parts causes seasonal differences in nutrient deposition between years (Callaway and Nadkarni 1991). Nutrient cycling is very efficient in woodlands and little inorganic nitrogen or phosphorus escapes from undisturbed habitats.

Although grassland productivity is often enhanced by nutrient inputs from litterfall and throughfall under trees, the shallow, fine roots of oaks can actually inhibit understory productivity (Callaway, Nadkarni, and Mahall 1991). Soils under oak canopies have higher nitrogen turnover and inorganic nitrogen availability than surrounding open grassland soils due to mineralization of oak leaf litter (Jackson et al. 1990). Even though aboveground production in grasslands does not benefit from the presence of oaks, soils under oak canopies harbor a reservoir of organic nitrogen that can be rapidly lost or redistributed if oaks are removed.

Maritime Chaparral and Sage Scrub

Chaparral, an assemblage of woody shrubs with hard, thick, evergreen leaves, covers 689 hectares within the watershed. GIS

Oak groves flank coastal grasslands above the slough and provide habitat for diverse terrestrial plants and animals. Photo credit: ESNERR.

mapping classifies chaparral into two vegetation communities: maritime chaparral is dominated by manzanita species, including *Arctosphylos pajaroensis* and *A. hookeri*. Other characteristic species include coyote brush, black sage (*Salvia mellifera*), and California lilac (*Ceanothus* spp.). Oak woodlands intermingle with chaparral in some parts of the watershed (Hanes 1981). This habitat remains relatively unaffected by introduced plants, perhaps because native chaparral species prevent invasion by producing allelopathic *phytotoxins* (Hanes 1981). A second type of shrub community, identified as sage scrub, covers 102 hectares; this includes a few areas of native coastal sage scrub. Many of the regions identified as sage scrub are old fields or other disturbed areas within maritime chaparral or oak woodland, and are often mostly poison oak, although coyote brush, sage (*Salvia* spp.), sagebrush (*Artemesia* spp.), and monkeyflower (*Mimulus* spp.) also occur.

As in woodland habitats, chaparral shrubs can grow year round, however, production in chaparral peaks in the spring when moisture availability is highest and day length increases. Chaparral species produce new leaves in fall, particularly the coastal sage shrubs such as *Artemisia* spp. and *Salvia* spp., which grow a complete new canopy after the first rains (Mooney 1981). Soils are characteristically sandy and have low capacity for water retention. After moisture, nitrogen appears to be the most limiting resource in these organically deprived soils. In addition, the steep, loosely consolidated slopes erode easily, producing severe mud slides during the rainy season. Thus, maintenance of chaparral vegetation on sandy slopes is critical for watershed protection. In areas where agricultural row-crop fields have replaced chaparral and oak woodland, erosion rates are estimated at 33 tons per acre per year, and can range higher in years of heavy rainfall (see chapter 5, "Soils").

Chaparral distribution and productivity are mostly determined by climate, slope aspect (direction), exposure, elevation, and fire (Hanes 1981). Fire is very important in the evolution and structuring of chaparral communities. Most chaparral plants show fire adaptations, such as production of seeds at an early age, seeds that are fire-resistant or that require fire to germinate, and sprouting from lignotubers or root-crown burls. Fire helps maintain species diversity by reducing the abundance of competitively dominant species. For example, grassland species can colonize immediately after fire (Hanes 1981). Fire also affects the dynamics of nutrient cycling by increasing levels of soil minerals and organic nutrients from ash; in addition,

Mixed community of maritime chaparral and coast live oaks on Elkhorn Highlands. Photo credit: Paul Zaretsky.

Table 8.2. Primary production of representative plant groups found within the Elkhorn Slough watershed.

Group	Habitat	Primary Production (gC $m^{-2}y^{-1}$)	Reference
phytoplankton	open water	420	Pace 1978
Salicornia virginica[1]	marsh	149	Reilly 1979
mixed grassland	upland	100	McNaughton 1968
chaparral	upland	185	Mooney 1981
eucalyptus	upland	224	Robles & Chapin 1995
strawberries[1,2]	upland-agricultural	1,200	Gliessman et al. 1996

[1] Aboveground biomass increase
[2] Includes berry production

Spring blooms on the coastal dune strand flanking Elkhorn Slough. Photo credit: Paul Zaretsky.

bacteria and fungi are more abundant in burned soils than unburned soils (Hanes 1981). On the other hand, fire decreases total nitrogen availability and redistributes nitrogen and phosphorus within soils that are highly susceptible to erosion (Weinhold and Klemmendson 1992). Mechanisms for replenishment of soil nitrogen involve fixation of atmospheric nitrogen (*Ceanothus* and some other chaparral shrubs are nitrogen-fixers; Hanes 1981) and deposition of aerosol NOx (oxidized forms of nitrogen). Phosphorus is replenished very slowly, primarily by weathering of parent material.

Dune Scrub/Beach

Dune scrub/beach habitat occurs along the western margin of Elkhorn Slough and covers only 58 hectares. This habitat supports a fragile and distinctive plant assemblage. As in the grassland community, the native plants in this habitat have been largely replaced by exotic species. Today, the most common dune inhabitants are ice plants (*Carpobrotus edulis*, *C. chilense*, and *Mesembryanthemum* spp.), which were introduced widely for beach stabilization (Bluestone 1970). Other common species include sea rocket (*Cakile maritima*), beach saltbush (*Atriplex leucophila*), beach burr (*Ambrosia chamissonis*), mock heather (*Ericameria ericoides*), and beach sagewort (*Artemisia pycnocephala*) (Bluestone 1970; Emory 1997).

Salinity (from salt spray), flooding, and moisture content are important determinants of species composition in dune habitats (Olff et al. 1993). In addition, interannual fluctuations of these factors seem to be responsible for variations in the success of many short-lived, annual species. Aboveground biomass production of all species is limited by nitrogen availability.

Inorganic soil nitrogen concentrations are very low, and only about 5% of nitrogen is present in the organically bound phase (Holton, Barbour, and Martens 1991). The most significant inputs of nitrogen to dunes appear to come from fog condensation, with much smaller contributions from bulk precipitation and nitrogen fixation (Holton, Barbour, and Martens 1991).

The dune scrub/beach habitat typically features a gradient of increasing substrate stability and vegetational complexity as one moves away from the ocean. The beach is relatively unstabilized sand with little or no vegetation; the dunes (usually subdivided into fore, middle, and rear dunes) are relatively stable sand with moderate vegetation; and the transitional zone is where dune and inland vegetation merge. At Moss Landing, for example, beach saltbush and sea rocket are the only species that grow on the beach (Bluestone 1970). In the fore dunes and intermediate dunes these colonizing species are joined by beach burr, sand verbena (*Abronia* spp.), and silky beach pea (*Lathyrus littoralis*). Ice plant, beach sagewort, and beach evening primrose (*Camissonia cheiranthifolia*) appear in the rear dunes. Development, such as road construction and agriculture, has eliminated most of the transitional zones in the Elkhorn Slough area (Bluestone 1970).

Dune habitats are also in a state of nearly continuous successional flux. The successional series in dune colonization is accompanied by an increase in biomass, a decrease in light penetration to the soil surface, a decrease in root/shoot ratio, an increase in dominance of long-lived tall species, and a decrease in abundance of small, short-lived species. Thus, light competition becomes increasingly important during succession, and tends to favor the accumulation of taller species at later successional stages.

Agricultural Fields

Agriculture occupies approximately 4,378 hectares within the Elkhorn Slough watershed. The dominant row crops are strawberries (1,450 ha in production in 1994), artichokes (520 ha), and raspberries (160 ha). Apple orchards account for 200 hectares, flowers and nurseries for 360. Broccoli, cauliflower, celery, lettuce, spinach, peas, and squash are also grown in the watershed. Of the area identified as agricultural land, 478 hectares were classified as currently fallow, but are known to have been recently under cultivation. The authors of the GIS report note that "many other fallow agricultural plots whose history is not known are probably scattered throughout the watershed; these successional areas would be classified as grassland where shrub cover (typically *Baccharis*) does not exceed 50%; otherwise they would be classified as sage scrub" (Van Dyke and Contreras 2001). An additional 61 hectares are classified as larger greenhouses, with 76 hectares of turf (non-agricultural irrigated grasslands, including golf courses) also occurring within the watershed. Land-use changes and the conversion of wildlands to agriculture are discussed in chapter 7, "History of Land Use."

With agricultural crops, moisture, nutrients, temperature, light, grazing, and pest infestation are carefully managed to ensure maximum plant productivity. Extensive use of groundwater for irrigation is necessary to maintain row crops in this region during the summer. Strawberries are planted either in the summer or fall; fall-planted strawberries start producing ripe fruit by March, although the peak production is usually in May and June (Gliessman et al. 1996). The plants can produce for two or more years, although most farmers in the watershed replant each year. The annual aboveground production for strawberries, including harvestable fruit and leaf production, is 1,200 gC m^{-2} y^{-1} (table 8.2). Lettuce, spinach, peas, and squash have shorter production cycles, yielding two and sometimes three crops in a single year.

Use of chemical fertilizers and pesticides is widespread in the watershed, particularly in strawberry production, which may use as many as 36 chemical fertilizers and pesticides (CDFA 1989). Prior to planting, strawberry fields are covered in plastic and fumigated with methyl bromide (categorized as a Class 1 ozone depleter by the Environmental Protection Agency) and chloropicrin to control soil pathogens and weeds. The phaseout of methyl bromide, initiated in 2001 and scheduled for completion in 2004, has prompted ongoing studies in the Elkhorn Slough watershed to find alternatives both to the soil fumigant and to other synthetic chemical inputs (Brown and Goldman 1999). The effects of nutrient and pesticide inputs from agricultural production on Elkhorn Slough are discussed in chapter 12, "Biogeochemical Cycling," and chapter 13, "Land Use and Contaminants."

Marsh Habitats

Salt marshes are the tidal wetlands between the permanently dry uplands and the permanently flooded slough. Salt marshes

Vegetated buffers planted between cultivated fields and wetlands help reduce damaging inputs of sediments and nutrients to the aquatic environment. Photo credit: Mark Silberstein.

Marsh dodder's orange filaments parasitize pickleweed in the salt marsh. Photo credit: Paul Zaretsky

in Elkhorn Slough occupy 1,147 hectares. Historically, before the diversion of the Salinas River and the dredging of Moss Landing Harbor to the Pacific Ocean in 1946, the marshes of the upper slough were much less saline (MacGinitie 1935). Today, marsh habitat throughout the slough is highly saline (25–35 psu) and dominated by pickleweed (*Salicornia virginica*). Cordgrass (*Spartina* spp.), an important component of many brackish wetlands, is conspicuously absent from Elkhorn Slough, perhaps due to the persistent high salinity. Salt stress reduces both photosynthesis and aboveground biomass of cordgrass, and saltwater intrusion has been identified as a major factor contributing to the rapid rate of cordgrass marsh deterioration on the Atlantic and Gulf Coasts (Pezeshki and DeLaune 1993; Ewing et al. 1995).

Flooded only during very high tides, the upper marsh is a transition area with high soil salinities. It supports species that are mildly tolerant of flooding and high salinity, including salt grass (*Distichlis spicata*), saltbush (*Atriplex* spp.), *Jaumea carnosa*, alkali heath (*Frankenia grandifolia*), and parasitic dodder (*Cuscuta salina*). The lower marsh is flooded on most high tides and is vegetated by extensive, monotypic stands of pickleweed; dodder can be present in the summer. A superior competitor for nitrogen, pickleweed may prevent the establishment and growth of cordgrass in the saline lower marshes of Elkhorn Slough (Covin and Zedler 1988).

Erosion within Elkhorn Slough is converting previously extensive tracts of pickleweed-dominated salt marsh into tidal mudflats. The loss of ecological functions provided by the salt marsh vegetation (e.g., sediment trapping, nutrient retention, denitrification) reduces the slough's ecological diversity and resistance to further change in biogeochemical dynamics. The loss of marsh vegetation may also facilitate further erosion. These problems are discussed further in chapters 4 ("Hydrography") and 14 ("Management Issues").

In contrast to the surrounding upland habitats, the salt marsh experiences a clear annual cycle of summer productivity and winter dormancy. The marsh remains green and productive all summer, long after drought-parched hillsides have turned golden brown. Winter dormancy turns the marsh brown just as the arrival of rain stimulates growth of new vegetation in upland habitats. Variation in rainfall promotes year-to-year changes in community composition and spatial distributions, particularly among annual species, and thus increases overall species diversity in the salt marsh (Callaway and Sabraw 1994).

Marsh vegetation traps sediments brought downstream during episodic winter floods. Sediment trapping stabilizes the marsh by maintaining surface elevation and enhances productivity through nutrient retention (Cahoon, Lynch, and Powell 1996). Productivity in the salt marsh may be nitrogen-limited, as experimental application of nutrients has enhanced standing crop, aboveground productivity, and nitrogen content of

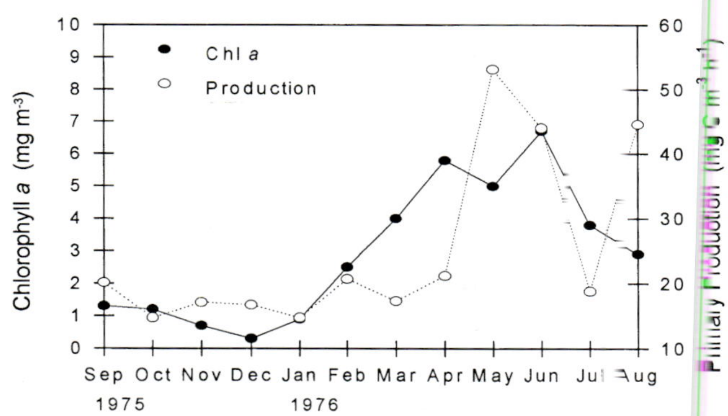

Figure 8.2. Seasonal distribution of phytoplankton chlorophyll *a* within Elkhorn Slough. From Pace 1978.

pickleweed in southern California (Page 1995). However, much of the inorganic nitrogen imported with groundwater and flood sediments may be released to the atmosphere via denitrification. Thus, the marsh also plays an essential role in filtering out dissolved nutrients and preventing eutrophication of aquatic habitats within the slough.

In the late 1920s, freshwater wetlands were abundant in the upper reaches of the Elkhorn Slough watershed and the nearby Pajaro River. Even then, however, the conversion of wetlands to agriculture was under way (Hayward 1931). The dominant species in freshwater marshes during this period were cattail (*Typha latifolia*), California bulrush (*Scirpus californicus*), and curly dock (*Rumex crispus*). Other common species (Hayward 1931) were:

Anthemis cotula	mayweed, stinkweed, dog-fennel
Baccharis douglasii	marsh baccharis
Carex abrupta	
Epilobium ciliatum ssp. *watsonii*	
Gnaphalium stramineum	cud weed, everlasting
Juncus effusus	
Lemna minuscula	duckweed
Oenanthe sarmentosa	
Picris echioides	bristly ox-tongue
Polygonum acre	
Polygonum muhlenbergii	
Polypogon monspeliensis	annual beard grass
Rorippa nasturtium-aquaticum	
Rubus ursinus	California blackberry
Rumex salicifolius	willow dock
Salix lasiolepis	arroyo willow

Today, freshwater habitat is limited to a relatively small area (108 hectares) of freshwater marsh, as well as 83 hectares of freshwater ponds on the National Estuarine Research Reserve and elsewhere in the watershed. These ponds are heavily colonized by the invasive, free-floating duckweeds *Lemna miniscula* and *L. minor*, and water fern, *Azolla filiculoides*. Light, nitrate, and potassium are necessary for optimal growth of duckweed (Schulthorpe 1967). Unidentified species of submerged aquatic vegetation have also been observed in some reserve ponds. Nothing is known about the pond phytoplankton communities or their relationships with the floating and submerged vascular plants. Six hectares of dry ponds, mostly agricultural in origin, also occur in the watershed. According to Van Dyke and Contreras (2001) these "generally fill with rainwater during the wet season and typically support various wetland species before drying out later in the year."

Marine Primary Producers

The marine habitats of Elkhorn Slough support a diverse flora of phytoplankton, macrophytic algae, and seagrasses.

Phytoplankton

Rapid tidal exchange between Monterey Bay and Elkhorn Slough helps support a rich community of phytoplankton within the slough. Diatoms (Bacillariophyceae) dominate the slough's highly turbid, green-brown waters.

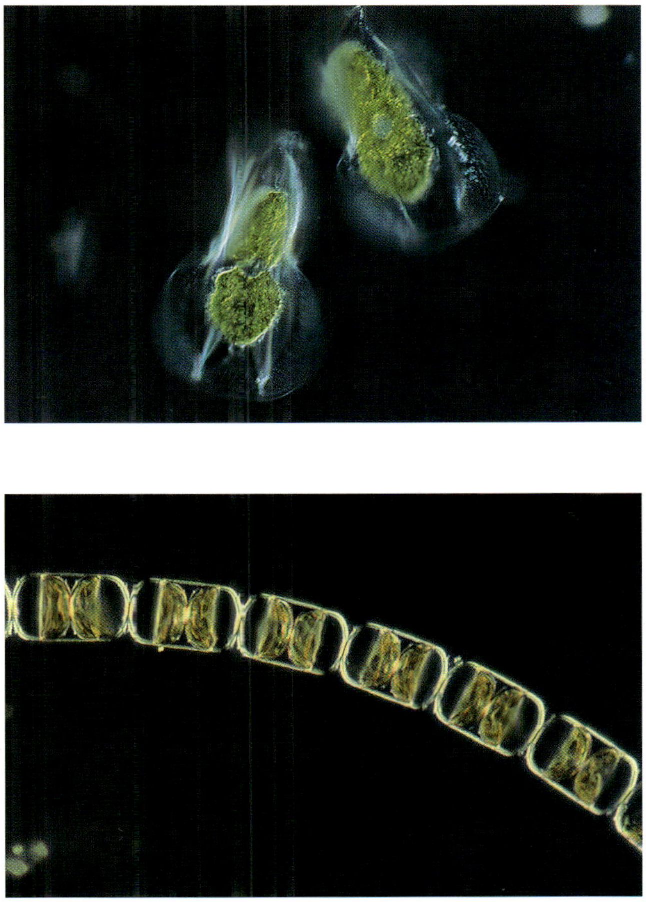

Diatoms are a key component of primary production in the slough's tidal waters. Upper photo shows individual diatoms; lower photo shows a diatom chain. Photo credit: Kenton Parker.

Phytoplankton abundance in the slough, as indicated by chlorophyll *a* concentrations, peaks in spring and summer (Pace 1978). Chlorophyll concentrations exceed 5 µg l^{-1} throughout the drought-dominated months of June to September (fig. 8.2). The extended summer abundance peak is supported primarily by advection into the slough of nutrient-rich waters and phytoplankton from upwelling centers north of Monterey Bay (Slager 1997). Although runoff may maintain high nutrient levels in the slough in winter, phytoplankton abundance may be limited by light availability and water column turbidity. The euphotic zone, the depth to which 1% of the surface irradiance penetrates, rarely exceeds 2 meters throughout the year (Nybakken, Cailliet, and Broenkow 1977; Zimmerman, Cabello-Passini, and Alberte 1994).

The factors controlling phytoplankton production in Elkhorn Slough are unclear, but light, nutrients, and grazing pressure are important in other estuarine systems (Boynton et al. 1982; Cloern 1982; Valiela 1995). Often carbon production by primary producers such as phytoplankton or macroalgae can be limited by nutrient concentrations. Most research has focused on nitrogen and phosphorus because these two elements often limit primary production (Dugdale and Goering 1967; Ryther and Dunstan 1971; Nixon and Pilson 1984; Howarth 1988), although silica (Conley et al. 1993; Conley and Malone 1992) and trace metals can sometimes be limiting.

The only seasonal study of chlorophyll *a* concentrations and primary productivity at Elkhorn Slough was conducted in 1975 and reported moderate chlorophyll *a* concentrations (0.2–15 µg chl *a* l^{-1}) and phytoplankton productivity (24 mg C m^{-3} h^{-1}) (Pace 1978). Recent measurements of chlorophyll *a* have indicated exceedingly high concentrations in some parts of the slough, sometimes in excess of 50 µg l^{-1} in areas with restricted circulation (Caffrey 1996), although in the well-flushed main channel, concentrations rarely exceeded 10 µg l^{-1} (Caffrey 1996).

Light availability is probably the most significant factor controlling phytoplankton production in the slough, given the high nutrient concentrations and low to moderate densities of benthic grazers (see chapter 9, "Invertebrates") in this very turbid estuary. Light limitation has been observed in other shallow, turbid estuaries with high nutrient levels (e.g., Four League Bay, La.; Randall and Day 1987) and in macrotidal estuaries (Monbet 1992). In nearby South San Francisco Bay, another shallow, turbid system with potential light limitation, benthic grazing also plays a significant role in controlling primary production (Cloern 1996).

Macrophytic Algae

Elkhorn Slough supports an abundant biomass assemblage of macrophytic algae. As is typical in estuarine habitats, macrophyte diversity is considerably lower in the slough than along the adjacent rocky shores of Monterey Bay. The most common macrophytes are the green algae *Enteromorpha* spp., *Ulva* spp., and *Derbesia marina* and the red algae *Gracilariopsis lemaneiformis*, *Gracilaria pacifica*, *Griffithsia* spp., *Polysiphonia* spp., and *Porphyra* spp. Brown algae, particularly the kelps so common in nearby Monterey Bay waters, are rare in Elkhorn Slough but occur occasionally on the riprap boulders below the Highway 1 Bridge.

Research by UC Santa Cruz phycologist Lynda Goff and coworkers in 1994 documented that another, unidentified species of *Gracilaria* has become established in Elkhorn Slough (Goff et al. 1994). Because this species is most closely (albeit distantly) related to species from the western Pacific (*G. tenuistipitata* from China and *G. chilensis* from New Zealand), it may have been introduced to the slough during the oyster farming period (see chapter 7) or from boats.

The distribution of aquatic and marsh macrophytes has changed significantly in the past 20 years due to a dramatic increase in erosion within Elkhorn Slough. Expansion of the mudflats and loss of pickleweed marsh have facilitated explosive growth of the green algae *Enteromorpha* to the

Researchers Alan Fukuyama and Jane King sample restored tidelands at the Elkhorn Slough reserve. Photo credit: Mark Silberstein.

point that many of the smaller tidal creeks and channels become clogged with floating mats in summer. *Enteromorpha* produce extensive mats throughout the estuary from spring through early fall, especially on mudflats. Suspended by rising tides, these mats often drift throughout the estuary.

The ability to float on the water surface gives these algae a significant competitive advantage over other macrophytes in the turbid, light-limited waters of the estuary. Of the free-floating forms, *Enteromorpha clathrata* var. *crinata* is the most common in estuaries (Abbott and Hollenberg 1976). A considerable but unquantified amount of *Enteromorpha* biomass is exported to Monterey Bay by the outgoing tide. *Ulva* spp. frequently colonize intertidal mudflats and subtidal habitats in the slough, particularly in summer. Although less visible than the floating mats of *Enteromorpha*, their abundance in subtidal channels suggests high productivity. The rhodophyte *Gracilariopsis lemaneiformis* becomes abundant on intertidal mudflats and subtidal habitats in fall and winter after the *Enteromorpha* bloom has subsided.

Controls on macroalgal productivity in Elkhorn Slough have not been investigated, but research in other systems suggests that desiccation, temperature, light, and nutrients (including dissolved inorganic carbon) can be important (Pregnall and Rudy 1985, Lee 1990). In addition, photosynthesis can be inhibited by high dissolved oxygen or the combination of high pH and low dissolved inorganic carbon (Gordon and Sand-Jensen 1990). In southern California estuaries, luxuriant macroalgal biomass and growth can be supported by high water column nutrient concentrations (Fong, Zedler, and Donohoe 1993).

Seagrass

Historically, the marine angiosperm eelgrass (*Zostera marina*) was very abundant in the central parts of Moss Landing Harbor and the main channel of Elkhorn Slough (MacGinnitie 1935). However, eelgrass beds have declined by more than 95% in Elkhorn Slough since the 1920s (MacGinnitie 1935). Dredging associated with harbor maintenance and high erosion in the main channel eliminated most of the shallow (2 meters or less) habitat that eelgrass requires for survival in these turbid waters. Eelgrass is particularly vulnerable to light limitation, and does not compete well with phytoplankton and weedy macrophytes like *Enteromorpha* in turbid, eutrophic estuaries (Zimmerman et al. 1991, 1995a, 1995b). By 1980, the only eelgrass in the slough was a few small patches near the Highway 1 bridge.

Recent broadening and shallowing of some parts of the main channel have allowed the eelgrass population to expand. Eelgrass has now colonized 10 hectares of shallow water in the Seal Bend area of the main channel, nearly a mile inland from the Highway 1 bridge. This increase of seagrasses may offset some of the slough's loss of ecological function associated with erosion of the pickleweed marsh, especially with regard to biomass production, nutrient retention, and denitrification processes. The ultimate impacts of erosion on overall ecosystem function cannot be determined with any degree of confidence at this time.

Worldwide losses of seagrass beds have proceeded at alarming rates in the last century due to disease outbreaks, physical disturbance (e.g., dredging, coastline development, fishing practices), and water quality deterioration (most often as water column light attenuation caused by particle loading, eutrophication, and nuisance algal blooms) (Thayer, Wolfe, and Williams 1975; Rasmussen 1977; Orth and Moore 1983; Wetzel and Penhale 1983; Cambridge and McComb 1984; Zimmerman et al. 1991). Thus there has been considerable research interest in determining prospects and techniques for restoration of eelgrass beds; some of this work has been done at Elkhorn Slough.

Exploratory eelgrass transplant studies were conducted in the Elkhorn Slough National Estuarine Research Reserve in 1988 and 1989 (Zimmerman and Alberte 1993). As expected, initial transplant mortality was high (>50%), particularly at the

Eelgrass exposed at low tide. Photo credit: Alan Fukuyama.

intertidal and stagnant subtidal sites. Eelgrass survival was better at subtidal sites with moderate flow, probably because the moving water prevented burial by drifting mats of *Enteromorpha* spp. and *Ulva* spp. These transplants survived and showed some evidence of vegetative expansion through the end of the study in 1993. This initial study and the recent appearance of naturally recruiting eelgrass populations on emerging shoals in the main channel (see below) demonstrate that the general environmental quality in Elkhorn Slough is adequate to support expanded eelgrass populations if substrate of appropriate depth (0 to –2 m *Mean Low Low Water* [MLLW]) and water flow (10–30 cm s^{-1} peak) is available.

Efforts have been made to predict eelgrass growth as a function of environmental conditions, particularly light availability. The relationships between light availability, leaf photosynthesis, translocation, and the carbon needs of leaves, roots, and rhizomes in eelgrass are now fairly well understood (Zimmerman et al. 1989, 1991, 1995a,b; Kraemer and Alberte 1993; Zimmerman and Alberte 1996; Zimmerman, Kohrs, and Alberte 1996; Zimmerman and Mobley 1997). In general, eelgrass requires the equivalent of 4–6 hours of photosynthesis-saturating irradiance each day to meet respiratory and growth needs for reduced carbon, but external factors such as epiphyte growth on the leaves and grazing can significantly increase light requirements (Neckles, Wetzel, and Orth 1993; Zimmerman, Kohrs, and Alberte 1996). However, modeling the requirements of individual shoots in isolation cannot be extrapolated to predict eelgrass production in larger areas. Another problem with predicting eelgrass performance in estuarine environments such as Elkhorn Slough is that light availability is highly variable on extremely short time scales (Dunton and Tomasko 1994; Zimmerman, Cabello-Passini, and Alberte 1994). The current models available are most useful for determining boundary conditions for the depth distribution and maximum densities of eelgrass populations.

The proximal causes for seagrass losses are usually evident, but it is also important to understand genetic diversity and gene flow among geographically isolated populations to assess population stability and develop sound management criteria. Unlike eelgrass populations elsewhere (McMillan 1982; Larkum and den Hartdog 1989; Heij and Nienhuis 1992), those in the northeast Pacific exhibit considerable genetic diversity (Alberte and Zimmerman 1993). Based on DNA fingerprint analysis of eelgrass populations between Alaska and the California-Mexico border, the Elkhorn Slough populations show a high genetic affinity to northern populations from Oregon, Washington, and Alaska. Eelgrass populations in Tomales Bay, however, show higher affinity to populations in southern California. Consequently, central California (Pt. Conception to Tomales Bay) may represent an important boundary and region of genetic mixing of northern and southern genotypes (Alberte and Zimmerman 1993).

Despite the reduction of eelgrass in the slough, its genetic diversity remains nearly as high as that of a nearby undisturbed population at Del Monte Beach, Monterey County (Alberte et al. 1994). In fact, there are several genetically distinct eelgrass populations within the slough. Studies in 1989 and 1991 (Alberte and Zimmerman 1993) found that these populations had different rates of light-saturated photosynthesis but similar rates of light-limited photosynthesis and shoot growth, and similar sucrose content. Thus, the extent to which genetically distinct populations possess features specific to their native habitats (ecotypic features) remains uncertain. All populations were capable of vigorous growth and vegetative reproduction in Elkhorn Slough. Consequently, genetic differences among the eight populations examined may be the result of nonselective (random) evolution among geographically isolated populations. Performance differences among these populations, however, may be demonstrated in habitat experiencing larger seasonal extremes of temperature or irradiance than are provided in Elkhorn Slough.

Management Issues and Research Recommendations

Although primary producers play a major role in structuring the appearance and function of Elkhorn Slough and the surrounding watershed, relatively little research has been done on these communities. Here we suggest research topics that would help develop a picture of how primary producers affect the functional health of Elkhorn Slough, the surrounding watershed, and adjacent marine communities. We also point out issues related to changes in primary production that must be addressed in order to limit eutrophication and the resultant impacts on the ecosystem.

System Productivity

Despite the importance of primary producers to slough health, very few studies of system productivity have been performed within the watershed. Rates of primary production, nutrient use, biomass production, and fate of plant biomass (whether grazed, exported, or buried) are poorly characterized for all plant communities. For example, how quickly can phytoplankton and macroalgae respond to pulses of nutrients from the watershed during rainy season? Residential developments and conversion to agriculture has led to increasingly fragmented plant communities. Is productivity within these fragmented communities the same as intact communities, and does fragmentation affect secondary productivity of herbivores such as insects and voles?

We also know little about the effects of grazing or exotic invasions on system productivity. Research comparing productivity of natural ecosystems versus those altered by grazing and nonnatives would help us understand the extent to which the area's original productivity has changed over time. Grasslands, oak woodlands, and riparian areas are being restored at a variety of sites within the watershed. How restoration efforts affect productivity, groundwater recharge, and nutrient and pesticide runoff is of critical importance and should be monitored as part of these projects.

Impacts of Erosion and Increased Nutrient Runoff

The loss of pickleweed marsh due to erosion is perhaps the most conspicuous challenge facing slough managers, and one that offers numerous research opportunities. For example:

- How does loss of marsh habitat affect ecosystem function?
- Does loss of marsh habitat further increase erosion rates?
- Does the loss of pickleweed marsh and its capacity for processing inorganic nutrients increase the eutrophic status of the slough?
- Will *Enteromorpha* populations increase to cover the expanding mudflat habitat? If so, how will the slough respond to increased abundance of drifting macroalgae that inundate beds of eelgrass, *Gracilaria*, benthic infauna (those living in sediments), and marsh habitat?
- Will loss of marsh vegetation increase turbidity in the slough, further restricting submerged macrophytes to depths shallower than 2 meters?

Addressing these questions would help quantify the impact of slough erosion on slough health and ecosystem functions. Human activities are also altering primary production via changes in nutrient levels. Intensive livestock grazing, row crop cultivation, and higher housing density all contribute to increased nutrient runoff. This runoff has caused serious eutrophication of Atlantic Coast estuaries, but effects of terrestrial nutrient inputs into Elkhorn Slough are poorly understood.

The slough's nutrient-rich, turbid condition suggests that aquatic primary production already may be light-limited. Increased algal abundance, particularly of floating *Enteromorpha* mats, could reduce water column oxygen during summer and severely restrict fish and invertebrate populations within the slough. Ratios of inorganic macronutrients indicate the slough is relatively low in phosphorus and silicate relative to nitrogen (see chapter 12, "Biogeochemical Cycling"). Further increases in nitrogen relative to other nutrients could shift the ecological balance within the water column away from diatoms and facilitate blooms of toxic and/or nuisance microalgae, as have occurred in other eutrophic estuaries (Burkholder et al. 1992; Holligam, et al. 1993; DeYoe and Suttle 1994). These issues are discussed in greater depth in the biogeochemistry chapter.

Links with Monterey Bay

The biogeochemical and ecological links between Elkhorn Slough and Monterey Bay are also poorly understood. Does the slough provide nutrients and biomass (phytoplankton, zooplankton, macroalgae, fish) to the bay, or does it require outside inputs to sustain itself? How will changes currently underway within Elkhorn Slough alter these relationships, and what are the long-term implications of such altered functional relationships? Research into these questions can help managers address the impacts of watershed and slough changes on the health of Monterey Bay.

References

Abbott, I. A., and G. J. Hollenberg. 1976. *Marine Algae of California*. Stanford: Stanford University Press.

Alberte, R. S., G. K. Suba, G. Procaccini, R. C. Zimmerman, and S. R. Fain 1994. Assessment of genetic diversity of seagrass populations using DNA fingerprinting: Implications for population stability and management. *Proceedings of the National Academy of Sciences* 91: 1049–1053.

Alberte, R. S., and R. C. Zimmerman. 1993. Restoration of eelgrass (*Zostera marina*) in estuarine environments: Evaluation of genetic and physiological performance variants. Final Report to National Estuarine Research Reserve Program, Elkhorn Slough Foundation.

Baker, H. G. 1989. Sources of naturalized grasses and herbs in California. In L. F. Huenneke and H. A. Mooney, eds., *Grassland Structure and Function: California Annual Grassland*, pp. 29–38. Dordrecht, Neth.: Kluwer Academic Publishers.

Bluestone, V. 1970. Distribution of selected coastal beach and dune plants along Monterey Bay, California. Master's thesis, Moss Landing Marine Laboratories. San Francisco State University.

Boynton, W. R., W. M. Kemp, and C. W. Keefe. 1982. A comparative analysis of nutrients and other factors influencing estuarine phytoplankton production. In V.S. Kennedy, ed., *Estuarine Comparisons*, pp. 69-90. New York: Academic Press.

Brown, M., and P. Goldman. 1999. BASIS promotes biological management in strawberries. *The Cultivar* (Center for Agroecology and Sustainable Food Systems, UC Santa Cruz) 17: 1–4.

Burkholder, J. M., E. J. Noga, C. H. Hobbs, and H. B. Glasgow, Jr. 1992. New "phantom" dinoflagellate is the causative agent of major estuarine fish kills. *Nature* 358: 407–410.

Cahoon, D. R., J. C. Lynch, and A. N. Powell. 1996. Marsh vertical accretion in a southern California estuary, U.S.A. *Estuarine Coastal and Shelf Science* 43: 19–32.

California Department of Food and Agriculture. 1989. *Annual Pesticide Use Report: Monterey County*. http://www.cdpr.ca.gov/docs/pur/pur89rep/comcnty/monter89.pdf

Callaway, R. M., and N. M. Nadkarni. 1991. Seasonal patterns of nutrient deposition in a *Quercus douglasii* woodland in central California. *Plant and Soil* 137: 209–222.

Callaway, R. M., N. M. Nadkarni, and B. E. Mahall. 1991. Facilitation and interference of *Quercus douglasii* on understory productivity in central California. *Ecology* 72: 1484–1499.

Callaway, R. M., and C. S. Sabraw. 1994. Effects of variable precipitation on the structure and diversity of a California salt marsh community. *Journal of Vegetation Science* 5: 433–438.

Cambridge, M. L., and A. J. McComb. 1984. The loss of seagrasses in Cockburn Sound, Western Australia. I. The time course and magnitude of seagrass decline in relation to industrial development. *Aquatic Botany* 20: 229–243.

Chiariello, N. R. 1989. Phenology of California grasslands. In L. F. Huenneke and H. A. Mooney, eds., *Grassland Structure and Function: California Annual Grassland*, pp. 47–58. Dordrecht, Neth.: Kluwer Academic Publishers.

Cloern, J. E. 1982. Does the benthos control phytoplankton biomass in south San Francisco Bay? *Marine Ecology Progress Series* 9: 191–202.

———. 1996. Phytoplankton bloom dynamics in coastal ecosystems: A review with some general lessons from sustained investigations of San Francisco Bay, California. *Reviews of Geophysics* 34: 127–168.

Conley, D. J. and T. C. Malone. 1992. Annual cycle of dissolved silicate in Chesapeake Bay: Implications for the production and fate of phytoplankton biomass. *Marine Ecology Progress Series* 81: 121-128.

Conley, D. J., C. L. Schelske and E. F. Stoermer. 1993. Modification of the biogeochemical cycle of silica with eutrophication. *Marine Ecology Progress Series* 101: 179-192.

Covin, J. D., and J. B. Zedler. 1988. Nitrogen effects of *Spartina foliosa* and *Salicornia virginica* in the salt marsh at Tijuana Estuary, California. *Wetlands* 8: 51–66.

DeVogelaere, A., J. Holte, M. Silberstein, and M. Jacobi. 1998. A species list for Elkhorn Slough and adjacent uplands. Unpublished report, Elkhorn Slough Foundation.

DeYoe, H. R., and C. A. Suttle 1994. The inability of the Texas "brown tide" alga to use nitrate and the role of nitrogen in the initiation of a persistent bloom of this organism. *Journal of Phycology* 30: 800–806.

Dugdale, R. C. and J. J. Goering. 1967. Uptake of new and regenerated forms of nitrogen in primary productivity. *Limnology and Oceanography* 12: 196–206.

Dunton, K. H.. and D. A. Tomasko. 1994. *In situ* photosynthesis in the seagrass *Halodule wrightii* in a hypersaline subtropical lagoon. *Marine Ecology Progress Series* 107: 281–293.

Emory, J. 1997. Beaches and dunes. In *A Natural History of Monterey Bay National Marine Sanctuary*, pp. 77–95. Monterey, CA: NOAA and Monterey Bay Aquarium.

Evans, R.A. and J. A. Young. 1989. Characterization and analysis of abiotic factors and their influences on vegetation. In L. F. Huenneke and H. A. Mooney, eds., *Grassland Structure and Function: California Annual Grassland*, pp. 13–28. Dordrecht, Neth.: Kluwer Academic Publishers.

Ewing, K. K. Mckee, I. Mendelssohn, and M. Hester. 1995. A comparison of indicators of sublethal salinity stress in the salt marsh grass, *Spartina patens* (Ait.) Muhl. *Aquatic Botany* 52: 59–74.

Fong, P., J. B. Zedler, and R.M Donohoe. 1993. Nitrogen vs. phosphorus limitation of algal biomass in shallow coastal lagoons. *Limnology and Oceanography* 38: 906-923.

Gliessman, S. R., M. R. Werner, S. L. Swezey, E. Caswell, J. Cochran, and F. Rosado-May. 1996. Conversion to organic strawberry management changes ecological processes. *California Agriculture* 50: 24–31.

Goff, L. J., D. A. Moon, and A. W. Coleman. 1994. Molecular delineation of species and species relationships in the red algal agarophytes *Gracilariopsis* and *Gracilaria* (Gracilariales). *Journal of Phycology* 30: 521–537.

Gordon, B. L. 1996. *Monterey Bay Area: Natural History and Cultural Imprints.* Pacific Grove, CA: Boxwood Press.

Gordon, D. M., and K. Sand-Jensen. 1990. Effects of O_2, pH, and DIC on photosynthetic net O_2 evolution by marine macroalgae. *Marine Biology* 106: 445–451.

Grove, T., R. Harris, M. Silberstein, R. Kvitek, D. Stow, R. Wright, L. Strnad, J. Van Coops, A. Hitt, J. Yap, and H. Gieben. 1997. Estuarine wetland and watershed inventory using NOAA's coastal change analysis project protocol in California's central coast. Final report to NOAA, Coastal Program Division (NA460Z0422).

Hanes, T. L. 1981. California chaparral. In F. di Castri, D.W. Goodall, and R. L Specht, eds., *Ecosystems of the World*, vol. 11: *Mediterranean-Type Shrublands*, pp. 139–174. Amsterdam: Elsevier Scientific Publishing Company.

Hayward, I. R. 1931. The marsh and aquatic plants of the Pajaro Valley. Master's thesis, Stanford University.

Heij, H. D., and P. H. Nienhuis. 1992. Intraspecific variation in isozyme patterns of phenotypically separated populations of *Zostera marina* L. in the south-western Netherlands. *Journal of Experimental Marine Biology Ecology* : 1-14.

Hickson, J. C. (editor). 1993. *The Jepson Manual: Higher Plants of California*. Berkeley: University of California Press.

Holligam, P. M., E. Fernandez, J. Aiken, W. M. Balch, P. Boyd, and H. Burkill. 1993. A biogeochemical study of the coccolithophorid *Emiliania huxleyi* in the North Atlantic. *Global Biogeochemical Cycles* 7: 879–900.

Holton, B., Jr., M. G. Barbour, and S. N. Martens. 1991. Some aspects of the nitrogen cycle in a California strand ecosystem. *Madroño* 38: 170–184.

Howarth, R. W. 1988. Nutrient limitation of net primary production in marine ecosystems. *Annual Review of Ecology* 19: 89-110.

Huenneke, L. F. 1989. Distribution and regional patterns of California grasslands. In L. F. Huenneke and H. A. Mooney, eds., *Grassland Structure and Function: California Annual Grassland*, pp. 1–12. Dordrecht, Neth.: Kluwer Academic Publishers.

Jackson, L. E., R. B. Strauss, M. K. Firestone, and J. W. Bartolome. 1990. Influence of tree canopies and grassland productivity and nitrogen dynamics in deciduous oak savanna. *Agriculture Ecosystems and Environment* 32: 89–106.

Kraemer, G. P., and R. S. Alberte. 1993. Age-related patterns of metabolism and biomass in subterranean tissues of *Zostera marina* L. (eelgrass). *Marine Ecology Progress Series* 95: 193–203.

Larkum, A., and C. den Hartdog. 1989. Evolution and biogeography of seagrasses. In A. Larkum, A. McComb and S. Shepherd, eds. *Biology of Seagrasses*. pp. 112-156. Amsterdam: Elsevier

Lee, S. Y. 1990. Primary productivity and particulate organic matter flow in an estuarine mangrove-wetland in Hong Kong. *Marine Biology* 106: 453–463.

MacGinnitie, G. E. 1935. Ecological aspects of a California marine estuary. *American Midland Naturalist* 16: 629–765.

McMillan, C. 1982. Isozymes in seagrasses. *Aquatic Botany* 14: 231-243.

McNaughton, S. J. 1968. Structure and function of California grasslands. *Ecology* 49: 962–972.

Mobley, C. 1989. A numerical model for the computation of radiance distribution in natural waters with wind-roughened surfaces. *Limnology and Oceanography* 34: 1473–1483.

Monbet, Y. 1992. Control of phytoplankton biomass in estuaries: A comparative analysis of microtidal and macrotidal estuaries. *Estuaries* 15: 563–571.

Mooney, H. A. 1981. Primary production in Mediterranean-climate regions. In F. di Castri, D. W. Goodall, and R. L Specht, eds., *Ecosystems of the World*, vol. 11: *Mediterranean-Type Shrublands*, pp. 249–255. Amsterdam: Elsevier Scientific Publishing Company.

Neckles, H. A., R. L. Wetzel, and R. J. Orth. 1993. Relative effects of nutrient enrichment and grazing on epiphyte-macrophyte (*Zostera marina* L.) dynamics. *Oecologia* (Heidelberg) 36: 265–274.

Nixon, S. W. and M. E. Q. Pilson. 1984. Nitrogen in estuarine and coastal marine systems. In E. J. Carpenter and D.G. Capone, eds., *Nitrogen in the Marine Environment*. pp. 565-648. New York: Academic Press

Nybakken, J., G. Cailliet, and W. Broenkow. 1977. *Ecologic and Hydrographic Studies of Elkhorn Slough Moss Landing Harbor and Nearshore Coastal Waters, July 1974–June 1976*, pp. 216–386. Moss Landing, CA: Moss Landing Marine Laboratories Technical Publication.

Olff, H., N. E. T. Wageningen, J. Huisman, and B. F. Van Tooren. 1993. Species dynamics and nutrient accumulation during early primary succession in coastal sand dunes. *Journal of Ecology* 81: 693–706.

Orth, R. J., and K. A. Moore 1983. Chesapeake Bay: An unprecedented decline in submerged aquatic vegetation. *Science* 222: 51–53.

Pace, S. D. 1978. Distribution, abundance, and rates of fecundity and growth of *Acartia tonsa* Dana and *Acartia californiensis* Trinast (Copepoda) in Elkhorn Slough California. Master's thesis, Moss Landing Marine Laboratories. San Jose State University.

Page, H. M. 1995. Variation in the natural abundance of 15N in the halophyte *Salicornia virginica* associated with groundwater subsidies of nitrogen in a southern California salt-marsh. *Oecologia* 104: 181–188.

Pezeshki, S. R., and R. D. DeLaune. 1993. Effects of soil hypoxia and salinity on gas exchange and growth of *Spartina patens*. *Marine Ecology Progress Series* 96: 75–81.

Pregnall, A. M., and P. P. Rudy. 1985. Contribution of green macroalgal mats (*Enteromorpha* spp.) to seasonal production in an estuary. *Marine Ecology Progress Series* 24: 167–176.

Randall, J. M., and J. W. Day Jr. 1987. Effects of river discharge and vertical circulation on aquatic primary production in a turbid Louisiana estuary. *Netherland Journal of Sea Research* 21: 231–242.

Rasmussen, E. 1977. The wasting disease of eelgrass (*Zostera marina*) and its effect on environmental factors and fauna. In C. P. McRoy and C. Helfferich, eds., *Seagrass Ecosystems: A Scientific Perspective*, pp. 1–51. New York: Marcel Dekker.

Reilly, M.E., 1979. The salt marsh vegetation of Elkhorn Slough. Western Society of Naturalists: Abstracts of Symposia and Contributed Papers: 3-4.

Robles, M., and F. S. Chapin III. 1995. Comparison of the influence of two exotic communities on ecosystem processes in the Berkeley hills. *Madroño* 42: 349–357.

Ruckleshaus, M. 1995. Estimates of outcrossing rates and of inbreeding depression in a population of the marine angiosperm *Zostera marina*. *Marine Biology* 123: 583–593.

Ryther, J. H. and W. M. Dunstan. 1971. Nitrogen, phosphorus and eutrophication in the coastal marine environment. *Science* 171:1008-1013.

Slager, C. 1997. Oceanography. In *A Natural History of Monterey Bay National Marine Sanctuary*, pp. 11–27. Monterey: NOAA and Monterey Bay Aquarium.

Schulthorpe, C. D. 1967. *The Biology of Aquatic Vascular Plants*. London: Edward Arnold.

Thayer, G., D. Wolfe, and R. Williams 1975. The impact of man on seagrass systems. *American Scientist* 63: 288–296.

Valiela, I. 1995. *Marine Ecological Processes*. New York: Springer-Verlag.

Van Dyke, E., and K. Contreras. 2001. Elkhorn Slough Watershed critical resource maps. Unpublished report, on file at Elkhorn Slough National Estuarine Research Reserve.

Weinhold, B. J., and J. O. Klemmedson. 1992. Effect of prescribed fire on nitrogen and phosphorus in Arizona chaparral soil-plant systems. *Arid Soil Research and Rehabilitation* 6: 285–296.

Wetzel, R. L., and P. A. Penhale. 1983. Production ecology of seagrass communities in the lower Chesapeake Bay *Marine Technology Society Journal* 17: 22–31.

Zimmerman, R., and R. Alberte. 1996. Effect of light/dark transition on carbon translocation in eelgrass *Zostera marina* seedlings. *Marine Ecology Progress Series* 136: 305–309.

Zimmerman, R. C., R. D. Smith, and R. S. Alberte. 1989. Thermal acclimation and whole-plant carbon balance in *Zostera marina* L. (eelgrass). *Journal of Experimental Marine Biology Ecology* 130: 93-109.

Zimmerman, R. C., A. Cabello-Passini, and R. S. Alberte. 1994. Modeling daily production of aquatic macrophytes from irradiance measurements: A comparative analysis. *Marine Ecology Progress Series* 114: 185–196.

Zimmerman, R., D. Kohrs, and R. Alberte. 1996. Top-down impact through a bottom-up mechanism: The effect of limpet grazing on growth, productivity, and carbon allocation of *Zostera marina*. *Oecologia* 107: 560–567.

Zimmerman, R. C., D. G. Kohrs, D. L. Steller, and R. Alberte. 1995a. Sucrose partitioning in *Zostera marina* L. in relation to photosynthesis and the daily light-dark cycle. *Plant Physiology* 108: 1665–1671.

———. 1995b. Carbon partitioning in the seagrass *Zostera marina* L.: Regulation by photosynthesis and the role of daily light-dark cycles. *Plant Physiology* 108: 1667–1671.

Zimmerman, R., and C. Mobley. 1997. Radiative transfer within seagrass canopies: Impact on carbon budgets and light requirements. In *Ocean Optics XIII, Proceedings*, 2963: 331–336. Bellingham: Society of Photo-optical Instrumentation Engineers.

Zimmerman R. C., J. L. Reguzzoni, and R. S. Alberte. 1995. Role of light availability on growth and metabolism of eelgrass (*Zostera marina* L.) transplants in San Francisco Bay. *Aquatic Botany* 51: 67–86.

Zimmerman R. C., J. L. Reguzzoni, S. Wyllie-Echeverria, M. Josselyn, and R. S. Alberte. 1991. Assessment of environmental suitability for growth of *Zostera marina* L. (eelgrass) in San Francisco Bay. *Aquatic Botany* 39: 345–352.

9

Chapter Nine

Invertebrates

Kerstin Wasson, James Nybakken, Rikk Kvitek, Caren Braby, Mark Silberstein

The diverse and sometimes bizarre marine invertebrates of Elkhorn Slough have long delighted locals, as tasty meals or subjects of study. Giant gaper clams and other delicious molluscs have been harvested by humans for thousands of years. Invertebrates also support tens of thousands of shorebirds and waterbirds that stop to feed in the slough as they migrate along the Pacific flyway, and help sustain local breeding populations of birds such as Great Egrets and Great Blue Herons. Many fish, including leopard sharks, prey on invertebrates, as do the sea otters that have recently become abundant in the slough.

Beginning in the 1920s, Elkhorn Slough's invertebrates caught the attention not only of predators but also of fine naturalists. Classic studies of the fat innkeeper worm, familiar to zoologists worldwide, were based at the slough. For decades, students and researchers have become mired in deep mud while attempting to acquaint themselves with the slough's invertebrate communities.

Invertebrates at the slough are also critical from a conservation perspective. Few extensive tidal wetlands remain on the Pacific Coast, and many mudflat species are nowhere else as abundant as in Elkhorn Slough. Drastic declines in the populations of some key species documented over the past decades are thus of serious concern. Various human activities in the Elkhorn Slough watershed have the potential to negatively affect invertebrates, and more research is needed to devise the best management strategies to protect them. Conservation efforts in the area should include explicit focus on these permanent slough residents, both for their own sake and to better protect more familiar but often transient visitors to the slough such as shorebirds, marine mammals, and sharks.

In this chapter, we begin by describing the different habitats with which invertebrates are associated at Elkhorn Slough and by reviewing the history of their study. Next we characterize invertebrate diversity and the natural history and ecology of prominent species. These early sections are not technical, and serve as a basic introduction to the slough's invertebrates. We then cover distributional patterns and changes over time. These middle sections are somewhat more technical, since they necessarily include mention of less commonly known invertebrate taxa; those readers without prior experience with invertebrates may want to skip or skim these parts. Finally, we discuss human influences on invertebrates at Elkhorn Slough, and recommend emphases for management and directions for future research. These sections, like the introductory ones, can be easily understood without any invertebrate background.

Habitat and Community Types

Each different marine, terrestrial, and freshwater habitat type at Elkhorn Slough supports a distinctive invertebrate community. However, only the marine communities have been well studied;

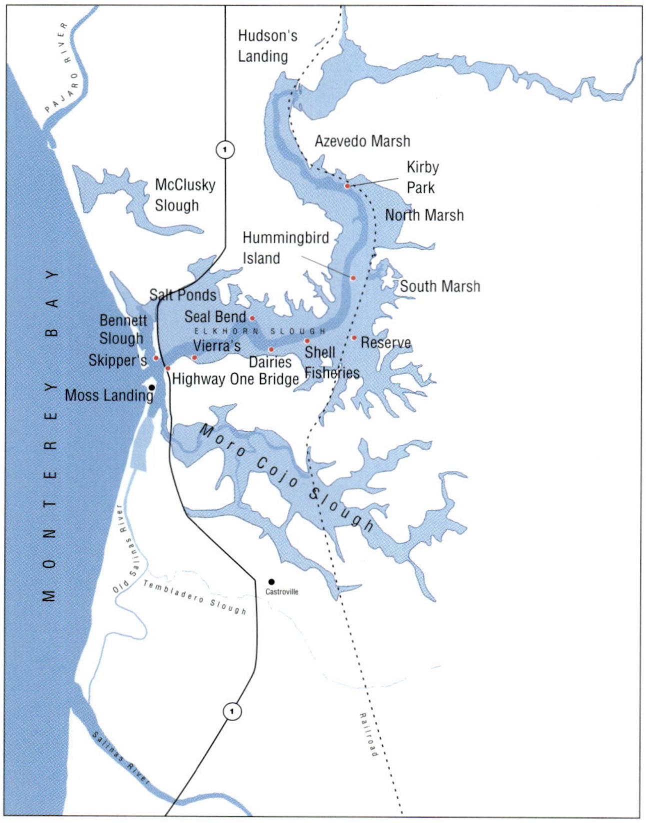

Figure 9.1. Sites of quantitative invertebrate surveys at Elkhorn Slough.

they are therefore the focus of this chapter. Many burrowing invertebrates, including countless clams and worms, live in the soft sediments of the extensive intertidal and subtidal mudflats. This *benthic infauna* includes the most prominent and intensely studied of the slough's marine invertebrates. Marine species that live on the surface of the sediments or on the occasional hard substrates found at the slough constitute the *benthic epifauna*, which has also been periodically studied. This group includes mobile animals such as snails or crabs, and sessile fouling organisms such as bryozoans and ascidians. Other marine invertebrates are part of the *plankton*, drifters in the water column for all or part of their lives. Planktonic species have received little attention, and are only briefly covered in the following pages.

Freshwater invertebrate communities are associated with riparian habitats in the Elkhorn Slough watershed, such as those along Carneros Creek. Upland habitats, ranging from marshes to grasslands to oak woodlands, also host a wealth of invertebrates, particularly spiders and insects. Because there has been little research on these communities, they are only briefly treated in this chapter.

History of Study

The rich marine invertebrate fauna of Elkhorn Slough has been the subject of intense, if somewhat intermittent, scientific study since the beginning of the last century. The first detailed investigation was the seminal work of George MacGinitie (1927, 1935), who described the invertebrate community near the current mouth of the slough (the area west of the Highway 1 bridge). His extensive monograph from 1935 is still the best source for the natural history of many slough species, especially the large, conspicuous, and charismatic worms, shrimps, snails, and bivalves of the mudflats. However, MacGinitie's report does not represent a comprehensive baseline, since no information is provided on the middle and upper slough, and since some taxa were better covered than others by his qualitative sampling.

In the 1970s, James Nybakken and other researchers from Moss Landing Marine Laboratories initiated a thorough characterization of the benthic infauna of the slough at four main sites (Skippers, Vierras, Dairies, and Kirby Park; fig. 9.1). The resulting report (Nybakken, Cailliet, and Broenkow 1977) was the first broad-scale quantitative assessment of the slough's marine invertebrates, and the first survey to cover the entire length of Elkhorn Slough. The focus of this investigation was free-living macroinvertebrates of intertidal mudflats. Fouling organisms such as bryozoans and ascidians, planktonic organisms such as copepods, parasitic organisms such as trematodes, and tiny animals such as nematodes, kinorhynchs, gastrotrichs, and rotifers were not covered.

In the early 1990s, Rikk Kvitek and his colleagues repeated the quantitative sampling regime of the 1970s, to assess changes in the intervening decades. Their unpublished findings (Kvitek et al. 1996) are included in this chapter. In the late 1990s, Kerstin Wasson and her collaborators (Wasson et al. 2001) thoroughly searched at ten intertidal slough sites for the presence of non-native invertebrate species.

Besides these four broad investigations of the marine invertebrate fauna of Elkhorn Slough, there have been many focused studies of particular taxa, especially by students and researchers from the nearby Moss Landing Marine Laboratories and the University of California, Santa Cruz. Most of these specific studies, as well as the four broad investigations, are summarized in this chapter.

Diversity of Marine Invertebrates

Over 550 species of marine invertebrates belonging to 16 phyla have been reported for Elkhorn Slough (appendix 9.1). Three taxa of invertebrates—polychaetes, crustaceans, and bivalves—dominate the slough's benthic infauna, as they do other soft-sediment communities around the world. Each of these three groups is represented by well over 100 species at the slough. However, many smaller taxa are also represented, and some of them (such as the echiuran fat innkeeper) are common. In addition to its rich native fauna, the slough hosts many non-native species introduced by humans, about 50 of which were reported by Wasson et al. (2001). Since microscopic species, parasites, and planktonic invertebrates have not received much attention, the true marine invertebrate diversity of the slough surely exceeds the number documented many times over.

Natural History and Ecology of Key Marine Species

Perhaps no locality offers a better opportunity for ecological study than does Elkhorn Slough. —MacGinitie 1935

The invertebrates of Elkhorn Slough have been the subject of many thoughtful ecological studies. In this section, we briefly and informally highlight some of the most common and charismatic characters of the invertebrate assemblage. We begin with the species that have been best studied at the slough, and close with a few taxa that have been neglected but would be ideal for focused investigations in the future. The best resources for more detailed background on these and other slough invertebrates are MacGinitie (1935), Morris, Abbott, and Haderlie (1980), and Ricketts et al. (1985).

Fat Innkeeper Worm If an invertebrate mascot had to be chosen for Elkhorn Slough, it would surely be the fat innkeeper or weenie worm, *Urechis caupo*, which can account for almost 80% of the subtidal invertebrate biomass in the main channel (Jolly 1997). Resembling a pink, squirming sausage with striking golden bristles adorning the anterior and encircling the posterior ends, this echiuran worm digs a semipermanent U-shaped burrow. The worm rhythmically contracts to pump water through the burrow and thereby through a mucus net it spins and deploys above its proboscis. When detritus or planktonic particles clog the net, it is reeled in and ingested. The worm earns its common name by hosting "as motley a crew of guests as one could hope to find" (Ricketts et al. 1985): a number of animals that take advantage of the safety of the burrows or of dining

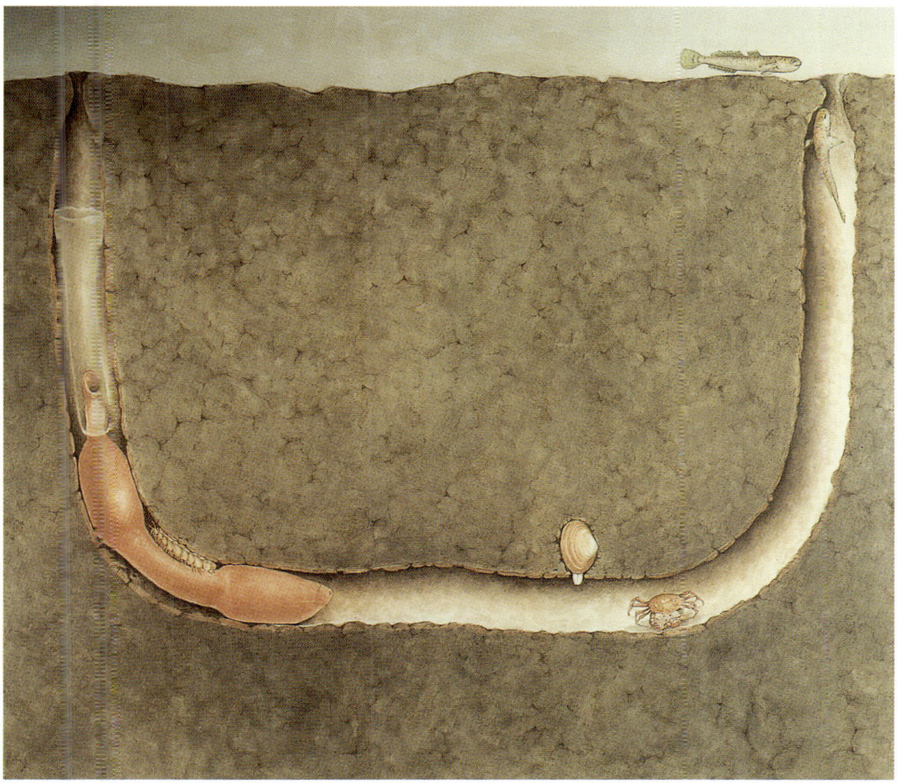

Fat innkeeper and guests in its burrow. Left to right: fat innkeeper, scaleworm, small clam, pea crab, goby. Illustration credit: Ann Caudle, © Monterey Bay Aquarium Foundation.

opportunities. An aggressive reddish scale worm, *Hesperonoe adventor*, pounces on discarded particles too big for the innkeeper, and sometimes steals food directly from the net. Two different species of pea crabs, *Scleroplax granulata* and *Pinnixa* sp., are found in the burrows, and a goby, *Clevelandia ios*, seeks shelter there between foraging bouts. A small clam, *Cryptomya californica*, protrudes its siphons into the burrow to take advantage of the worm's feeding current.

The behavior of the fat innkeeper was first investigated by Fisher and MacGinitie (1928), who, in a classic study, placed animals in clear glass tubes in aquaria to observe daily activities such as respiratory movements, feeding, cleaning of the burrow, and resting. The slough population of fat innkeepers has supplied gametes and larvae for embryologists across the country since the 1930s, and these worms continue to be popular subjects of developmental and physiological studies. For instance, Arp, Hansen, and Julian (1995) have examined their physiological tolerance to high sulfide conditions, while Toomey and Epel (1995) found that fat innkeepers gain protection from some environmental toxins through a special immunological defense system.

Phoronopsis viridis This phoronid is another odd worm characteristic of Elkhorn Slough. A member of a very small phylum with only a dozen or so extant species, *P. viridis* individuals live in stiff, sand-encrusted tubes from which they extend their green tentacles to suspension feed. MacGinitie (1935) described such an abundance of these worms that parts of the seafloor in Elkhorn Slough appeared solid green due to their tentacles, and Nybakken et al. (1977) found them to be very common as well, with thousands per square meter in some areas. (The green-tentacled *P. viridis* has been considered by some to be synonymous with a white-tentacled species, *P.*

Common slough clams, with characteristic siphon lengths. Left to right: gaper, Washington, bent-nosed, littleneck, cockle, geoduck. Illustration credit: Ann Caudle, © Monterey Bay Aquarium Foundation.

harmeri, we maintain the distinction for clarity: only green-tentacled phoronids have been observed in the slough.) The high densities of these worms in central California are unique in the world; phoronids are typically so rare that few zoologists have ever seen them alive (Morris, Abbott, and Haderlie 1980). Unfortunately, as we discuss later, the slough's phoronid populations are in imminent danger of local extinction.

Gaper Clam The gaper clam (*Tresus nuttallii*) is one of the largest American clams, with a shell length up to 25 centimeters. The openings to the deep burrows of this clam are a common sight in the mudflats of the lower slough. Young clams burrow actively, gradually moving deeper into the sediment; as adults, they inhabit burrows that may be more than one meter deep. When disturbed (for instance by the vibrations of approaching footsteps), they retract their long siphons and squirt forceful geysers of water. Since the siphon must be long enough to reach the surface, it is too big to fit into the shell when retracted, hence the common name "gaper." The siphon tip has distinctive hard plates that help to protect it from the nibbles of passing predators. These little bits of hard substrate in an otherwise soft mud habitat support the growth of at least fifty species of invertebrates and algae. Gapers also serve as hosts for pea crabs, which nestle in their spacious mantle cavities.

Gapers eat suspended planktonic particles drawn down to their gills through the siphon. In turn, gapers are eaten by bat rays, leopard sharks, sea otters, and humans. The clams are often heavily infested with larval tapeworms, whose definitive hosts are bat rays. Laurent (1971) investigated reproduction and growth of gapers in Elkhorn Slough and found nearly continuous spawning year round and very rapid growth of juveniles, features that are perhaps responsible for the persistence of this species at moderate densities despite high predation levels. In a related study, Clark, Nybakken, and Laurent (1975) confirmed these results for reproduction and growth, and developed growth curves useful for estimating the ages of clams of known sizes. They also detected twofold differences in juvenile recruitment between years, but what is responsible for this variation remains unknown, an intriguing topic for future studies.

Bent-nosed Clam While the gaper is the largest clam, *Macoma nasuta*, the bent-nosed clam, is the most abundant of the larger bivalves of the slough. Addicott (1952) carefully investigated distributional patterns, burrowing, and feeding in the bent-nosed clam and three other *Macoma* species in the slough. Together these four species accounted for nearly 900 individuals per square meter in areas of the lower slough.

Moon snail cruising over a mudflat on its large foot. Photo credit: Christopher Richard.

The bent-nosed clam, so called because the valves are bent to the right at the siphon end, is an active burrower for its entire life. When buried, it lies on its left side and extends the incurrent siphon out over the surface of the sediment. The incurrent siphon gyrates in a circular motion, sweeping organic particles into the feeding current of the clam. The excurrent siphon stays below the sediment, but produces a visible volcanolike mound on the surface. A large bent-nosed clam emits about one million platter-shaped fecal pellets a year, and pellet formation by this species visibly alters the sedimentary environment in areas where individuals are abundant (Addicott 1952).

Other clams often encountered at the slough include the Washington, *Saxidomus nuttallii*; and the littleneck, both the native *Protothaca staminea* and the nonnative *Venerupis philippinarum*.

Moon Snail Exploring the mudflats of Elkhorn Slough on a low tide, one is likely to encounter sand-encrusted, rubbery collars reminiscent of the plunger of a "plumber's friend." These collars are the egg cases of the moon snail, *Polinices lewisii*. From them hatch larvae, which settle on sea lettuce

Sea hare. Illustration credit: Ann Caudle, © Monterey Bay Aquarium Foundation.

(*Ulva* spp.), eating diatoms on the alga and the alga itself, for their first months of life. As an adult, however, the moon snail is a voracious carnivore, preying mostly on clams. It plows through the mud on a foot that becomes improbably large when inflated with water; surprisingly, the foot can fit within the shell when the water is expelled. When a clam is located, the snail either uses its foot to surround and suffocate its victim, or it drills a hole into the shell, using its toothlike radula and chemical secretions. Empty clamshells bearing this telltale sign of a fatal encounter with a moon snail are commonly washed up on slough mudflats.

Sea Hare The California sea hare, *Aplysia californica*, is a massive (up to 3 kg) herbivore on algae and eelgrass. When irritated, it exudes a cloud of ink, colored dark purple by pigments from red algae it has consumed. Sea hares are seasonally abundant in the slough, where they congregate to mate. Simultaneous hermaphrodites, they often copulate in chains or rings, with each individual receiving sperm from one animal while donating sperm to another. Afterward s/he lays a huge egg mass, resembling a tangled bunch of yellow yarn. In addition to entertaining field biologists with its interesting habits in the wild, the sea hare has been a popular subject of laboratory studies, particularly neurobiological investigations, due to its large, accessible neurons. Indeed, for many years, a sea hare "ranch" at Elkhorn Slough supplied neurobiologists worldwide with research material.

Ghost Shrimp The ghost shrimp, *Callianassa californiensis*, is an extremely active burrower, using its second and third pairs of legs as shovels for digging. Despite its seemingly ceaseless activity, the respiratory rate of the ghost shrimp is low, and it can survive without oxygen for six days. The ghost shrimp is a selective deposit feeder, picking desirable organic particles from the sediment it digs through and from the surface near the openings of its burrows. By constantly excavating tunnels, the ghost shrimp loosens the packing of the sediments and allows oxygenated water to percolate much deeper than it would otherwise. MacGinitie (1935) said of it, "Perhaps no animal exerts more influence on mudflat associations than does this shrimp." His assertion, based on the sheer abundance of the ghost shrimp, and on their alteration of the physical environment, was never tested. Nowadays large ghost shrimp have become quite rare in the slough, but their very absence may likewise be shaping the species composition of the mudflats, since some animals may have depended on the disturbance and oxygenation of the sediments that the shrimps' burrowing activities provided.

Ghost shrimp. Photo credit: James Nybakken.

Shore Crabs The striped shore crab (*Pachygrapsus crassipes*) and yellow shore crab (*Hemigrapsus oregonensis*) are the most conspicuous crabs of Elkhorn Slough, scurrying about in almost every bank and pickleweed marsh. Sliger (1982) found distinct distributional differences between these two common grapsids in the slough, based on different preferences for substrate type and tidal elevation. Striped shore crabs are found in the upper parts of banks that are firmly consolidated by pickleweed roots, where the water flowing through their burrows remains quite

Shore crab. Photo credit: Greg Hoffman, Elkhorn Slough Foundation.

clear; they have gills that clog easily and cannot survive long in muddy water. Yellow shore crabs are found in the lower parts of banks, where the water is often clouded with sediments; they can tolerate muddy water, and perhaps avoid higher banks due to predation by striped shore crabs.

Both of these rambunctious crabs are semiterrestrial, spending much of their time out of water. Like many crabs, they are quite omnivorous, eating macroalgae and diatom films, but also killing small animals and scavenging remains. In turn, shore crabs are preyed upon by many shorebirds. Ramer, Page, and Yokel-vich (1991) found that shore crabs were the most important prey of the willet, one of the most numerous wintering shorebirds at the slough.

Pea Crabs Elkhorn Slough is a hotbed of pea crab (pinnotherid) diversity. At least eight species are present, all of them commensal with other mudflat invertebrates. For instance, *Pinnixa faba* often inhabits the mantle cavity of the gaper clam, while *P. franciscana* frequently shares the burrows of fat innkeeper worms (see illustration, p. 137). Shore crabs may be more conspicuous, but pea crabs are probably more abundant in the slough, and theirs are the most common crab larvae in the plankton (Hsueh 1991). Pea crabs typically eat organic particles scraped from the host's body or taken from the host's feeding current. Experiments could be carried out to determine whether the relationship should be characterized as parasitism rather than commensalism. These ubiquitous little crabs would also make rich systems for a variety of other investigations. For instance, there is often strong size dimorphism, with large females and small males, which suggests an interesting mating system. Host preference and settlement behavior would also lend themselves readily to study.

Amphipods and Tanaids The small amphipod and tanaid crustaceans of the slough, while so diverse and abundant that they are often numerically dominant in benthic cores, have received little focused attention. They certainly merit further study at the slough. Amphipods and tanaids are ideal for laboratory experiments, for instance examining mate-guarding behavior, or for toxicological studies. In the field, population dynamics, settlement behavior, or foraging strategies could be fruitfully explored.

Polychaetes Oddly, there have been very few studies examining the natural history or ecology of the slough's segmented worms; like the amphipods, these common inhabitants have been comparatively neglected by local researchers. MacGinitie (1935) stated that he spent little time with them because they were too hard to identify. Surely these diverse and abundant members of slough communities merit focused studies. For instance, vermicelli worms, *Notomastus tenuis*, resembling red vermicelli and dense in every shovelful of mud in the lower slough, would be well suited to biomechanical experiments examining breakage threshold when stretched, or burrowing capability. Likewise, the ubiquitous *Streblospio benedicti* of the upper slough has been shown elsewhere to have both planktonic and benthic larvae, and would lend itself to investigations of the ecological causes and consequences of this unusual developmental flexibility.

A dense handful of vermicelli worms in mud.
Photo credit: Kenton Parker.

Distribution of Dominant Species along the Length of Elkhorn Slough

Benthic Intertidal Infauna

Some benthic infaunal species occur throughout the slough. For instance, a few small crustacean species (the tanaid *Leptochelia dubia*, cumaceans of the genera *Cyclaspis* and *Cumella*, and amphipods of the genus *Corophium*) are common at sites along the length of the slough. Nonetheless, the benthic invertebrate communities of the lower and upper slough are fundamentally different. Some widely distributed species are abundant only near the mouth, while others are completely restricted to either the lower or the upper slough.

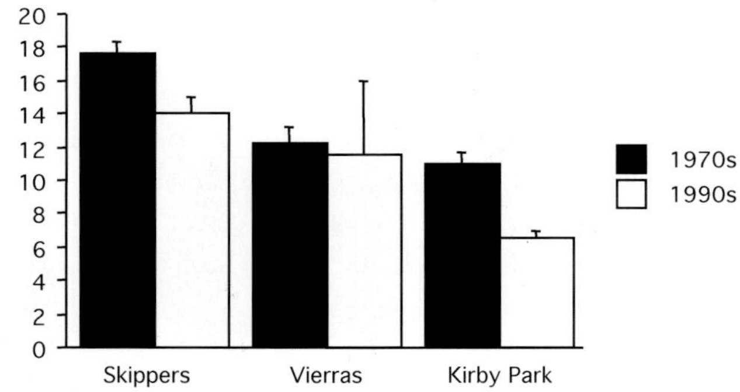

Figure 9.2. Invertebrate species diversity in benthic cores. Diversity declined with distance from the mouth, and declined between the 1970s and 1990s. Based on Nybakken et al. (1977) and Kvitek et al. (1996).

Quantitative studies in the 1970s (Nybakken et al. 1977) and 1990s (Kvitek et al. 1996) revealed that infaunal invertebrate diversity is greatest near the mouth and decreases towards the head (fig. 9.2). There is a gradient of sediment size along the length of the slough, with predominantly coarse beach sands at the mouth and finer silt and clay deposits at the head (Addicott 1952; Nybakken et al. 1977). This gradient affects the distribution of burrowing organisms that are specialized for certain substrate types. Turbidity is typically higher in areas with fine sediments, and this may prevent the upper slough from being settled by those suspension feeders whose capture devices easily become clogged. Current speed is generally higher near the mouth; conversely, residence time of the water is longer at the head (see chapter 4, "Hydrography"). Consequently, salinity near the mouth is consistently fairly similar to adjacent oceanic waters, while water in the upper reaches can become quite hyposaline during winter rains or hypersaline during hot dry periods. Likewise, other parameters to which organisms are sensitive (e.g., temperature, dissolved oxygen, pH) show much stronger daily and seasonal variation in the upper reaches of the slough than at the mouth. The environment therefore generally becomes more stressful for animals in a gradient from the mouth to the head. Only species with broad tolerances are found in the upper slough, and this limits the species diversity there.

In the lower slough, the most abundant bivalve is the bent-nosed clam, *Macoma nasuta*, which is absent from the upper slough. The larger clam species, such as gaper (*Tresus nutallii*), littleneck (*Protothaca staminea*), and Washington (*Saxidomus nutalli*) clams and basket cockles (*Clinocardium nutallii*), all of which are collected by humans as food, are also confined to the more sandy, coarse sediments of the lower slough, perhaps because the muddier water of the upper slough makes their suspension feeding too inefficient. Blue mud shrimp (*Upogebia pugettensis*) and ghost shrimp (*Callianassa californiensis*) are apparently limited mostly to the lower slough, probably because they need more sandy sediments to construct permanent burrows; oddly, however, a new population of ghost shrimp has recently become established in the mid-slough, in finer sediments. Among polychaetes, the capitellids *Capitella capitata* and *Notomastus tenuis* and the opheliid *Armandia brevis* are common in the lower and middle slough.

The calmer waters and finer sediments of the upper slough are dominated by different species, many of them nonnative. A small suspension-feeding venerid, the Atlantic gem clam (*Gemma gemma*), is by far the most abundant bivalve in the upper slough, and the Japanese littleneck clam, *Venerupis philippinarum*, is also common in the middle and upper slough. The native razor clam, *Tagelus californianus*, while rarer, is also characteristic of the upper slough. The polychaete fauna of the upper slough is dominated by two spionid species, *Exogone lourei*, probably a native, and *Streblospio benedicti*, an introduced Atlantic species. Since Pacific Coast estuaries are geologically young (see chapter 2, "Geology"), they may have relatively few and less specialized species, and thus be vulnerable to invasion from other regions, such as the Atlantic, where estuaries and their endemic faunas are much older. The more oceanic and more diverse fauna of the lower slough may thus be more resistant to invasion than the truly estuarine (strongly influenced by freshwater) communities of the upper slough.

Benthic Intertidal Epifauna

The distribution and abundance of the benthic epifauna have not been rigorously studied at Elkhorn Slough, but appear also to be related to distance from the mouth. Various large mobile species, such as the moon snail (*Polinices lewisii*), sea hare (*Aplysia californica* and *Chelidonura inermis*), and the ocher sea star (*Pisaster ochraceus*) occur mostly in the lower slough. In contrast, the Japanese mud snail, *Batillaria attramentaria* occurs everywhere, but is most abundant in the upper slough. The two grapsid shore crabs (*Hemigrapsus oregonensis* and *Pachygrapsus crassipes*) co-occur along much of the length of the slough. *Cancer* crabs are found mostly in the lower slough, while the non-native green crab (*Carcinus maenas*) has been collected mostly in the upper slough (Wasson et al. 2001). Some epifaunal species are associated with plants or algae. MacGinitie (1935) examined invertebrates associated with eelgrass beds near the mouth of the slough. This characteristic epifauna has not received much attention in past decades, perhaps because of the decrease in eelgrass after the opening of the harbor mouth in 1946. In recent years, eelgrass beds in the mid-slough have begun to expand, and brief surveys have again revealed a diverse fauna, including the slug *Phyllaplysia taylori* in abundance (Silberstein, pers. obsv.). While eelgrass has generally declined in the last decades, mats of macroalgae (especially *Enteromorpha* and *Ulva* spp.) have probably increased during this period due to high levels of nutrients; algal mats are particularly dense in the upper slough (see chapter 8, "Primary Producers"). A recent study examined the fauna associated with macroalgal mats and found that amphipod densities are high in the mats, which provide these crustaceans food as well as shelter (Allen 1992).

Pilings and gravel bars in the slough host a mostly sessile, suspension-feeding community that has not been well studied. In the lower slough, for instance on the harbor jetty and the Highway 1 bridge, there are many species typical of the sheltered rocky intertidal, including various native anemones, sea stars, barnacles, and hydroids. Some nonnative species (such the bay mussel, *Mytilus galloprovincialis* and the bryozoan *Watersipora subtorquata*) are also common here. In the mid-slough (Elkhorn Slough National Estuarine Research Reserve) and upper slough (Kirby Park and Hudson Landing), most of the fouling species are introduced; there is a diverse assemblage of non-native sponges, bryozoans, hydroids, anemones, and ascidians (Wasson et al. 2001). However, one native species is common and conspicuous on rocks and pilings of the upper slough: the oyster *Ostrea conchaphala* (formerly *O. lurida*) forms extensive beds, providing additional hard substrate that hosts many nonnative fouling species.

Nudibranch gastropods, which consume fouling species such as hydroids, ascidians, and bryozoans, are found in the slough near their preferred prey (McDonald and Nybakken 1978). Some open coast nudibranchs seem to venture only into the lower slough to lay egg masses, while others, including two nonnative species (Wasson et al. 2001) are found in the uppermost reaches of the slough.

Subtidal Communities

In 1975, Nybakken and colleagues conducted a series of observational dives at five stations (shown in fig. 9.1), which resulted in a qualitative description of the slough's subtidal communities (detailed in Nybakken et al. 1977 and summarized below). As with the intertidal communities, they found distinctive differences in dominant species between the lower and upper slough.

At the Highway 1 bridge, the current and old bridge pilings and the surrounding rocks were covered with the anemone *Metridium senile*, with occasional *Anthopleura* anemones and mussels (*Mytilus* sp.) interspersed. In the soft substrate, the siphons of large gaper clams (*Tresus nuttallii*) and the burrows of fat innkeepers (*Urechis caupo*) were fairly abundant. The rock crab *Cancer antennarius* was also observed.

Near Vierras and the former outfall of the power plant, the sandy mud bottom was densely populated with the siphons of the rough piddock, *Zirfaea pilsbryi*, and with lesser numbers of gaper clam siphons and fat innkeeper burrows. Sabellid and terebellid polychaetes were also observed, as were the moon snail *Polinices* and the sea hare *Chelidonura*. The sea hare *Aplysia californica* and its egg masses were also especially abundant.

Farther up the slough near the intertidal sampling station known as the Dairies, the softer bottom sediments were dominated by rough piddock. Also seen here was the tube-dwelling anemone *Pachycerianthus fimbriatus*, various terebellid and sabellid polychaetes, and large moon snails. Even farther up, across from Hummingbird Island, the bottom was dominated by large empty shells of the Japanese oyster, *Crassostrea gigas*, remnants from past oyster-culturing efforts in the slough. These shells harbored a rich epifauna of bryozoans, anemones, and colonial tunicates. The nudibranch *Aeolidia papillosa* was also common.

Near Kirby Park, the bottom was of soft mud, with scattered cement blocks and rocks. Few large clams were present and the station was instead characterized by sponges, burrowing anemones, and nudibranchs.

Planktonic Invertebrates

Much less is known about planktonic invertebrates of the slough than about their benthic counterparts. Nybakken et al. (1977) carried out zooplankton tows, sampling many different

Planktonic larvae of a spionid polychaete worm. Photo credit: Kenton Parker.

species, of which calanoid copepods of the genus *Acartia* were most abundant. Zooplankton densities were highest in the upper slough, near Kirby Park. Pace (1978) examined the distribution, abundance, and fecundity of three copepods in the slough. He found that the upper slough was dominated by resident populations of *Acartia californiensis*, whereas the lower slough was dominated by *A. tonsa* and *A. clausii*, species also common on the open coast. This suggests the planktonic communities of the upper and lower slough may differ, as do their benthic counterparts.

Planktonic invertebrate larvae have received some attention at Elkhorn Slough. An examination of settlement of polychaete larvae into collecting jars anchored near the benthic intertidal sampling sites (of Nybakken et al. 1977) revealed that the upper slough harbored a distinct assemblage, apparently retained by a relatively isolated water mass (Oliver and Jong 1981; Kvitek et al. 1996). The spatial distribution of settling polychaete larvae closely matched that of adults for dominant species. Abundance of juveniles of the broadcast spawners *Armandia brevis* and *Capitella capitata* increased toward the slough mouth, and peak abundances occurred during late fall and winter. Juveniles of the brooders *Exogone lourei* and *Streblospio benedicti* increased in abundance toward the head of the slough, with peak abundances occurring during the summer.

Larval stages of brachyuran crabs are perhaps the best-studied components of the plankton in Elkhorn Slough. Hsueh (1988, 1991) examined seasonal patterns of abundance of crab larvae at four stations in the slough. He found 11 species, of which he could identify 7: *Cancer gracilis*, *C. productus*, *Hemigrapsus oregonensis*, *Lophopanopeus bellus*, *Pachygrapsus crassipes*, *Pinnixia franciscana*, *P. weymouthi*. The most abundant crab larvae were those of a pea crab, *Pinnixa franciscana*, commensal with *Urechis* and other burrowers. All stages of pea crab larvae (from newly hatched to competent to settle) were found in the slough, which suggests that they were released by resident females and retained within the slough throughout their development into adults. Larvae of *Hemigrapsus oregonensis* were second most common. Early zoeal stages were abundant, especially in the upper slough, which suggests that resident females released them from this area; later stages were scarce in the slough, so these grapsid larvae likely undergo the rest of their planktonic period elsewhere. Similarly, Von Thun et al. (2000) found that the larvae of *Hemigrapsus*, *Pachygrapsus*, and *Cancer gracilis* were exported from the slough, while larvae of *Pinnixa franciscana* and another unidentified pea crab were retained. Additionally, they found that larvae of *Emerita analoga*, the mole crab, were actually imported into the slough, presumably from nearby sandy beaches. In another recent study, *Cancer* spp. larvae were identified and enumerated from a number of sites in and around the slough to assess possible impacts of expansion of the Moss Landing power plant (Tenera 2000). Three species of *Cancer* crab megalops were found in the Elkhorn Slough area: *C. jordani* (50% of *Cancer* larvae), with peak concentrations in September; *C. productus* (25%), peaking in December; and *C. anthonyi* (25%), with high concentrations in both September and January.

Changes in Invertebrate Abundance and Diversity Over Time

Intertidal Benthic Invertebrate Communities, 1920s–1970s

Conditions at Elkhorn Slough changed dramatically between the 1920s and the 1970s, the interval between thorough studies by MacGinitie (1935) and Nybakken et al. (1977). During this period, the boat harbor was developed in the lower part of the slough, a new entrance to Monterey Bay was created, the lower slough was continually dredged to facilitate vessel traffic, and a large power plant was erected with its intake and outfall situated in the slough. These anthropogenic changes undoubtedly had an effect on the slough's invertebrate fauna.

It is difficult, however, to compare slough invertebrate communities of these eras. The study by MacGinitie (1935) was qualitative, limited to a few sites near the mouth, and included both epifauna and infauna, while that by Nybakken et al. (1977) was quantitative, sloughwide, and limited to benthic infauna. Nybakken et al. found many species of polychaetes, including members of eight families (ctenodrillids, dorvilleids, goniadids, hesionids, magelonids, orbiniids, phyllocodids, and spionids) not reported by MacGinitie. This difference is almost certainly due to MacGinitie's less thorough search effort and species identifications rather than actual changes. Conversely, MacGinitie recorded more species of decapod crustaceans, probably because Nybakken et al. used small core samples that do not collect larger decapods. Among the molluscs, species attached to hard substrates are better represented by MacGinitie, who searched such habitats, while many more nudibranchs were found by Nybakken et al. because of particular interest in this taxon. In summary, comparison between the species lists of MacGinitie (1935) and Nybakken et al. (1977) reveals numerous differences in species richness for different phyla, but the differing methods and focus of the studies make it challenging to separate real differences from those due to sampling bias.

While comparisons of diversity and species composition between the two eras are difficult owing to differences in search effort, radical changes in abundance are more reliable indicators of significant change in the invertebrate communities. Some species (indicated with a plus in appendix 9.1) that were described as abundant by MacGinitie were only rarely encountered, or not found at all, by Nybakken et al. or subsequent researchers, despite efforts in the appropriate habitats. The following are the most notable examples of such species.

Scyphistomae (jellyfish polyps) once carpeted rocks near the mouth of the slough, forming vast aggregations during all seasons (Galigher 1925; MacGinitie 1935). However, none of these polyps were ever found by Nybakken et al. or subsequent researchers. The timing and cause of their disappearance remains mysterious. Likewise, another cnidarian, the burrowing sand anemone (*Zaolutus actius* = *Harenactis attenuata*), was described as common by MacGinitie, but has not been reported since.

Among bivalves, one nonnative species disappeared. MacGinitie (1935) reported that *Mya arenaria*, the Atlantic soft-shell clam, was common, and it was still present in the 1950s (Addicott 1952). By the 1970s (Nybakken et al. 1977), the species was apparently absent; directed searches in the 1990s (Wasson et al. 2001) also failed to uncover it.

Three native bivalve species (*Clinocardium nutallii*, *Protothaca staminea*, and *Saxidomus nuttalli*) noted as abundant by MacGinitie were no longer considered common by Clark, Nybakken, and Laurent (1975), perhaps due to overharvesting by humans.

One gastropod mollusc, the horn snail (*Cerithidea californica*), apparently went locally extinct during this time period. Though it was not reported by MacGinitie, early published records and museum specimens confirm its presence at Elkhorn Slough until about 1940 (Byers 1999). This native snail was likely displaced by the ecologically similar but competitively superior Japanese mud snail, *Batillaria attramentaria*, which became abundant in the slough in the latter part of the twentieth century (Byers 1999). About 18 species of trematode parasites are known to be associated with *Cerithidea*; these species have presumably also disappeared from the slough along with their host.

Among crustaceans, the blue mud shrimp (*Upogebia pugettensis*) was abundant in MacGinitie's day, but had become rare by the 1970s (Nybakken et al. 1977) and remains so today, perhaps owing to overharvesting as bait or to high levels of pollutants.

Among worms, the giant nemertean *Cerebratulus californiensis* (up to 4 meters long!) was abundant in areas with large polychaetes in the 1930s (MacGinitie 1935) but rarely encountered in the 1970s (Nybakken et al. 1977). Other worms reported by MacGinitie as abundant but only very rarely found by Nybakken et al. and subsequent investigators include the polychaetes *Glycera rugosa* and *Eudistyllia polymorpha* and the sipunculid *Themiste perimeces*. The apparent decline of these three species is perhaps related to reduction in suitable habitat; they are associated with eelgrass beds, which decreased in extent after the opening of the harbor mouth in 1946.

Infaunal Intertidal Communities, 1970s–1990s

Physical conditions at Elkhorn Slough continued to change between the 1970s and 1990s. Nutrient levels increased (see chapter 12, "Biogeochemical Cycling") and tidal scour became stronger due to increased tidal volume following the breaching of levees in 1983 to reflood the South Marsh area (see chapter 4). There were also significant differences in the benthic infaunal community between the 1970s (Nybakken et al. 1977) and the 1990s (Kvitek et al. 1996). Since sampling in both periods spanned multiple seasons and years, differences probably reflect long-term trends rather than short-term fluctuations attributable to climatic variation. Overall, comparison of the Nybakken et al. and Kvitek et al. studies reveals a significant decline in total invertebrate species diversity in benthic intertidal cores (fig. 9.2). For instance, at the Skippers Restaurant site near the mouth of the slough, about 20 species were found per core in the 1970s, but only about 15 in the 1990s. In terms of abundance, Kvitek et al. found no overall trends: some species decreased in abundance between the mid-1970s and the mid-1990s, while others increased (see below). In contrast, Lindquist (1998) documented an overall decrease in benthic invertebrate densities along the main channel during this twenty-odd-year period.

Among polychaetes, Kvitek et al. found that the common capitellid species, *Capitella capitata* and *Notomastus tenuis*, became more abundant near the mouth of the slough, while the introduced spionid *Streblospio benedicti* became more abundant at its head. The most striking change among crustaceans was the large-scale invasion of an Asian amphipod, *Grandidierella japonica*. This species was not present in the 1970s, but is now one of the most abundant crustaceans along the length of the slough (Kvitek et al. 1996). Lindquist (1998) found that amphipod and tanaid crustaceans increased in abundance between the 1970s and 1990s. The fat innkeeper, *Urechis caupo*, also dramatically increased in abundance. In the 1960s and 1970s it was uncommon and restricted to a few areas in the lower slough, where MacGinitie had also reported it. But by the 1990s *Urechis* was abundant throughout many areas of the lower slough (Kvitek et al. 1996).

Other species declined significantly in abundance between the 1970s and 1990s. Most dramatic has been the virtual disappearance of the phoronid worm *Phoronopsis viridis*. This species was abundant on mudflats near the mouth of the slough in the 1930s (MacGinitie 1935) and in the 1970s (Nybakken et al. 1977) but was not detected during quantitative sampling in the 1990s (Kvitek et al. 1996). Isolated patches of this unique green-tentacled creature are still occasionally sighted after hours of searching (K. Wasson, unpublished data), but it is surely in real danger of becoming locally extinct. The ghost shrimp, *Callianassa californiensis*, was common in the lower slough in the 1930s (MacGinitie 1935) and the 1970s (Nybakken et al. 1977), but rare in the 1990s (Kvitek et al. 1996). Popular with fishermen as bait, these shrimp may have been overharvested, or they may be sensitive to certain pollutants. Likewise, the gaper clam, *Tresus nuttallii*, has decreased significantly in abundance. In 1969 there were about ten siphons per square meter in sampling at Skippers; in 1979, only about four; and in 1992, none at all (Kvitek et al. 1996). The gaper is harvested by both people and sea otters. However, sea otters probably do not account for most of the observed decline, since much of the reduction in abundance occurred prior to otter occupation of the slough in the early 1980s. Clam digging is therefore more likely to be the cause of the decline. In terms of benthic polychaetes, Lindquist (1998) found a general decrease between the 1970s and 1990s. One final species that decreased dramatically during this period was the cephalochordate *Branchiostoma californiense*. This lancelet was abundant in the lower slough in the 1970s (J. Pearse, pers. comm.), but now is virtually never encountered.

Subtidal Bivalves, Crabs, and Fat Innkeepers, 1980s–1990s

In 1984, a few young male sea otters (*Enhydra lutris*) began feeding near the mouth of the slough, mostly on large infaunal clams. Initially, the otters had no detectable effect on the abundance or size distribution of infaunal bivalves (Kvitek et al. 1988). They preferentially foraged in areas where bivalves were prevented from burrowing deeply due to a clay layer close to the surface.

In the following decade, a dozen or so otters used the lower slough area seasonally. In 1995, a large group (up to 54 animals) began living in the slough year round (see chapter 11, "Birds and Mammals"). To determine whether their presence was affecting subtidal bivalve populations, the 1986 sampling regime of Anderson and Kvitek (1987) was repeated in 1996 by Jolly (1997). She found that the mean shell length of the two largest clams, *Saxidomus nuttalli* and *Tresus nuttallii*, had decreased significantly and that only 8% of the gaper clams were of reproductive size. As a result of this change in size distribution, there was also a greater than 60% decrease in the biomass per square meter of each species. However, she did not find a decrease in the density of the clams, perhaps because there had been a recent recruitment event or perhaps due to sampling error (she found that the siphon count method used by Anderson and Kvitek underestimated the true number of clams, as determined by excavation). Her work also revealed that sea otters commonly eat fat innkeeper worms and crabs in the slough. Overall, the otters fit predictions of optimal foraging theory, selecting prey items that provide the greatest energetic benefit for the least time spent foraging. For all prey species, profitability correlated with size. With their preference for large individuals, sea otters will likely continue to alter the size distribution of their preferred prey, particularly clams. If the abundance of reproductive-sized adults declines dramatically because of otter foraging, populations of these invertebrates could be strongly affected.

A recent study (Kao 2000) documented a shift in diet of slough leopard sharks over the last twenty years, perhaps owing to changes in availability of prey items. Subtidal clams and crabs accounted for most of the leopard shark diet in the 1970s (Talent 1976), while fish and fat innkeeper worms were the main prey in the 1990s (Kao 2000). This change in diet may be attributable to declines in the abundance of clams and crabs during this period, in recent years perhaps because of increases in the number of sea otters.

Freshwater and Terrestrial Invertebrates of the Elkhorn Slough Area

Very little is known about the freshwater and terrestrial invertebrates of Elkhorn Slough. Not even a list of locally occurring species is available. Taking these animals into account would surely double the invertebrate species richness of the slough area. Here we will just briefly mention important freshwater and terrestrial habitats and what little is known about their invertebrate communities.

A key freshwater habitat is Carneros Creek, which flows into the upper slough. A student recently searched for aquatic arthropods there and found a variety of species, including crayfish, beetles, mayflies, and dragonflies, water boatmen, whirligigs, backswimmers, and midges (J. Jones, pers. comm.). Other freshwater areas near the mouth of Elkhorn Slough, such as the old Salinas River channel and McClusky Slough, have yet to be surveyed for invertebrates.

The slough's upper marshes support a diverse community of terrestrial invertebrates. As part of a pickleweed trampling study, Woolfolk (1999) assessed the diversity of terrestrial invertebrates in experimental plots as an indicator of marsh health. Further work on the insects and other invertebrates of the slough's marshes and salt pans would surely be fruitful.

The grasslands, oak woodlands, and other uplands of the slough certainly host a rich fauna of invertebrates, including insects, spiders, isopods, earthworms, and snails. However, virtually nothing is known about the occurrence or ecology of these species at the slough. Some 23 species of butterflies occur in the Elkhorn Slough area (P. Johnson, pers. comm.; J. Lane, pers. comm.) and over 50 species of moths have been collected (S. Fork, pers. comm.). Over 100 additional species of other insects have been collected on the ESNERR and are curated in the reserve's collection (S. Fork, per. comm.). Representatives of eight spider families are known from the area (F. Sala, pers. comm.). Four species of earthworms have been identified from areas around Elkhorn Slough, one of them native and three introduced species common in disturbed soils (Silberstein et al. 1997).

Human Influences on Invertebrates at Elkhorn Slough

Invertebrate distributions, abundances, and behaviors at Elkhorn Slough are shaped by the activities of humans as well as by other, more natural processes. For thousands of years, indigenous people directly influenced species targeted for food and shaped invertebrate communities by altering disturbance regimes, for instance through burning of grasslands. However, human impacts on invertebrates at the slough have surely increased dramatically in the last few hundred years, and particularly in the last century. In the absence of baseline studies, we will never know what invertebrate communities at Elkhorn Slough looked like before intense human colonization and use of the area.

In this section, we will discuss some of the major anthropogenic factors that influence invertebrate communities at the slough. Again, we focus primarily on the marine realm, since little is known about the response of the slough's freshwater and terrestrial species to human influences. Since invertebrates are simultaneously influenced by multiple human activities, it is difficult to sort out the role each individual factor plays in shaping community dynamics. Future experimental studies are essential to untangle these complex processes, and to test different strategies for ameliorating negative effects.

Diversion of Freshwater

Until the mid-1800s, freshwater or brackish wetlands dominated the lowlands of the central Monterey Bay area, including Elkhorn Slough. The diking of wetlands to create farmlands and the diversion of the Salinas River in the early 1900s to prevent flooding deprived the slough of most of its freshwater input (see chapter 4). These fundamental changes certainly altered the composition of the slough's invertebrate communities, but we cannot describe these alterations, since scientific study of the slough's invertebrates did not begin until decades later.

Strong Tidal Flushing Due to Opening of Harbor Mouth

The opening of a new wide mouth to Monterey Bay for the newly created Moss Landing Harbor in 1946 further changed the character of the slough (see chapter 4). Virtually overnight, it became a strongly flushed tidal embayment, rather than a sluggish tidal estuary. The presence of an artificially large entrance continues to shape the slough's communities and habitats today. Biologically, strong tidal flushing caused communities throughout the slough to become composed entirely of marine, rather than brackish or freshwater organisms. These marine invertebrate communities in turn have been affected by vertebrate predators that took advantage of the large permanent mouth and the deeper channels to establish large populations in the slough. Abundant leopard sharks and other elasmobranchs, and more recently, sea otters, influence invertebrate communities in the slough.

In addition to directly shaping the biological communities, the harbor mouth has marked physical influences on habitat structure. Strong tidal currents continue to widen and deepen both the main channel and smaller tidal creeks. Intertidal mudflats are being eroded to lower tidal heights, and salt marshes are being cut back and converted to mudflats. In the intertidal, Kvitek et al. (1996) found that there has been a decrease in fine unconsolidated sediments along the main channel since the 1970s (cf. Nybakken et al. 1977), probably as a result of tidal scour. Similarly, in the subtidal between Hummingbird Island and Kirby Park, fine sediments along the middle of the channel have been scoured away, leaving hard polished clay with a patchy surface veneer of coarse rubble (Kvitek et al. 1996).

Doubtless, tidal erosion is influencing invertebrates by changing the nature of the available habitats. Species that cannot withstand strong currents or that require very fine sediments for burrowing or feeding may not persist in the lower slough. As the channels are further carved away, community structure will necessarily continue to be altered. To date, no studies have attempted to examine the impacts of erosion on invertebrates in the slough. Such work is necessary for predicting future community changes.

Dredging

Ever since the harbor mouth was opened in 1946, harbor channels have been regularly dredged to facilitate boat traffic. The direct effects of dredging on subtidal communities in Moss Landing Harbor were studied in the 1970s and are summarized by Nybakken et al. (1977). Dredging apparently favored two polychaete species, *Capitella capitata* and *Armandia brevis*, which settled rapidly and became abundant immediately following this disturbance of the sediments. Eventually, these species declined and others, including several oligochaetes and bivalves of the genus *Macoma*, again became established. Dredging also may indirectly affect slough communities. Contaminants such as DDT transported into the Moss Landing Harbor area via agricultural runoff in the old Salinas River channel may be mobilized from the sediments and swept into the slough during incoming tides.

Introduction of Nonnative Species

Two anthropogenic transport mechanisms are responsible for the introduction of most nonnative marine species to Elkhorn Slough (Wasson et al. 2001). The first is oyster culturing. Atlantic (*Crassostrea virginica*) and Japanese (*C. gigas*) oysters were farmed in the slough for nearly a century (see chapter 7),

Massive mounds of a bright orange, nonnative sponge, Hymeniacidon sinapium. *Photo credit: Kerstin Wasson.*

The nonnative Japanese mud snail. Photo credit: © Monterey Bay Aquarium Foundation.

with activity peaking in the 1930s and 1940s and ending in the 1970s owing to concern about water quality. Many species growing on or among oysters were inadvertently introduced with them. Ironically, while the oysters failed to establish a permanent population after farming ceased, some of the species introduced with them continue to thrive in the slough today. For instance, a bright orange sponge (*Hymeniacidon sinapium*) associated with oysters and probably originally from the Atlantic forms massive aggregations in upper slough channels with high flow. The Japanese mud snail (*Batillaria attramentaria*) was also likely transported to the slough with oysters. This snail is found at incredible densities (2,000–10,000 per square meter) in upper intertidal areas (Byers 1999), and upward of a billion individuals likely inhabit Elkhorn Slough (Wasson et al. 2001). In experiments at Bolinas Lagoon, McDermott (1996) and Byers (2000) demonstrated the potential for *Batillaria* to competitively exclude a native horn snail, *Cerithidea californica*, by having lower mortality (in part because it is infected with fewer parasites) and by more efficiently converting food resources into growth. As a result, *Batillaria* has replaced *Cerithidea* in various California marshes, including Elkhorn Slough (Byers 1999, 2000).

The second key means by which nonindigenous species reach Elkhorn Slough is via San Francisco Bay or other bays with major ports (Wasson et al. 2001). International shipping has transported hundreds of nonnative invertebrate species to places like San Francisco Bay, mostly on fouled hulls or in ballast water (Cohen and Carlton 1995). Once they become established, populations of nonnative species then spread to nearby estuaries. Larvae or mobile adults may travel on currents from San Francisco Bay (or Los Angeles or other regional harbors) to Elkhorn Slough. Alternatively, local boat traffic may inadvertently transport the invaders from bays with international shipping to smaller harbors like Moss Landing Harbor.

Since oyster culturing ceased nearly thirty years ago, most recent invaders arriving at Elkhorn Slough (as well as some earlier ones) are probably organisms transported from San Francisco Bay and other large bays (Wasson et al. 2001). For instance, tortellini slugs (*Philine* spp.) from the western Pacific arrived in San Francisco Bay in 1992, probably in ballast water. This mollusc now is very abundant in the slough and elsewhere in Monterey Bay (Wasson et al. 2001). Similarly, the European green crab (*Carcinus maenas*) first appeared on the West Coast in San Francisco Bay in 1989, and within a few years had spread hundreds of miles to the north and south; the species has been in Elkhorn Slough since 1994 (Grosholz and Ruiz 1995). While these two species apparently spread from San Francisco Bay to other estuaries as larvae transported by currents, other species were spread on the hulls of boats. For example, a burrowing Australian isopod (*Sphaeroma quoyanum*) has apparently spread throughout California bays and harbors via boat-fouling. The burrows of this species riddle almost every bank in the slough today. An Australian tubeworm (*Ficopomatus enigmaticus*) also probably arrived at the slough from San Francisco Bay via shipping. First identified from the slough in 1994, it now forms massive reefs in areas with freshwater input (Wasson et al. 2001).

The effects of invasive invertebrates on native communities or habitat structure at the slough have not been studied. Given their sheer abundance, some of the invaders are surely

Tube worms native to Australia encrust the railroad trestle at Hudson's Landing. Photo credit: Kerstin Wasson.

significantly influencing native communities. By consuming high levels of some prey items, the invaders may alter many links in the food web of the community they have invaded. Filterers such as the bright orange sponge *Hymeniacidon sinapium* affect the planktonic community, scrapers such as the Japanese mud snail affect benthic diatoms, and voracious predators such as green crabs and tortellini slugs affect a number of native invertebrate prey species (especially small bivalves). Both direct and indirect effects of invaders on natives remain poorly understood in estuarine systems. However, a recent study (Grosholz et al. 2000) demonstrated remarkably sharp declines in native shore crabs and small clams, and concurrent increases in some opportunistic polychaetes, following the invasion of the European green crab in Bodega Bay. Some invasive invertebrates may also alter the physical structure of the community. For instance, Australian isopods probably weaken banks with their burrows, and thereby may increase the already high erosion rates in the slough.

Moss Landing Power Plant

The Moss Landing power plant, operated by PG&E from the 1940s to the 1990s and by Duke Energy since then (see chapter 7), may significantly impact invertebrate communities in Elkhorn Slough. An enormous volume of water (equivalent to 6–28% of the volume of the slough when operation is at full capacity) is pumped into the cooling system daily, killing many planktonic organisms that pass through the intake filters and potentially harming bigger animals that become trapped on them (Tenera 2000). The only invertebrates quantified in recent studies of entrained organisms (organisms drawn into the cooling system) were *Cancer* crabs, deemed important apparently because they are harvested commercially (Tenera 2000). Large numbers of *Cancer* larvae were captured in plankton tows near the intake pipes, but calculations suggest that given the extremely high fecundity of crabs, loss of these larvae may not translate into loss of adults (Tenera 2000). Other planktonic invertebrates (such as copepods and clam and worm larvae) as well as the phytoplankton eaten by many slough suspension feeders are killed in large numbers too, but the extent and consequence of their loss to invertebrate communities has not been examined. Mitigation for expansion of the power plant will result in restoration of wetlands and adjacent uplands in the area to compensate for the loss of these planktonic organisms (see chapter 14, "Management Issues"), and new monitoring programs funded by the power plant will help to shed light on its effects on the slough's communities.

Harvesting by clammers and anglers has affected some of the slough's invertebrate populations. Photo credit: ESNERR.

Harvesting

Humans have been harvesting invertebrates in Elkhorn Slough for millennia. Local Ohlone (Costanoan) middens 8,000 years old contain clam and cockle shells (see chapter 6), revealing the importance of slough invertebrates in ancient diets. The slough continued to be an important source of shellfish after European colonization (see chapter 7).

Today, people still use the areas around the Highway 1 bridge (see fig. 9.1) to dig for clams, particularly gapers, and for ghost shrimp and worms to use as fishing bait. Changes in benthic community structure between the 1930s and 1990s (described previously) may be attributed in part to these activities. A recent investigation (Gardner and Kvitek 1997) compared density and size structure of three commonly harvested species (gaper clams, ghost shrimp, and fat innkeeper worms) in tidal flats temporarily designated as either harvest or no-harvest zones. After two years, there was no difference in density or size of the three species between the two zones. It is difficult, however, to draw conclusions from this result, because the populations may require longer time periods for recovery from harvesting, and because restrictions were not well enforced. Further experimental studies are needed to better characterize the effects of human harvesting on invertebrate communities near the mouth of the slough. Such studies may reveal that human harvest is negligible compared to other influences such as sea otter foraging. If, however, harvesting is shown to have significant effects, alternative management strategies should be considered.

Trampling

Woolfolk (1999) demonstrated that human and cattle trampling reduced pickleweed (*Salicornia virginica*) height and flower production and increased susceptibility to invasion by nonnative plant species. She also assessed impacts of human trampling on marsh invertebrates. Homopterans (plant bugs) and spiders were reduced following trampling, while dipterans (flies) were most abundant in heavily trampled areas with high levels of damage to pickleweed. Trampling thus has consequences for marsh invertebrate community dynamics.

Pollutants

High levels of nutrients and various contaminants occur in Elkhorn Slough (see chapters 12, "Biogeochemical Cycling," and 13, "Land Use and Contaminants"), largely washed in with sediments eroded from surrounding farmlands during rains or periods of heavy irrigation. Some of the highest levels of DDT in the state have been detected in Elkhorn Slough by the Department of Fish and Game's Mussel Watch program (see chapter 13). The impacts of these pollutants on slough communities are largely unknown, but may be profound. For instance, the extensive mats of green algae (*Ulva* and *Enteromorpha*) may be the result of high nutrient levels. These algal mats may favor the amphipods or leptostracans living among them (Allen 1992), but may decrease recruitment, survival, or reproductive success of invertebrates dwelling beneath them. Indeed, such algal mats may be responsible for the near extinction of phoronids at Elkhorn Slough, which have elsewhere been shown to be sensitive to coverage by macroalgae.

Research following cohorts of the ghost shrimp, *Callianassa californiensis*, revealed periodic die-offs in the north Moss Landing Harbor area, which receives runoff directly from adjacent agriculture (Gardner and Kvitek 1997). The possibility that this mortality is due to periodically high pesticide levels should be explored with detailed correlative studies linking ghost shrimp size structure and abundance to pesticide levels.

Although not well studied, invertebrate distributions in the marshes and uplands of the Elkhorn Slough watershed are also undoubtedly strongly influenced by pollutants, especially in areas adjacent to farmed lands. For instance, earthworms are much more common in grasslands on the Elkhorn Slough Reserve than near strawberry fields at the Azevedo Ranch, possibly because the fumigant methyl bromide is lethal to them (Silberstein et al. 1997). Not surprisingly, differences in insect communities between the reserve and the Azevedo Ranch were also detected (Silberstein et al. 1997); the soil surface fauna in the farmed areas is less diverse and less abundant due in part to the influence of pesticides, as well as to other effects of farming, such as soil disturbance.

Management Issues and Research Recommendations

Conservation Strategies

Based on the preceding discussion of human influences, we present five recommendations to reduce negative anthropogenic effects and protect native invertebrate biodiversity at Elkhorn Slough:

Minimize future invasions by nonnative species. Requirements for regular boat-hull cleaning and appropriate disposal of non-native fishing bait and seaweed packing material would reduce the likelihood of new marine invasions. Regular monitoring for potential invaders would also allow for early detection and eradication, limiting impacts of future invasions.

Reduce pollution. Efforts already underway to reduce amounts of polluted sediments entering Elkhorn Slough from surrounding agricultural and residential uplands should be supported and expanded to better protect invertebrate communities from negative effects of eutrophication and contaminants. In addition, preventing flow of the Salinas River into the old river channel would reduce levels of pollutants in the Moss Landing Harbor area, and perhaps also in the slough as a whole, since disturbances such as dredging likely transport sediments from the harbor up into the slough proper.

Decrease tidal erosion. Measures to mute tidal flow should be evaluated. If projects such as constructing tide gates at the Parson's Slough bridge or slowing flow with shoals or vegetation enhancement are judged to be useful in decreasing tidal scour, they should be implemented.

Establish no-take zones in the lower slough. While the greatest native diversity of invertebrates occurs near the mouth of the slough, this region is mostly unprotected from human exploitation. By establishing no-take zones adjacent to areas in which harvesting is permitted, the impact of human harvesting on benthic intertidal invertebrates can be examined in rigorous long-term studies.

Support applied research and monitoring. More research is urgently needed to develop and test science-based, adaptive management strategies for ameliorating human threats to native invertebrates at the slough, and to design appropriate long-term monitoring programs (see below).

Research Priorities

Despite a wealth of studies on invertebrates at Elkhorn Slough, too little is known about individual species and about community dynamics to predict changes resulting from human activities and to confidently design appropriate conservation and management strategies. We hope that in future decades there will be an increase in applied research on invertebrates at the slough, to complement past and ongoing basic scientific studies. We divide research priorities for invertebrates into three categories:

Investigations of human influences and tests of strategies for reducing negative human impacts. Experiments, correlational studies, and models are needed to better understand the effects of particular human activities on slough invertebrates. Such investigations will also help to identify activities with the biggest negative impacts so that they can receive priority for management. For instance, some pesticides may exert stronger influences than others on invertebrates; some types of human harvesting may disturb invertebrate communities more than others do; some non-native species may have broader negative influences than others. Once effects of human activities are better understood, experimental tests and predictive modeling will be required to provide strong scientific background for management decisions about how threats to slough communities can best be diminished.

Identification of vulnerable habitats, communities, or species, and development of adequate conservation strategies for them. Research is needed to identify critical invertebrate habitats at Elkhorn Slough. Comparisons with other embayments should be undertaken to determine whether there are habitat types at the slough that are rare elsewhere on this coast; these should receive high priority for conservation and enhancement. Further, a geographical information system approach should be used to identify and map habitats at the slough that are particularly valuable to invertebrates, in terms of supporting the greatest number of native species or particularly vulnerable or rare species.

Marine invertebrates have rarely been the focus of conservation research, but such work is necessary if we are to protect them. For instance, there is a pressing need for conservation research on the phoronid worm *Phoronopsis viridis*. Elkhorn Slough was one of only a few places in the world where this worm was extremely abundant, as noted in both the 1930s and the early 1970s (MacGinitie 1935; Nybakken et al. 1977). In recent years, however, it has virtually disappeared from the slough (Kvitek et al. 1996; K. Wasson, unpublished data), and it has also undergone a sharp decline at Bodega Bay (Grosholz et al. 2000). There are only a dozen or so extant species in this phylum worldwide, so efforts should be taken to determine and, if possible, reverse the factors that have led to this precipitous decline. Timely research may be the only way to prevent extinction of this unique species.

The California brackish-water snail, *Tryonia imitator,* is another threatened species that urgently requires scientific attention. Once widespread in California, this species is now limited to small isolated populations, including some marshes in the Elkhorn Slough area (Kellogg 1980). To adequately protect this and other species, we must better understand their ecology.

Characterization of poorly known invertebrate communities. We know very little about some invertebrate communities at the slough. Without at least knowing which species are present, we will be unable to detect losses or invasions over time. Baseline characterization of these poorly known communities is necessary; more detailed studies would be desirable.

Planktonic invertebrates. Very little is known about either planktonic adults or larvae. Planktonic forms should be studied for their own sake, but also to shed light on benthic community dynamics, which are influenced by larval abundance and distribution. Particularly obvious systems for focused studies are ubiquitous mudflat species, including various clams and the fat innkeeper worm; nothing is known about the larval ecology of these species in Elkhorn Slough.

Parasitic invertebrates. Very little is known about invertebrate parasites in the slough, yet these may be essential in regulating population dynamics. For instance, parasites such as trematodes or rhizocephalans may help control nonnative invertebrates such as Japanese mud snails and European green crabs.

Microscopic invertebrates. We do not even have a species list for tiny taxa such as gastrotrichs and rotifers. Indeed, nematodes are likely as dominant in terms of abundance and diversity as polychaetes in the slough, but have received no attention. The diversity of these communities should be characterized as part of a fuller understanding of invertebrate community dynamics at the slough.

Freshwater and terrestrial invertebrates. These groups deserve significantly more attention than they have received. Freshwater species may be useful indicators of water quality in local riparian habitats. A better understanding of terrestrial insects, especially pollinators and seed dispersers, would complement upland vegetation restoration efforts.

Long-Term Monitoring Recommendations

There has been no consistent long-term monitoring of invertebrate communities in Elkhorn Slough. We recommend development and implementation of the following monitoring program as the minimum needed to understand and protect the slough's invertebrate communities; coverage of more taxa and habitats is desirable.

The sampling regime for benthic infaunal communities carried out at four stations along the length of the slough by Nybakken et al. (1977) and Kvitek et al. (1996) should be repeated annually. This includes benthic cores to quantify intertidal mudflat invertebrates and estimation of numbers and sizes of common species of intertidal and subtidal bivalves. At the same times and stations, baited minnow traps should be set to capture snails and crabs, and plankton samples should be taken. Settlement plates should be deployed at these sites to assess recruitment of native vs. nonnative fouling organisms. Besides labor-intensive field collection of samples and laboratory identifications, this will require resources for data management, analysis, and dissemination.

The above sampling will primarily detect changes in distribution and abundance. It is also necessary to track changes in diversity over time, to detect new invasions or impending local extinctions early enough for successful intervention. The species list (appendix 9.1) should be continually updated using data from various research projects as well as the quantitative monitoring. Directed searches for suspected declining native species or new invaders should be carried out as necessary to improve the accuracy of the list. In order to better monitor diversity of invertebrates, which are often taxonomically challenging, a comprehensive voucher collection for Elkhorn Slough invertebrates should be established and curated so that it can readily be accessed by interested researchers. Images of each species should also be included in a searchable, Web-based digital library of the voucher collection.

Development and implementation of such a monitoring program will allow us to better characterize the rich invertebrate fauna of Elkhorn Slough and to detect changes in community dynamics over time.

Acknowledgments

We thank the mud-loving students, staff, and faculty at Moss Landing Marine Laboratories—J. Oliver, P. Slattery, C. Jong, B. Anderson, K. Finn, and many others—for their continued help with field collections and sample processing over the years. We also gratefully acknowledge T. Newberry, J. Pearse, and their many students at the University of California, Santa Cruz for the attention they have focused on the diversity and natural history of slough invertebrates over the past decades. M. Brown, J. Caffrey, B. Christensen, T. Newberry, J. Pearse, J. Thompson, and B. Tyler thoughtfully reviewed and thereby much improved this chapter.

Appendix 9.1. Species list of marine and brackish water invertebrates in Elkhorn Slough. The first reference to the presence of the species at Elkhorn Slough is listed after its name. Additionally, references 1, 2, and/or 3—the only thorough surveys of the slough's invertebrates—are listed if those authors reported the presence of the species. The number of species is listed with each higher taxon name.

Key for References: 1 = MacGinitie 1935; 2 = Nybakken, Cailliet, and Broenkow 1977; 3 = Kvitek et al. 1996 and R. Kvitek, pers. comm.; 4 = Wasson et al. 2001; 5 = DeVogelaere et al. 1998; 6 = J. Pearse, pers. comm.; 7 = Ricketts et al. 1985; 8 = Pace 1978; 9 = M. Silberstein, pers. comm.; 10 = Hsueh 1988; 11 = Wasson, pers. comm.; 12 = Nybakken, pers. comm.; 13 = Kudenov and Blake 1985; 14 = Addicott 1952; 15 = Kellogg 1980; 16 = Byers 1999; 17 = J. Brown, pers. comm.

*indicates species is not native to this bioregion
+means this species is locally extinct, or so rare it has not been found in recent searches for it

Taxon	References	Number of species
PHYLUM PORIFERA (sponges)		**9**
*Cliona ? celata**	1, 2	
*Halichondria bowerbanki**	4	
Haliclona cinera	1	
*Haliclona loosanoffi**	4	
Haliclona permollis	2	
Halisarca sacra	1	
*Hymeniacidon sinapium**	4	
Mycale macginitiei	1	
Topsentia sp.	4	
PHYLUM CNIDARIA		**25**
CLASS HYDROZOA (hydroids)		11
Abietinaria filicula	1	
Aglaophenia struthionides	1	
Bougainvillia mertensi	1	
Campanularia sp.	1	
*Cordylophora caspia**	4	
*Ectopleura crocea**	1	
Obelia gracilis	1	
Obelia longissima	1, 2	
Opercularella lacerata	1	
Polyorchis penicillatus	2	
Syncoryne mirabilis	1	
CLASS SCYPHOZOA (jellyfish)		4
*Aurelia aurita**	1	
Chrysaora melanaster	17	
Pelagia colorata	2	
Pelagia noctiluca	2	
CLASS ANTHOZOA (sea anemones)		10
Anthopleura elegantissima	1, 2	
Anthopleura sola	6	
Anthopleura xanthogrammica	1, 2	
*Diadumene franciscana**	4	
*Diadumene leucolena**	4	
*Diadumene lineata**	4	
Edwardsiella sp.	3	
Metridium senile	1, 2	
Pachycerianthus fimbriatus	2	
Zaolutus actius+	1	
PHYLUM CTENOPHORA		**1**
Pleurobrachia bachei	1	
PHYLUM PLATYHELMINTHES (flatworms)		**9**
CLASS TREMATODA		4
*Cercaria batillariae**	4	
Epibdella pacifica	1	
Probolitrema californiense	1	
Udonella myliobati	1	
CLASS POLYCLADIDA		4
Eurylepta aurantiaca	1	
Kaburakia excelsa	12	
Notoplana sp.	6	
Stylochoplana heathi	1	
CLASS CESTODA		1
Phyllobothrium sp.	1	
PHYLUM NEMERTEA		**9**
Carcinonemertes sp.	12	
Carinoma mutabilis	1, 12	
Cerebratulus californiensis	6, 12	
Cerebratulus sp. 1	1	
Cerebratulus sp. 2	1	
Malacobdella grossa	14	
Micrura sp.	1	
Pantinonemertes californiensis	12	
Paranemertes peregrina	1	
PHYLUM ECHIURA		**2**
Listriolobus pelodes	9	
Urechis caupo	1, 3	
PHYLUM SIPUNCULA		**3**
Phascolosoma gouldi	1	
Siphonosoma ingens	6	
Themiste perimeces	1, 12	
PHYLUM ANNELIDA		**144**
CLASS POLYCHAETA		143
Amaeana occidentalis	2	
Ampharete labrops	2	
Anaitides williamsi	2	
Armandia brevis	1, 2, 3	
Audouinia tentaculata	1	
Axiothella sp.	3	
Boccardia columbiana	2	
Boccardia hamata	2	
Boccardia proboscidea	2	
Boccardia redeki	12	
Boccardia rou	3	
Boccardia sp.	3	
Capitella capitata	1, 2, 3	
Chaetozone setosa	2	
Chone gracilis	2	

Chone infundibuliformis	1
Cirratulus cirratus	2
Cirriformia spirabrancha	2
Cistenides brevicoma	7
Ctenodrilus serratus	2
Diopatra ornata	2
Diopatra splendidissima	2
Dispio uncinata	2
Dorvillea articulata	2
Eteone californica	2
Eteone dilatae	2
Eteone sp.	3
Eudistylia polymorpha	1
Eulalia quadrioculata	2
Eumida bifoliata	2
Eumida sp.	3
Exogone lourei	2, 3
Exogone sp.	3
*Ficopomatus enigmaticus**	4
Glycera alba macrobranchiata	1
Glycera convoluta	2
Glycera robusta	1, 2
Glycera rugosa	1
Glycera sp.	3
Glycinde sp.	2, 3
Gyptis brevipalpa	2
Halosydna brevisetosa	1
Haploscoloplos pugettensis	2
Harmothoe lunulata	2
Harmothoe priops	2
Hemipodus borealis	1, 2
Hesionura sp.	2
Hesperonoe adventor	1, 6
Hesperonoe complanata	1
Hesperonoe sp.	1
*Heteromastus filiformis**	4
Heteromastus filobranchus	2
Leitoscoloplos elongatus	5
Leodice longicirrata	1
Loimia medusa	1
Lumbrineris cruzensis	2
Lumbrineris japonica index	1
Lumbrineris limicola	2
Lumbrineris luti	2
Lumbrineris sp.	3
Lumbrineris tetraura	1, 2
Lumbrineris zonata	2
Magelona sacculata	2
Malacoceros glutaeus	2
Mediomastus californiensis	2, 3
Mesochaetopterus taylori	7
Microphthalmus sp.	2
Naineris dendritica	2
Neanthes brandti	7
Neanthes virens	1, 2
Neoamphitrite robusta	1
Nephtys caeca ciliata	1
Nephtys caecoides	1, 2
Nephtys californiensis	3
Nephtys cornuta franciscana	2, 3
Nereis dumerilii agassizi	1
Nereis grubei	2
Nereis procera	1, 2
Nereis sp.	3
Nereis vexillosa	1, 6
Nereis virens	1
Nereis virens brandti	1
Nerinides acuta	2
Northia elegans	1, 2
Notomastus giganteus	1
Notomastus magnus	2
Notomastus sp.	3
Notomastus tenuis	1, 2, 3
Onuphis eremita	1, 2
Owenia collaris	2
Owenia sp.	2
Paleanotus bellis	2
Pareurythoe californica	1, 2
Pectinaria auricoma	1
Pectinaria californiensis	2
Perinereis monterea	1
Pherusa plumosa	5
Pholoe glabra	2
Phyllodoce muscosa	2
Pilargis berkeleyi	1
Pilargis maculata	2
Pista elongata	1
Platynereis bicanaliculata	2
Platynereis sp.	3
Polycirrus sp.	2
Polydora brachycephala	2
Polydora citrona	2
Polydora ligni *	2
Polydora socialis	2
Polydora sp.	3
Prionospio cirrifera	2
Prionospio pinnata	2
Prionospio pygmaea	2
Protodorvillea gracilis	2
Pseudeurythose reducta	13
*Pseudopolydora paucibranchiata**	2
Pseudopolydora sp.	3
Pygospio elegans	2
Rhynchospio glutaea	5
Schistomeringos longicornis	5
Schistomeringos rudolphi	2
Scololepis tridentata	2
Scoloplos armiger	2
Scoloplos sp.	2
Serpula vermicularis	6
Sigambra tentaculata	2
Spiochaetopterus costarum	2
Spiophanes berkeleyorum	2
Spiophanes bombyx	2, 3
Spiophanes missionensis	2, 3
Spriorbis sp.	6
Sthenelais fusca	1
Sthenelais tertiaglabra	5
Sthenelais verruculosa	2
*Streblospio benedicti**	2, 3
Styllaroides plumosa	1
Syllides sp.	2
Tharyx monilaris	2
Tharyx parvus	2
Tharyx sp.	3
Travisia sp.	3
Trochochaeta multisetosum	2
Typosyllis armillaris	2

CLASS HIRUDINEA (leeches)	1
Branchellion sp.	1

PHYLUM ARTHROPODA	**161**
CLASS PYCNOGONIDA (sea spiders)	3
Ammothea hilgendorfi	2
Phoxichilidium femoratum	2
Pycnogonum stearnsi	2

CLASS CRUSTACEA	158
Subclass Cirripedia (barnacles)	11
Balanus crenatus	11

Balanus glandula	6
*Balanus improvisus**	4
Balanus nubilis	1
Chthamalus sp.	6
Lepas hilli	1
Megabalanus californicus	1, 6
Pollicipes polymerus	6
Sacculina sp.	1
Semibalanus cariosus	5
Tetraclita rubescens	6

Subclass Branchiura (fish lice)	1
Argulus melanostictus	1

Subclass Branchiopoda	2
Evadne nordmanni	2
Podon leuckarti	2

Subclass Copepoda	15
Acartia californiensis	8
Acartia clausii	8
Acartia longiremis	8
Acartia tonsa	8
Calanus pacificus	2
Eucalanus bungi	2
Eurytemora hirundoides	2
Hemicyclops callianassae	1
Hemicyclops thysanotus	1
Microcalanus sp.	2
Modiolicola gracilis	1
*Mytilicola orientalis**	4
Oithona spinifera	2
Tortanus discaudatus	2
Trebius caudatus	1

Subclass Ostracoda	3
Euphilomedes carcharodonta	2
Euphilomedes longiseta	2
Euphilomedes oblonga	2

Subclass Malacostraca	126
Order Isopoda (Pill Bugs)	15
Austrosignum tillerae	12
Exosphaeroma inornata	5
Exosphaeroma media	2
Gnorimosphaeroma oregone	3
*Iais californica**	4
Ianiropsis montereyensis	2
Idotea resecata	1
Idotea wosnesenskii	2
Limnoria lignorum	1, 2
*Limnoria quadripunctata**	2
Lironeca vulgaris	1
Munna ubiquita	2
Phyllodurus abdominalis	1
Portunion conformis	2
*Sphaeroma quoyanum**	4

Order Tanaidacea	6
Anatanais normani	2
Leptochelia dubia	2, 3
Leptognathia sp.	12
Sinelobus sp.	4
Tanais carolinii	2
Tanais sp.	12

Order Cumacea	5
Cumella vulgaris	3
Cyclaspis sp.	2, 3
Cyclaspis nubila	2
Hemilamprops californiensis	2
Lamprops sp.	2

Order Amphipoda (sand hoppers)	54
Allorchestes angusta	2
Allorchestes sp.	3
Ampithoe lacertosa	1
*Ampithoe valida**	4
Anisogammarus confervicolus	1, 2
Aoroides columbiae	1, 2
Aoroides sp.	3
Argissa hamatipes	2
Atylus tridens	2
Caprella acutifrons	1
Caprella equilibria	1
Caprella brevirostris	2
Caprella californica	2, 3
Caprella ferrea	2
Caprella gracilior	2
Caprella mendax	2
*Caprella mutica**	4
Caprella natalensis	2
Caprella penantis	2
Caprella scaura	1
Caprella sp.	3
Caprella verrucosa	5
*Corophium acherusicum**	2
*Corophium insidiosum**	2
Corophium salmonis	1
Corophium sp.	3
Corophium spinicorne	1, 2
*Corophium uenoi**	2
Dulichia sp.	2
Eohaustorius sencillus	2
Erichthonius sp.	3
Gammaropsis sp.	3
*Grandidierella japonica**	3
Grandidierella sp.	3
Ischyrocerus pelagops	2
*Jassa marmorata**	4
Jassa sp.	2
Listriella diffusa	2
Maera sp.	2
*Melita nitida**	4
Melita sp.	2
Metopa sp.	2
Monoculodes spinipes	2
Orchestia traskiana	2
*Parapleustes derzhavini**	4
Photis sp.	2
Podocerus sp.	2
Pontogeneia sp.	3
Protomedeia articulata	2
Rhepoxynius daboius	2
Rhepoxynius variatus	2
Synchelidium shoemakeri	2
Tiron biocellata	2
Tritella laevis	2

Order Mysidacea (opossum shrimps)	1
Acanthomysis sp.	2

Order Leptostraca	1
Nebalia pugettensis	2

Order Decapoda (shrimps and crabs)	43
Betaeus longidactylus	1
Callianassa californiensis	1, 2
Callianassa gigas	1
Callianassa sp.	3
Cancer antennarius	1, 2

Cancer anthonyi 1, 2
Cancer gibbosulus 1
Cancer gracilis 1, 2
Cancer jordani 1, 2
Cancer magister 1
Cancer productus 1, 2
*Carcinus maenas** 4
Crangon nigricauda 1, 2
Crangon nigromaculata 17
Hemigrapsus nudus 1
Hemigrapsus oregonensis 1, 2
Heptacarpus paludicola 1, 2
Heptacarpus stichensis 1
Hippolyte californiensis 1
Isocheles pilosus 2
Lophopanopeus bellus 10
Loxorhynchus grandis 2
Opisthopus transversus 1
Pachycheles rudis 1
Pachygrapsus crassipes 1, 2
Pagurus granosimanus 5
Pagurus hirsutiusculus 1
Pagurus samuelis 1
*Palaemon macrodactylus** 7
Petrolisthes cinctipes 1
Pinnixa faba 1
Pinnixa franciscana 1, 2
Pinnixa longipes 1
Pinnixa schmitti 1
Pinnixa sp. 3
Pinnixa tomentosa 1
Pinnixa tubicola 1, 2
Pinnixa weymouthi 10
Pugettia producta 1, 2
Randallia ornata 2
Scleroplax granulata 1, 2
Synalpheus lockingtoni 2
Upogebia pugettensis 1, 2

Order Euphausiacea (krill) 1
Euphausia pacifica 2

PHYLUM BRACHIOPODA 1
Glottidia albida 2

PHYLUM PHORONIDA 1
Phoronopsis viridis+ 1, 2

PHYLUM KAMPTOZOA (=ENTOPROCTA) 2
*Barentsia benedeni** 4
*Barentsia gracilis** 1, 12

PHYLUM BRYOZOA (=ECTOPROCTA) 18
*Amathia vidovici** 4
*Bowerbankia gracilis** 1, 4
Bugula "*neritina*"* 4
Bugula californica 12
Bugula sp. 6
*Bugula stolonifera** 4
Buskia seriata 4
Celleporaria brunnea 6
*Conopeum tenuissimum** 4
Crisia occidentalis 12
*Cryptosula pallasiana** 4
Dendrobeania lichenoides 12
Membranipora membranacea 1
Reginella hippocrepis 11
*Schizoporella unicornis** 4
Tricellaria occidentalis 12
Tricellaria sp. 12
Watersipora "*subtorquata*"* 4

PHYLUM MOLLUSCA 160
CLASS POLYPLACOPHORA (chitons) 7
Lepidochitona dentiens 5
Lepidochitona raymondi 1
Lepidozona sinudentata 1
Mopalia ciliata 1
Mopalia hindsii 1, 6
Mopalia muscosa 1, 6
Nuttallina californica 6

CLASS BIVALVIA 62
Adula diegensis 1
Bankia setacea 1, 2
Chaceia ovoidea 1
Clinocardium nuttallii 1, 2
Cooperella sp. 3
Cooperella subdiaphana 2
Crassadoma giganteus 1, 6
Crassostrea virginica 1
Cryptomya californica 1, 2, 3
*Gemma gemma** 2, 3
Hiatella arctica 1, 2
Leptopecten latiauratus 12
Lithophaga plumula 1
Lyonsia californica 2
*Lyrodus pedicellatus** 1, 4
Macoma acolasta 2
Macoma balthica 2, 3
Macoma dolabriformis 1
Macoma inquinata 1, 2, 14
Macoma nasuta 1, 2, 3
Macoma secta 1, 2
Mactra sp. 2
Modiolus capax 1
Modiolus rectus 1, 2
*Musculista senhousia** 4
Musculus sp. 2
*Mya arenaria**+ 1, 2
Mysella aleutica 2
Mysella sp. 2, 3
Mytilus californianus 6
*Mytilus trossulus/galloprovincialis** 1, 2
Naeromya compressa 5
*Nuttallia nuttallii** 1
Orobitella rugifera 1
Ostrea conchaphila 1, 2
Panopea generosa 1, 2
Pecten hindsii 1
Penitella penita 1
Petricola carditoides 1
Platyodon cancellatus 1
Pododesmus cepio 1
Protothaca sp. 3
Protothaca staminea 1, 2, 3
Protothaca tenerrima 1, 2
Pseudochama exogyra 1
Pseudopythina compressa 1
Saxidomus nuttalli 1, 2, 3
Siliqua lucida 1, 2
Solen sicarius 1, 2
Tagelus californianus 1, 6
Tellina bodegensis 1
Tellina meropsis 2
Tellina modesta 1, 2, 3
Tellina nuculoides 2
Tivela stultorum 1
Trachycardium quadragenarium 2
Transennella tantilla 2
Tresus capax 5

Taxon	
Tresus molli	3
Tresus nuttallii	1, 2
*Venerupis philippinarum**	4
Zirfaea pilsbryi	1, 2
CLASS GASTROPODA	88
Subclass Prosobranchia	40
Acanthina spirata	1, 2
Alia carinata	2
Alia gausapata	5
Alia gouldii	2
Alia sp.	3
Alvinia compacta	2
Asperiscala bellastriata	5
Assiminea californica	2
*Batillaria attramentaria**	2
Carinaria sp.	2
Cerithidea californica+	16
Crepidula nummaria	1
Diodora aspersa	1
Epitonium bellastriatum	2
Kellia laperousii	1, 2
Kurtziella plumbea	2
Lacuna porrecta	1
Lacuna unifasciata	1
Littorina plena/scutulata	1, 6
Lottia digitalis/austrodigitalis	6
Lottia gigantea	6
Lottia limatula	1, 6
Lottia paradigitalis	6
Macclintockia scabra	1, 2
Marseniopsis sharonae	6
Nassarius fossatus	1, 6
Nassarius mendicus	2
Nassarius rhinetes	2, 3
Nucella emarginata	2
Olivella biplicata	1, 2
Olivella pycna	2, 3
Olivella sp.	3
Polinices draconis	1
Polinices lewisii	1, 2
Polinices pallidus	5
Tectura persona	1
Tegula funebralis	1, 2
Tryonia imitator	15
*Urosalpinx cinerea**	4
Vitrinella sp.	2
Subclass Pulmonata	2
*Myosotella myosotis**	2
Onchidella borealis	2
Subclass Opisthobranchia	46
Acanthodoris lutea	2
Acanthodoris pilosa	2
Acanthodoris rhodoceras	2
Aeolidia papillosa	1, 2
Aglaja diomedea	2
Alderia modesta	2
Ancula lentiginosa	2
Ancula pacifica	2
Anisodoris nobilis	17
Aplysia californica	1, 2
Aplysiopsis smithi	2
Archidoris montereyensis	2
Bulla gouldiana	2
Catriona alpha	2
Chelidonura inermis	1, 2
Coryphella cooperi	2
Coryphella sp.	1, 2
Cumanotus beaumonti	2
Cylichna attonsa	2
Dendronotus frondosus	2
Dendronotus iris	2
Dendronotus sp.	1
Dendronotus subramosus	17
Diaulula sandiegensis	2
Doto amyra	2
Doto sp.	1
Elysia hedgpethi	2
Emarcusia morroensis	2
Eubranchus rustyus	2
Flabellina trilineata	2
Galvina sp.	1
Haminoea vesicula	1, 2
Melibe leonina	2
Okenia angelensis	2
*Okenia plana**	4
Onchidoris bilamellata	2
Onchidoris hystricina	2
Phidiana crassicornis	1, 2
*Philine ?auriformis**	4
Philine sp.	1
Phyllaplysia taylori	1, 2
Polycera atra	2
Polycera hedgpethi	2
Stiliger fuscovittatus	2
*Tenellia adspersa**	4
Trinchesia albocrusta	2
CLASS CEPHALOPODA	3
Loligo opalescens	2
Octopus rubescens	7
Paroctopus apollyon	1
PHYLUM ECHINODERMATA	**7**
Amphiodia occidentalis	1
Dendraster excentricus	1, 2
Leptosynapta albicans	1
Ophiothrix spiculata	1
Paracaudina chilensis	1
Pisaster ochraceus	1, 2
Strongylocentrotus purpuratus	2
PHYLUM CHORDATA	**7**
SUBPHYLUM UROCHORDATA (tunicates)	6
*Botrylloides violaceus**	4
Didemnum carnulentum	4
Halocynthia johnsoni	1
*Molgula manhattensis**	4
Pyura haustor johnsoni	7
*Styela clava**	4
SUBPHYLUM CEPHALOCHORDATA (lancelets)	1
Branchiostoma californiense	2
TOTAL SPECIES:	**559**

References

Addicott, W. O. 1952. Ecological and natural history studies of the pelecypod genus *Macoma* in Elkhorn Slough, California. M.A. thesis, Stanford University.

Allen, J. K. 1992. Benthic invertebrates living in macroalgal mats on intertidal mudflats of Elkhorn Slough, California. M.S. thesis. Moss Landing Marine Laboratories. San Jose State University.

Anderson, B. S., and R. G. Kvitek. 1987. Sea otter predation and the distribution of bivalve prey in the Elkhorn Slough National Estuarine Research Reserve. Technical memorandum, NOAA.

Arp, A. J., Hansen, B. M., and Julian, D. 1992. Burrow environment and coelomic fluid characteristics of the echiuran worm *Urechis caupo* from populations at three sites in northern California. *Marine Biology* 113: 613–623.

Byers, J. 1999. The distribution of an introduced mollusc and its role in the long-term demise of a native confamilial species. *Biological Invasions* 1: 339–353.

———. 2000. Mechanisms of competition between two estuarine snails: Implications for exotic species invasions. *Ecology* 81: 1225–1239.

Clark, P., J. Nybakken, and L. Laurent. 1975. Aspects of the life history of *Tresus nuttallii* in Elkhorn Slough. *California Fish and Game* 61: 215–227.

Cohen, A. N., and J. T. Carlton. 1995. Nonindigenous aquatic species in a United States estuary: A case study of the biological invasions of the San Francisco Bay and Delta. A report for the United States Fish and Wildlife Service, Washington, D.C., and the National Sea Grant College Program, Connecticut Sea Grant, NTIS Report Number PB96-166525.

De Vogelaere, A., J. Holte, M. Silberstein, and M. Jacobi. 1998. A species list for Elkhorn Slough and adjacent uplands. Elkhorn Slough National Estuarine Research Reserve.

Eissinger, R. A. 1970. Density distributions of economically important species of clams in Elkhorn Slough, Monterey Bay. Unpublished report in the collection of the Elkhorn Slough Foundation, Moss Landing Marine Laboratories.

Fisher, W., and G. E. MacGinitie. 1928. The natural history of an echiuroid worm. *The Annals and Magazine of Natural History* 10: 204–213.

Folk, R. L., and W. C. Ward. 1957. Brazos River Bar: A study in the significance of sediment grain-size parameters. *Journal of Sedimentary Petrology* 27: 3–26.

Galigher, A. E. 1925. On the occurrence of the larval stages of scyphozoa in the Elkhorn Slough, Monterey Bay, California. *American Naturalist* 59: 94–96.

Gardner, M. E., and R. Kvitek. 1997. Tide flat restoration and management: implementation of a novel recovery program in Elkhorn Slough. NOAA Report, Moss Landing Marine Laboratories.

Grosholz, E. D., and G. M. Ruiz. 1995. Spread and potential impact of the recently introduced European Green Crab, *Carcinus maenas*, in central California. *Marine Biology* 122: 239–247.

Grosholz, E. D., G. M. Ruiz, C.A. Dean, K. A. Shirley, J. L. Maron, and P. G. Connors. 2000. The impacts of a nonindigenous marine predator in a California bay. *Ecology* 81: 1206–1224.

Hsueh, P. 1988. Seasonal occurrence and abundance of brachyuran larvae in Elkhorn Slough and nearshore waters of Monterey Bay, California, September 1985–October 1986. M.A. thesis, Moss Landing Marine Laboratories. California State University, Fresno.

———. 1991. Seasonal occurrence and abundance of brachyuran larvae in a coastal embayment of central California. *Journal of Crustacean Biology* 11: 546–552.

Jolly, J. M. 1997. Foraging ecology of the sea otter, *Enhydra lutris*, in a soft-sediment community. M.S. thesis, University of California, Santa Cruz.

Kao, J. 2000. Diet, daily ration, and gastric evacuation of the leopard shark (*Triakis semifasciata*). M.S. thesis, Moss Landing Marine Laboratories. California State University, Hayward.

Kellogg, M. G. 1980. Status of the California brackish water snail *Tryonia imitator* in central California. Special Publication 80-3, California Department of Fish and Game, Inland Fisheries Endangered Species Program.

Kudenov, J. D., and J. A. Blake. 1985. *Pseudeurythoe reducta*, new species (Polychaeta: Amphinomidae) from central California. *Bulletin Southern California Academy of Sciences* 84: 38–40.

Kvitek, R. G., A. K. Fukayama, B. S. Anderson, and B. K. Grimm. 1988. Sea otter foraging on deep-burrowing bivalves in a California, USA, coastal lagoon. *Marine Biology* 98: 157–168.

Kvitek, R. G., M. A. Silberstein, J. S. Oliver, P. N. Slattery, C. Jong, J. Nybakken, B. S. Anderson, and K. Finn. 1996. Persistence and change of infaunal invertebrate communities in the Elkhorn Slough. Unpublished manuscript, Elkhorn Slough National Estuarine Research Reserve.

Laurent, L. L. 1971. The spawning cycle and juvenile growth rate of the gaper clam, *Tresus nuttallii*, of Elkhorn Slough, California. M.A. thesis, Moss Landing Marine Laboratories. San Francisco State University.

Lindquist, D. C. 1998. Assessing the effects of erosion on the trophic relationships of fishes in Elkhorn Slough, CA. M.S. thesis, Moss Landing Marine Laboratories. California State University, Hayward.

MacGinitie, G. E. 1927. Ecological aspects of Elkhorn Slough. M.A. thesis, Stanford University.

———. 1935. Ecological aspects of a California marine estuary. *American Midland Naturalist* 16: 629–765.

McDermott, S. P. 1996. Parasites, density, and disturbance: factors influencing coexistence of *Cerithidea californica* and *Batillaria attramentaria*. M.S. Thesis, Moss Landing Marine Laboratories. California State University, Fresno.

McDonald, G. R., and J. W. Nybakken. 1978. Additional notes on the food of some California nudibranchs with a summary of known food habits of California species. *Veliger* 21: 110–119.

Morris, R. H., D. P. Abbott, and E. C. Haderlie. 1980. *Intertidal Invertebrates of California*. Stanford: Stanford University Press.

Munson, D. A. 1976. An analysis of some of the factors affecting the distribution of the metazoan fauna in a California estuarine sulfide ecosystem. Ph.D. diss., Stanford University.

Nybakken, J. W., G. M. Cailliet, and W. W. Broenkow. 1977. *Ecologic and Hydrographic Studies of Elkhorn Slough, Moss Landing Harbor, and Nearshore Coastal Waters, July 1974 to June 1976.* Moss Landing, CA: Moss Landing Marine Laboratories Technical Publication.

Oliver, J. S., and C. Jong. 1981. Polychaete larval retention, flushing, and invasion in the Elkhorn Slough, California. (Abstract). *Estuaries* 4: 278.

Pace, S. D. 1978. Distribution, abundance and rates of fecundity and growth of *Acartia tonsa* Dana and *Acartia californiensis* Trinast (Copepoda) in Elkhorn Slough, California. M.A. thesis, Moss Landing Marine Laboratories. San Jose State University.

Ramer, B. A., G. W. Page, and M. M. Yoklavich. 1991. Seasonal abundance, habitat use, and diet of shorebirds in Elkhorn Slough, California. *Western Birds* 22:157–174.

Ricketts, E., J. Calvin, J. Hedgpeth, and W. Phillips. 1985. *Between Pacific Tides*. Stanford: Stanford University Press.

Silberstein, M., M. Werner, L. Wunderlich, M. Los Huertos, S. Swezey, S. Gliessman, R. Tjeerdema, A. De Vogelaere, and S. Zembsch. 1997. Report on lands owned by the Nature Conservancy in Elkhorn Slough. Azevedo Ranch, Elkhorn Slough, California. Elkhorn Slough Foundation Report.

Sliger, M. C. 1982. Distribution and microhabitat selection of *Hemigrapsus oregonensis* (Dana) and *Pachygrapsus crassipes* Randall in Elkhorn Slough, Monterey County, California. M.S. thesis, Moss Landing Marine Laboratories. California State University, Hayward.

Talent, L. G. 1976. Food habits of the leopard shark, *Triakis semifasciata*, in Elkhorn Slough, Monterey Bay, California. *California Fish and Game* 62: 286–298.

Tenera, E. S. 2000. Moss Landing Power Plant Modernization Project 316(b) Resource Assessment. Report available from Duke Energy, 655 3d Street, PMB 49, Oakland, Calif. 94607.

Toomey, B. H., and D. Epel. 1995. A multixenobiotic transporter in *Urechis caupo* embryos : Protection from pesticides. *Marine Environmental Research* 39: 299–302.

Von Thun, S., K. Puglise, A. Culley, L. Kuhnz, S. Morgan, and S. McAfee. 2000. Larval transport in Elkhorn Slough, California. (Abstract). In *4th International Larval Biology Meeting, Santa Cruz, California.*

Wasson, K., C. J. Zabin, L. Bedinger, M. C. Diaz, and J. S. Pearse. 2001. Biological invasions of estuaries without international shipping: The importance of intraregional transport. *Biological Conservation* 102(2): 143-153.

Woolfolk, A. M. 1999. Effects of human trampling and cattle grazing on salt marsh assemblages in Elkhorn Slough, California. M.S. thesis, Moss Landing Marine Laboratories. California State University, Sacramento.

10

CHAPTER TEN

Fishes

Mary M. Yoklavich, Gregor M. Cailliet, Dion S. Oxman, James P. Barry, David C. Lindquist

Fishes are among the most conspicuous and best-studied inhabitants of Elkhorn Slough. Long a source of food and recreation, they are of special interest to humans and—as both predator and prey—play a critical role in the slough ecosystem.

The fish fauna in Elkhorn Slough is abundant, diverse, and dominated by marine and estuarine species (Cailliet et al. 1977; Yoklavich et al. 1991; Yoklavich, Stevenson, and Cailliet 1992). The slough provides critical habitat not only for year-round residents, but also for marine species from nearshore waters that enter sloughs to feed, mate, and spawn. Many marine fishes, including a number of economically important species, inhabit the slough's relatively warm, calm waters as juveniles before moving to nearshore coastal waters.

Early studies on the slough's fish populations provided baseline information critical to long-term monitoring of Elkhorn Slough's fish *assemblages*, their response to potential environmental changes, and their contribution to the nearshore fishery resources of Monterey Bay. More recent studies reveal that human impacts on Elkhorn Slough have fundamentally changed available fish habitat, resulting in changes to fish assemblages and an overall decline in diversity of both fishes and their prey.

In this chapter we review the results of all studies of Elkhorn Slough's fish assemblages and prey items, with particular reflection on their response to significant habitat modifications (e.g., increased erosion and tidal scour) that have occurred during the last twenty years. We also offer recommendations for further research to monitor the continued impact of human and natural changes on the slough's fish populations, and to understand the contribution that the slough makes to fish resources in the Monterey Bay area.

Historical Perspective

The ichthyofauna of Elkhorn Slough was described first by George MacGinitie (1935), who documented the general natural history of the coastal end of the embayment in the 1930s. In the 1950s and 1960s, Earl S. Herald and colleagues from the California Academy of Sciences collected data on elasmobranch (shark and ray) populations by monitoring sportfishing derbies. Using data from Gary Kukowski (1972), a graduate student at Moss Landing Marine Laboratories (MLML), Bruce Browning (1972) produced the first checklist of the slough's fishes as part of a series on California's wetlands.

Detailed scientific studies of fish assemblages in Elkhorn Slough began in the mid-1970s and continue today. The faculty and students of MLML have conducted the majority of these studies. Investigations have focused on distribution, abundance, diversity, seasonality, and feeding habits and energetics of adult and juvenile fishes in various habitats throughout the slough system and adjacent marine coastal areas (Ambrose 1976; Ruagh 1976; Appiah 1977; Cailliet et al. 1977; Antrim 1981;

Barry and Cailliet 1981; Yoklavich 1982a, b; Barry 1983; ABA Consultants 1989; Oxman 1995; Lindquist 1998; Cailliet et al., unpublished data). Taxonomic and life history data, including seasonal and spatial patterns of abundance, also have been collected on egg and larval stages of slough fishes (Wang 1981; Yoklavich, Stevenson, and Cailliet 1992).

The elasmobranchs are especially well studied. Background information comes from angling derbies in the 1940s. More recently, researchers have studied feeding ecology (Talent 1976, 1982), reproduction (Martin 1982; Talent 1985; Martin and Cailliet 1988a; Kusher, Smith, and Cailliet 1992), and age and growth (Kusher 1987; Yudin 1987; Martin and Cailliet 1988b; Yudin and Cailliet 1990; Cailliet 1992) of slough elasmobranchs. Field and laboratory experiments on the feeding energetics of the leopard shark are among the most recent projects (San Filippo 1995; Kao 2000).

Interest in the fish fauna was stimulated further by the designation of Elkhorn Slough as a National Estuarine Research Reserve (ESNERR) in 1979 (U.S. Dept. of Commerce 1979; Elkhorn Slough Estuarine Sanctuary Advisory Committee 1985). The reserve's goals include:

- initiation of long-term monitoring of fish populations in Elkhorn Slough;
- assessment of fish responses to natural and human-induced environmental changes; and
- determination of the slough's contribution to the nearshore fishery resources of Monterey Bay.

With support from ESNERR, the results of several of the past studies on fish assemblages in the slough have received recent attention, especially in terms of slough dependence or opportunism, culminating in scientific publications with broad distribution (Yoklavich et al. 1991; Yoklavich, Stevenson, and Cailliet 1992; and Barry et al. 1996).

Distribution, Abundance, and Diversity

Overview

A minimum of 102 species of fishes from 43 families has been identified in Elkhorn Slough and adjacent waters (appendix 10.1). The vast majority (82 species) are marine fishes from Monterey Bay. Sixteen of these marine species use Elkhorn Slough as a spawning or nursery ground; known as immigrant marine fishes (or slough opportunists), this group includes the northern anchovy, Pacific herring, cabezon, and 6 species of flatfish (halibut, sole, sanddab, etc.). Eight fish species are permanent residents that spawn and complete their entire life cycle in Elkhorn Slough; residents include the Pacific staghorn sculpin, black surfperch, bay pipefish, and 5 species of gobies. Six other species are partial residents—they primarily live and reproduce in the slough but move to the ocean during some seasons or life stages. The partial residents are topsmelt, jacksmelt, shiner and white surfperches, leopard shark, and bat ray. Six species are primarily associated with freshwater; these are American and threadfin shad, mosquitofish, prickly sculpin, threespine stickleback, and striped bass. Notably, only 4 species of fish (yellowfin goby, mosquitofish, American shad, and striped bass) are considered to be nonnatives to the Elkhorn Slough system. There seems to have been little opportunity for the introduction of exotic fish species. The slough's narrow opening may isolate it from tanker traffic and mariculture, the activities usually implicated in the introduction of nonnative fish species.

Common and Best Recognized Species

Among the many species collected in Elkhorn Slough, several have been consistently common as adults and juveniles in surveys over the past twenty-five years.

Shiner surfperch.
Photo credit: © Monterey Bay Aquarium Foundation.

Surfperches (Embiotocids) are the most diverse (14 species) and abundant group of fishes in the slough. Shiner surfperch seasonally are common in many slough habitats, including the main channel, tidal creeks, and Bennett Slough. Adult shiners are relatively small (maximum total length is 18 cm) and silvery, with three yellow vertical bars on the mid-body. This partial resident gives birth to live young (usually 7–10) in the shallow parts of the slough in summer months and leaves the slough for deeper water in winter. Shiners eat small crustacea living on the slough floor, and are important forage food for birds, pinnipeds, and fish. Shiners commonly are caught by anglers both at the mouth and upper reaches of the slough.

Able to tolerate a wide range of salinities, the **Pacific staghorn sculpin** is most abundant in upper slough areas and in Bennett Slough. This relatively large sculpin (total length to 46 cm) is the color of mud (green-brown to gray), has a flattened head, no scales, and large antlerlike spines forward on the gill cover.

Pacific staghorn sculpin. Illustration credit: Ann Caudle, © Monterey Bay Aquarium Foundation.

Staghorns spawn in the slough in winter months. They eat small crustacea living on the slough floor, and are themselves preyed upon by birds (notably blue herons and cormorants) and harbor seals. Anglers commonly take staghorns throughout the slough.

Northern anchovy is a marine immigrant that visits the slough to spawn in summer and fall. Large numbers of young anchovies dominate the shallow inland areas of the slough in fall and winter. Anchovies feed on zooplankton and diatoms in the water column. They are an important forage fish for coastal marine birds, fish, and mammals. Rarely caught by anglers in the slough, anchovies have been the target of a significant bait and reduction commercial fishery within the Monterey Bay National Marine Sanctuary.

Northern anchovy. Photo credit: Bruce Stewart.

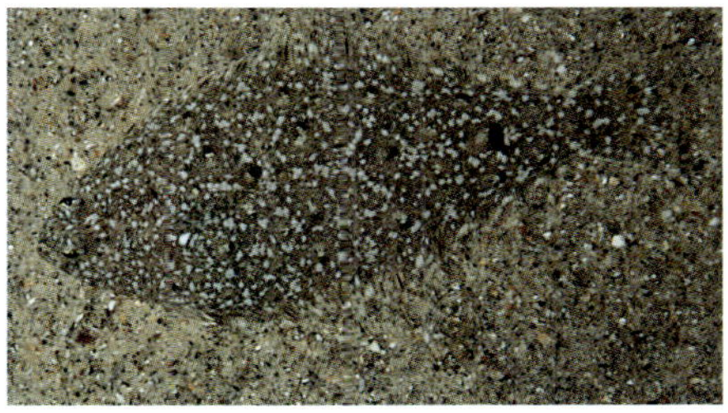

Speckled sanddab. Photo credit: © Monterey Bay Aquarium Foundation.

The speckled sanddab is a relatively small flatfish (maximum total length to 17 cm) that commonly occurs in summer and fall during the egg, larval, and juvenile stages, primarily in areas closest to the slough entrance. Juvenile sanddabs feed on small crustacea near the slough floor, and are prey for marine birds, fish, and mammals, including harbor seals and otters. They also are taken by anglers fishing off the jetty at the slough entrance.

Part-time slough residents, adult **leopard sharks** are abundant in spring and summer, when they give birth to live young in the warm tidal creeks. These sharks are about 20 to 25 centimeters long at birth and grow to a maximum of 2 meters. Leopard sharks feed along the slough's floor on mobile prey such as small fishes, crabs, and fat innkeeper worms. They are taken occasionally by recreational anglers in the slough.

Leopard shark. Illustration credit: Ann Caudle, © Monterey Bay Aquarium Foundation.

Starry flounder are easy to recognize by their diamond shape and alternating black and yellow bands on their dorsal and anal fins. This marine immigrant commonly was found in both Elkhorn and Bennett Sloughs during surveys conducted in the 1970s and 1980s, but currently is not abundant. Although

Starry flounder. Photo credit: © Monterey Bay Aquarium Foundation.

most of the starry flounders taken in slough surveys have been juveniles (<30 cm long), they can reach up to 90 centimeters in length. They mostly prey on molluscs (e.g., bivalve siphons) and infaunal worms. Starry flounders have been caught by sportfishers throughout the slough.

Distribution by Habitat

The Elkhorn Slough system traditionally has been divided into several distinct fish habitats; these are the Moss Landing Harbor and adjacent nearshore sandflats of Monterey Bay, the lower and upper main channel, tidal creeks, Bennett Slough, pickleweed marshes, and salt evaporation ponds. All are connected by the exchange of tidal water, but differ in water depth, tidal influence (primarily salinity and current flow), and biological components such as plants that provide refuge for prey and spawning. As a result, each habitat supports a different assemblage of fish species.

Overall, fish distribution patterns vary with distance from the mouth of the slough. For example, marine species typically reside in the lower slough and harbor, where waters are strongly influenced by ocean and bay hydrographic properties such as higher salinity, lower water temperature, and variable turbulence compared with upper reaches of the slough. Resident fishes are distributed widely but are most abundant in the upper slough. Freshwater species occupy the middle and upper slough including tidal creeks, ponds, and salt marshes. Dominant species of the upper slough and tidal creeks are best characterized as *euryhaline*, with affinities toward higher temperature. In the following sections, we detail assemblages found in the slough's various habitats and discuss changes in these assemblages over time.

Moss Landing Harbor and Adjacent Coastal Waters Assemblages of fishes in the harbor have not been thoroughly surveyed, but sporadic sampling indicates that the species composition resembles a combination of fishes found in the lower channel of the slough and in coastal nearshore waters. Much of this information was gathered from fishes collected by impingement in the Pacific Gas and Electric (now Duke Energy North America) cooling water intake structures (PGE Co. 1983).

Nearshore juvenile and adult fishes were surveyed systematically at least monthly from August 1974 to June 1976 with a small otter trawl fished along 1 kilometer of the sandy shelf (water depth = 5–10 m) to the north and south of the harbor entrance. Abundance consistently was low throughout all seasons, with an average of 13 fish per 10-minute tow. Species diversity (expressed as the number of species) was high (34 species) compared to upper areas of the slough. The four most abundant species in nearshore waters have not been collected anywhere in the slough and yet represented nearly 50% of all fishes taken in the nearshore ocean habitat. These were sand sole, barred and spotfin surfperch, and curlfin turbot. Two abundant nearshore species (speckled sanddab and white surfperch) are marine immigrants into lower slough habitats.

Elkhorn Slough: Main Channel and Tidal Creeks The main channel covers approximately 1.4 square kilometers and extends inland about 10 kilometers from the bay. Depth and width vary considerably with tidal height, but range from nearly 200 meters wide and 7.5 meters deep at the mouth to 15 meters across and 1.5 meters deep at the head of the slough near Hudson's Landing (Crampton 1994). Extensive mudflats fringe the main channel. Lying about 1.3 meters above mean low low water (MLLW), the mudflats are flooded during higher *semidiurnal tides*, at which time several fish species move into this habitat from the main channel. A network of tidal creeks meanders through the salt marshes and, together with the main channel, covers nearly 10 square kilometers. Key features that could affect fish distribution include water depth, turbulence and turbidity, salinity, temperature, food, and predation.

Species composition, abundance, and distribution of fishes in different areas of the main channel and selected tidal creeks are best known from detailed surveys conducted monthly from the mid-1970s to 1980, and repeated twice in the 1990s (Yoklavich et al. 1991; Oxman 1995; Cailliet and Oxman, unpubl. data). Small otter trawls were used to collect adult and juvenile fishes at six sites (fig. 10.1): the Bridge and Dairy stations in the deeper main channel of the lower slough; Rubis and Long Canyon, tidal creeks midway up the slough (beach seine and channel nets also were used at these sites); and Kirby Park and Hudson's Landing, in the shallower waters of the upper slough.

In the 1970s, mean fish abundance was consistently higher in the main slough than in the tidal creeks (table 10.1). Abundance (209 fish per tow) and diversity (37 species) were highest at the Bridge, with shiner surfperch (43%), white surfperch (17%), black surfperch (15%), and speckled sanddab (10%) accounting for almost 85% of the catch

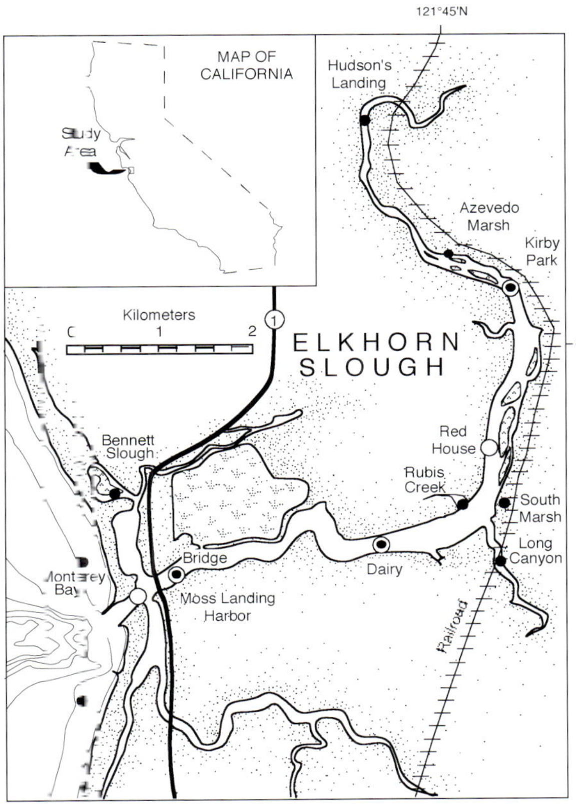

Figure 10.1. Juvenile and adult (●) and larval (○) fish sampling locations in Elkhorn and Bennett Sloughs and the adjacent nearshore ocean.

(table 10.2). The Dairy featured the same diversity but much lower abundance (66 fish per tow); the three surfperches also were predominant here. Fish assemblages at Kirby Park and Hudson's Landing featured intermediate levels of diversity and abundance. Shiner surfperch (54%), Pacific staghorn sculpin (13%), and English sole (11%) were the three most abundant species at Kirby Park. The sculpin (58%), northern anchovy (18%), and Pacific herring (15%) predominated at Hudson's Landing. The tidal creeks (Long Canyon and Rubis Creek stations) had comparable diversity but the lowest mean abundance (12 and 39 fish per tow). The Pacific staghorn sculpin predominated (>40%) at both sites.

Three stations (Bridge, Dairy, and Kirby Park) were resurveyed in the early 1990s. Abundance was significantly lower (table 10.1), but surfperches, flatfishes, and the Pacific staghorn sculpin remained the predominant species at all sites (table 10.2). Cabezon (17%) was the second most abundant fish at the Bridge. California tonguefish (12%) were common at Kirby Park.

Surveys in the mid-1990s again found low abundance (>70% lower than 1970s levels) at the deep channel sites (Bridge and Dairy). Surfperches remained predominant at both sites, but northern anchovy and Pacific herring were common at the Dairy. Conversely, abundance was two to four times higher (than 1970s levels) at three of the four sites farther up the slough. This largely was due to seasonally high numbers of northern anchovy, Pacific herring, and shiner surfperch.

Species diversity seems to have decreased throughout the slough in these twenty years of studies (table 10.1). While the deep main channel stations once had higher numbers of species than the shallow upper slough stations, diversity is now similar and relatively low at all sites. The Bridge station has maintained the highest diversity (24 species in 1995–1996), probably due to the site's diverse habitats (including submerged rocks and pier pilings) and its proximity to the ocean. Marine fishes typical of rocky coastal habitats have been collected in low numbers at the Bridge, but rarely at other stations; these include the scalyhead sculpin, blackeye goby, kelp greenling, rainbow surfperch, onespot fringehead, lingcod, cabezon, and 11 species of juvenile rockfishes. The decline in species richness and change in species composition with increasing distance inland has been noted in other estuarine systems as well. For example, the number of fish species in estuaries of temperate southwestern and tropical northern Australia was highest in the mouth and mid-estuary, largely reflecting the occurrence of marine species that rarely were found in the upper areas (Loneragan et al. 1986) and greater diversity of habitats in lower areas (Blaber, Brewer, and Salini 1989).

Graduate students from Moss Landing Marine Laboratories retrieve a small otter trawl in the main channel during the 1974–1976 Elkhorn Slough fish survey. Left to right: Doug Vaughan, Ed Osada, Dave Ambrose, and Brooke Antrim. Photo credit: Gregor Cailliet.

Table 10.1. Summary of diversity, abundance, and dominance of fishes captured with otter trawls at six locations in Elkhorn Slough during 1974–1980 (from Yoklavich et al. 1991), 1991–1992 (from Oxman 1995) and 1995–1996 (Cailliet and Oxman unpubl. data). Five-minute tows conducted at Rubis Creek and Long Canyon were adjusted to 10-minute tows.

1974–1980	**Bridge**	**Dairy**	**Kirby Park**	**Hudson's Landing**	**Rubis Creek**	**Long Canyon**
# of species	37	37	26	24	32	20
# of tows	35	49	49	44	50	52
Total # of fish	7,326	3,244	3,807	4,479	1,933	596
Mean # / tow (SE)	209.3 (82.4)	66.2 (2.0)	77.7 (15.3)	101.8 (21.8)	38.7 (9.6)	11.5 (1.7)
Dominance index	0.26	0.19	0.32	0.37	0.22	0.29

1991–1992	**Bridge**	**Dairy**	**Kirby Park**	**Hudson's Landing**	**Rubis Creek**	**Long Canyon**
# of species	25	22	30	–	–	–
# of tows	27	28	28	–	–	–
Total # of fish	452	312	1,191	–	–	–
Mean # / tow (SE)	16.7 (5.1)	11.1 (2.9)	42.5 (10.7)	–	–	–
Dominance index	0.12	0.14	0.26	–	–	–

1995–1996	**Bridge**	**Dairy**	**Kirby Park**	**Hudson's Landing**	**Rubis Creek**	**Long Canyon**
# of species	24	15	21	19	17	16
# of tows	33	32	30	22	28	31
Total # of fish	1,801	626	7,973	2,285	2,258	924
Mean # / tow (SE)	54.6 (10.7)	19.6 (5.6)	265.8 (96.8)	103.9 (34.1)	80.6 (39.4)	29.8 (8.2)
Dominance index	0.15	0.22	0.70 *	0.32	0.29	0.30

* = Elevated dominance index was due to the great numbers of northern anchovy caught between April and September at Kirby Park.

Overall similarity of fish assemblages at different sites can be compared by means of a percent similarity index (PSI), which is the sum of the smallest percentage by number of each pair of species. This index ranges from zero (no similarity) to one (identical species arrays). PSI values greater than 0.60 were interpreted as significant. In the 1970s, the fish assemblages near the mouth of the slough (e.g., Bridge and Dairy stations) were very different from those of the tidal creeks (table 10.3; Barry 1983; Yoklavich et al. 1991). The assemblage at the eastern end of the slough (e.g., Hudson's Landing) was more similar in composition to the tidal creek fauna, while the assemblage at Kirby Park was more like those from the main channel to the west.

Twenty years later, however, these geographical differences have disappeared (table 10.3). Fish assemblages in the lower main channel are mostly unchanged, but assemblages in the tidal creeks now resemble those of the lower slough. In fact, assemblages from the Dairy and Long Canyon stations have the highest PSI (0.76) of all 1990s comparisons, due to three dominant species (shiner surfperch, northern anchovy and Pacific herring). These changes in fish assemblages coincide with the continued erosion and scouring of the slough during the last twenty-five years, which has resulted in a geomorphology of the tidal creeks that is now more similar to that of the main channel (Malzone and Kvitek 1994)

Bennett Slough Bennett Slough is a shallow (0.3–2.5 m) tidally influenced embayment (figure 10.1), with a soft mud bottom often covered by sea lettuce. It originally was connected to Elkhorn Slough via a single small culvert (0.6 m diameter) at the north end of Moss Landing Harbor. This culvert was destroyed by the 1989 earthquake and subsequently replaced with six larger culverts (each 1.2 m diameter). This likely has increased tidal current velocities and water depth adjacent to the culverts.

The fish assemblage in Bennett Slough generally is similar to those surveyed at the tidal creek and Hudson's Landing stations of Elkhorn Slough (Yoklavich et al. 1991). Species composition is least similar to that of the Moss Landing Harbor and adjacent

Table 10.2. Relative abundance (%) of dominant species totaling 80% or greater of fishes collected by small otter trawl during the day in Elkhorn Slough, 1974–1980 (from Yoklavich et al. 1991), 1991-1992 (from Oxman 1995), and 1995–1996 (from Cailliet and Oxman, unpubl. data). See legend of appendix 10.1 for life-style categories (in parentheses).

	1974–1980						1991–1992			1995–1996					
Species	Bridge	Dairy	Kirby Park	Hudson Landing	Rubis Creek	Long Canyon	Bridge	Dairy	Kirby Park	Bridge	Dairy	Kirby Park	Hudson Landing	Rubis Creek	Long Canyon
speckled sanddab (MI)	10.3	4.4					18.4	14.4		7.0					
white surfperch (PR)	16.6	28.4					7.1	14.1		15.1	12.1				
English sole (MI)		4.0	10.7				9.1	23.4	21.2						
starry flounder (MI)		3.9	5.9			5.9									
shiner surfperch (PR)	43.4	28.8	53.7		15.5		10.6	16.3	42.9	26.9	35.5		9.8	31.4	48.1
black surfperch (R)	14.6	13.4			7.7	3.7	15.3			18.0					
Pacific staghorn sculpin (R)			13.1	57.6	41.1	51.2	4.4	9.9	13.4				44.1	9.1	
northern anchovy (MI)				18.3	9.2					9.2	16.1	83.1	32.3	41.5	17.3
Pacific herring (MI)				15.0	3.8						21.2				16.2
queenfish (MI)						9.5									
topsmelt (PR)						4.9									
arrow goby (R)						3.7									
leopard shark (PR)						3.7									
lingcod (MI)								5.8							
cabezon (MI)							17.3								
bay pipefish (R)										6.3					
California tonguefish (MI)									12.0						
Total (%)	84.9	82.9	83.4	90.9	81.3	82.6	82.2	83.9	89.5	82.5	84.9	83.1	86.2	82.0	81.6
No. of dominant species	4	6	4	3	6	7	7	6	4	6	4	1	3	3	3

coastal waters. From 86 samples collected with beach seines in the mid-1970s, Appiah (1977) described a relatively low-diversity fish assemblage (20 species) that had a high average abundance (59.1 fish per sample) dominated by 3 euryhaline species (Pacific staghorn sculpin, starry flounder, and arrow goby). Jacksmelt, topsmelt, and three-spined stickleback also were numerous. Most of these species are slough residents or partial residents.

The fishes in Bennett Slough were resurveyed with beach seines in April 1996 (J. Field and S. Sundberg, MLML unpubl. report), well after the earthquake and culvert modifications. This survey of more than 1,300 fishes (about 25% of the total number of fishes from the earlier survey) was dominated by jacksmelt, staghorn sculpin, topsmelt, arrow goby, and bay goby. Most of these fishes were young-of-the-year, suggesting the likely role of Bennett Slough as a nursery. Although limited in scope, this study indicates that the fish assemblage in Bennett Slough has not changed significantly, despite the physiographic changes resulting from post-earthquake construction.

South Marsh The South Marsh (fig.10.1) restoration site, located within the ESNERR, represents approximately 20% of the wetted area and 30% of the total volume of Elkhorn Slough (Malzone and Kvitek 1994). Once an active salt marsh, this area was almost completely cut off from the main slough with the construction of the Southern Pacific Railway in the late 1800s, resulting in a brackish-water habitat. Tidal flow was restored in 1983.

Small (1986) surveyed the fish and macroinvertebrate fauna of South Marsh before and after the restoration. Prerestoration surveys in April and August 1983 resulted in fairly high abundances (677 to 1,240 fish per tow) of 9 small species. The euryhaline three-spined stickleback was predominant, but species of greater marine affinity, such as topsmelt, Pacific staghorn sculpin, Pacific herring, California tonguefish, arrow goby, northern anchovy, shiner surfperch, and jacksmelt, also were present.

Samples taken in April and August 1984, after South Marsh was opened to tidal influence, had higher diversity (16 species) but much lower abundances (42 to 73 fish per tow). Marine and brackish-water fishes were more prevalent, with Pacific staghorn sculpin and northern anchovy the dominant species. Other common species included the arrow goby, California tonguefish, Pacific herring, longjaw mudsucker, plainfin

Table 10.3. Comparison of fish species composition in otter trawl collections from six stations in Elkhorn Slough during 1974–1980 (top half of matrix and bold) and 1995–1996 (bottom half of matrix and not bold) based on percent similarity index (PSI).

	Bridge	Dairy	Kirby Park	Hudson's Landing	Rubis Creek	Long Canyon
Bridge	---	**0.77**	**0.60**	**0.07**	**0.32**	**0.16**
Dairy	0.59	---	**0.48**	**0.11**	**0.36**	**0.18**
Kirby Park	0.19	0.29	---	**0.27**	**0.45**	**0.32**
Hudson's Landing	0.26	0.41	0.47	---	**0.61**	**0.62**
Rubis Creek	0.52	0.66	0.56	0.56	---	**0.66**
Long Canyon	0.44	0.76	0.31	0.66	0.63	---

midshipman, English sole, shiner surfperch, yellowfin goby, and bay goby (Small 1986). Small numbers of California halibut, bay pipefish, Pacific sardine, starry flounder, diamond turbot, and bat ray also were collected. The restoration process apparently increased saltwater mixing and allowed a higher diversity of marine and estuarine fishes to occupy this formerly brackish-water marsh.

In subsequent monthly surveys of South Marsh through 1985 (Small 1986), 10 species were consistently abundant (Pacific staghorn sculpin, shiner surfperch, northern anchovy, arrow goby, Pacific herring, California tonguefish, English sole, California halibut, starry flounder, and bat ray). Strong peaks in abundance were detected in late spring and early summer, primarily due to the influx of juveniles. Thus, the fish assemblage in this newly developed area of the South Slough is now quite similar to that of the adjacent slough waters.

Salt Evaporation Ponds: Azevedo Marsh The Azevedo Marsh system comprises several salt evaporation ponds, covering about 4.7 square kilometers that were separated from the main slough by Southern Pacific Railway dikes in the late 1800s (see ABA Consultants 1989). Commonly called pocket marshes, these ponds are connected to the slough by culverts under the railroad dikes. They often are filled by freshwater discharge, but there is significant tidal exchange (especially since 1982, when the dikes were partially breached by erosion), so the ponds sometimes become brackish or even marine. These ponds probably represent one of the most extreme habitats occupied by fishes in the slough system. They are characterized by highly variable water quality and are dominated by seasonally changing plant assemblages, including pickleweed and drift algae. They are shallow, very muddy, offer restricted access and egress, and they receive heavy runoff of nutrients and pesticides from adjacent agricultural fields.

The fishes in the saltponds of Azevedo Marsh were surveyed by beach seine during spring 1998 (A. Spotswood and A. Dewitt, MLML unpubl. report). In general, the species composition was quite similar to that of Bennett Slough, dominated by Pacific staghorn sculpin, arrow goby, and topsmelt; large numbers of northern anchovy were collected at high tide in the largest pond. Other fishes included shiner surfperch, mosquitofish, three-spined stickleback, and yellowfin goby.

Overall, this is an interesting mixture of freshwater, brackish-water, and saltwater fishes. However, much higher than normal rainfall in 1998 (an El Niño year) might have influenced these results. Species composition in these ponds might differ considerably in other years or seasons. The longjaw mudsucker was common in North Azevedo Pond during summer 1998 (S. Ross, North Carolina NERR unpubl. data). This goby seems particularly well adapted to the pond's harsh environment, which includes extreme variations in temperature and dissolved oxygen (ESNERR unpubl. data). Small leopard sharks also have been observed in the Azevedo ponds (M. Silberstein, pers. comm.).

Seasonality and Spawning

The abundance and distribution of several fish species in Elkhorn Slough vary significantly from season to season primarily as a function of the fishes' reproductive habits. Like other estuarine systems, Elkhorn Slough serves as a spawning and nursery ground for many fishes, including some commercially and recreationally important species (Barry 1983; Yoklavich et al. 1991; Yoklavich, Stevenson, and Cailliet 1992). Its protected waters (relative to the open ocean) provide abundant food resources, perhaps lower predation pressure, and a variety of shallow habitats with elevated water temperature for juveniles and spawning adults.

The overall abundance and diversity of fishes in Elkhorn Slough peak during summer (Cailliet et al. 1977; Barry 1983; Small 1986; Yoklavich et al. 1991). Many marine species enter the slough as young juveniles or reproductively active adults during this season. Also, locally spawned juveniles grow large enough by summer to be collected in the survey nets.

Abundance and diversity decline during fall, when many of the partial-resident and marine-immigrant juveniles leave the slough and return to the ocean. Fishes overwintering in the slough are mainly resident species capable of coping with the wide range in salinity that is typical of the rainy season (see chapter 4, "Hydrography").

Seasonal variations in abundance and distribution of adult and juvenile fishes in the Elkhorn Slough system are similar to patterns described for other estuarine systems. Higher abundance and species richness during the summer invasion of young-of-the-year marine species also have been documented for fish assemblages in other temperate bays and estuaries, and were positively correlated with temperature (Allen and Horn 1975; Hoff and Ibara 1977; Allen 1982; Onuf and Quammen 1983).

Seasonal distribution and abundance of adults, juveniles, and larvae vary significantly from species to species, depending on their spawning cycles. For example, increased springtime abundance of juvenile Pacific herring in the upper slough follows winter spawning in eelgrass beds of central California bays and estuaries. The high abundance of the three dominant species of surfperches (shiner, black, and white) in summer and early fall reflects their summer spawning (Antrim 1981). Shiner surfperch enter upper areas of the slough to spawn (Barry 1983), while white surfperch spawn at the lower stations (Antrim 1981).

The English sole, a commercially important marine immigrant, relies heavily on large northern estuaries as nursery grounds (Myers 1974; Krygier and Pearcy 1986; Pearcy and Toole 1980; Gunderson et al. 1990). Young benthic (bottom-dwelling) juveniles that are newly settled from the plankton enter Elkhorn Slough in early spring, increase in abundance in the main channel, and reach peak numbers at Kirby Park in summer. Elkhorn Slough is thought to be the southernmost major nursery for juvenile English sole, which are conspicuously absent from fish surveys of southern California bays (Horn and Allen 1985). Their occurrence in shallow inshore areas may be limited by specific thermal tolerances during various stages in development.

Although categorized as partial residents, two species of elasmobranchs, the leopard shark and bat ray, occur nearly year-round and spawn in Elkhorn Slough (Talent 1985). Other elasmobranch species appear to be more seasonal in occurrence. Patterns in occurrence may be related to food abundance, breeding areas, and nursery grounds (Barry and Cailliet 1981; Barry 1983; Martin and Cailliet 1988a; Talent 1976, 1982, 1985).

Fish Eggs and Larvae

Some fishes lay eggs that are free floating; others attach their eggs to plants or other substrates. Floating fish eggs and newly hatched larvae are part of the ichthyoplankton that is moved about by wind and water currents. The distribution of young fishes at these developmental stages provides evidence for the occurrence, timing, and location of spawning within the slough.

Patterns of seasonal and spatial abundance of free-floating fish eggs and larvae in Elkhorn Slough are known from ichthyoplankton samples collected monthly from 1974 to 1976 at five stations (Harbor, Bridge, Dairy, Red House, and Kirby Park) along the main channel (fig. 10.1; Yoklavich, Stevenson, and Cailliet 1992). Twenty-nine taxa of fishes were collected in this study (appendix 10.1). Gobiid (gobies) and clupeoid (herrings and anchovies) larvae were numerically predominant. Seven species accounted for 94% of the total larval fish catch: in order of abundance, longjaw mudsucker, northern anchovy, arrow goby, unidentified osmerid (probably surf smelt), Pacific staghorn sculpin, white croaker, and Pacific herring. Among the eggs, northern anchovy, unidentified sanddab, and white croaker accounted for 83% of the catch.

Larval fish assemblages varied seasonally in Elkhorn Slough. Northern anchovy and two gobies (longjaw mudsucker and arrow goby) were most abundant during summer and fall.

California halibut. Left: Three larval stages of development, ranging from 7.3 mm long (top) to 14.0 mm long (bottom). Photo credit: Bruce Stewart. Right: Adult about 1.2 meters long, held by former Moss Landing Marine Laboratories graduate student Brooke Antrim. Photo credit: Gregor Cailliet.

Both gobies have protracted spawning seasons from February to September (Weisel 1947; Prasad 1948) that peak during late summer when water is warm (18–23°C). Northern anchovies spawn primarily in the lower slough in summer and early fall, in water temperatures of 14–18°C. Newly hatched northern anchovies are distributed progressively eastward from the harbor to at least Kirby Park. The shallow upper slough waters provide a nursery for these larvae during fall and winter.

A different assemblage of young larvae is found during winter and early spring. This group is more diverse but less abundant than the summer-fall assemblage. Pacific staghorn sculpin, surf smelt, two species of Atherinids (topsmelt and jacksmelt), and Pacific sand lance are the dominant larvae in this season. This group also is more variable in occurrence, with pulses of high numbers of young surf smelt and Pacific sand lance collected in single samples.

There are at least two spatially distinct larval assemblages in the slough, one located in the most inland areas and another at the near-ocean stations. These distributions are attributed to reproductive specialization (egg type and spawning origin of adults) and hydrographic conditions.

Upper slough waters featured fewer taxa but greater numbers of individual larvae. Resident gobies produced the most abundant larvae in the slough, and their numbers increased with distance from the harbor. Adults and juveniles of longjaw mudsucker and arrow goby were collected almost entirely in tidal creeks and inland slough waters (Barry 1983; Yoklavich et al. 1991). The extensive pickleweed beds bordering these regions provide suitable habitat for spawning gobies.

Pacific herring larvae were abundant in spring in the upper slough where pickleweed beds occupy the intertidal marsh zone. Submerged aquatic vegetation is relatively sparse elsewhere in Elkhorn Slough. The lack of adequate spawning substrate, such as seagrass beds, potentially limits use of the lower slough as nursery habitat. Restoration of seagrass beds in Elkhorn Slough could modify distribution and enhance recruitment for species with adhesive, *demersal* eggs, such as Pacific herring.

Along with localized spawning habits of adults, significant differences between larval assemblages in the lower and upper slough suggest that the tidal prism (a wedge of marine water flowing along the bottom of the main channel on a rising tide to about 4.8 km from the harbor entrance) assists in retaining young fishes within the slough. Mid-slough species composition represents a transition between lower and upper groups. Water is exchanged daily by tidal currents, and seasonal variations in temperature and visibility are similar at harbor entrance, Bridge, and Dairy stations. The most abundant larvae at these stations were very small northern anchovy and surf smelt, species that are classified as slough immigrants.

The water mass to the east of the tidal prism fluctuates widely in temperature and salinity, and has lower visibility. The high water *residence time* and increased summer water temperatures and available substrate for adhesive demersal eggs in shallow inland marsh habitats of the slough could enhance reproductive success, larval retention, and survival.

There are similarities in species composition among Elkhorn Slough larval fish assemblages and those described in the few comparable studies of other West Coast estuaries. Gobiid and clupeoid larvae dominate most systems, including almost all southern California bays and estuaries (McGowen 1977; Leithiser 1981; Nordby 1982; Horn and Allen 1985), northern California's Humboldt Bay (Eldridge and Bryan 1972), and Yaquina Bay, Oregon (Pearcy and Myers 1974). Seasonal and spatial patterns of larval fish abundance also were similar in many of the West Coast studies of inshore larval fish assemblages.

Feeding Habits

The diets of the dominant fishes in Elkhorn Slough are known from analysis of the stomach contents of juvenile and adult fishes collected at eight stations in 1974–1980 (Ambrose 1976; Cailliet et al. 1977; Antrim 1981; Barry and Cailliet 1981; Barry 1983; Barry et al. 1996). Diets at four of these stations were reexamined in 1996–1997 to assess potential effects of slough erosion and scour on the fishes' trophic ecology (Lindquist 1998).

Based on these studies, four *trophic guilds* have been described for the slough (fig. 10.2). The most distinct guild, which includes Pacific staghorn sculpin, arrow goby, shiner surfperch, and speckled sanddab, feeds principally on *epifaunal* crustacea (especially gammarid amphipods and harpacticoid copepods). A second guild, including English sole, starry flounder, white surfperch, and bat ray, eats mostly molluscs and *infaunal* worms (especially during the 1970s). A third guild consists of fishes that primarily prey on mobile crustaceans (such as mysids, *Crangon* shrimp, and grapsid crabs); these are sand sole (at near ocean stations), leopard shark (in the 1970s), and white surfperch (in the 1990s). The fourth guild comprises fishes that feed primarily on zooplankton and diatoms in the water column; topsmelt, Pacific herring, and northern anchovy are among the species in this group.

Prey diversity was noticeably lower in the 1990s than in the 1970s. This trend was evident at all four stations and for all of the nine fish species found in both studies. The main difference was the reduced use of infaunal worms and molluscs (Lindquist 1998). For example, the English sole had the most diverse diet (44 prey taxa, including many infaunal worms) in the 1970s, but ate only 14 prey items in the 1990s. Infaunal prey were largely replaced with epifaunal crustacea in the English sole's diet, and with mobile crustacea (specifically *Hemigrapsus* crabs) in the diet of the white surfperch.

Fishes of estuaries and shallow marsh habitats often are categorized as trophic generalists that use abundant, but highly variable, prey resources. Diets of the benthic foraging fishes (such as sanddab, starry flounder, and shiner surfperch) in Elkhorn Slough reflected prey availability in core samples of bottom sediments (Barry et al. 1996; Lindquist 1998). Prey availability in the main channel, however, changed during the twenty years of studies. Overall densities of benthic invertebrates have decreased (Lindquist 1998). Notable were the increased relative abundance of epifaunal crustacea (amphipods and tanaids) and the dramatic decrease in infaunal worms.

Increased rates of erosion, water current velocity, and scour of the slough (Williams et al. 1992; Crampton 1994; see chapter 4), which were initially caused by the construction of the entrance to Moss Landing Harbor in 1947, have resulted in significant changes to the vegetated salt marshes, mudflats, main channel, and tidal creeks. In the late 1970s, Long Canyon was 1–1.5 kilometers long, 5–10 meters wide, and 2 meters deep at its intersection with the main channel (about 4 km east of Moss Landing Harbor entrance) and received freshwater runoff from the adjacent watershed. Rubis Creek, which intersects the main channel across from Long Canyon, was about the same width and depth as Long Canyon in the late 1970s, but received little freshwater runoff. Long Canyon increased 62% in width from 1980 to 1987, and is continuing to erode. Rubis Creek increased 8% in width during this time. The main channel also has broadened and deepened (J. Oliver et al., MLML unpubl. report; Malzone and Kvitek 1994). Mudflat habitat that fringes the main channel and harbors dense assemblages of infaunal and epifaunal invertebrates likely is diminishing (ABA Consultants 1989).

As the main channel and tidal creeks of the slough continue to broaden and deepen, the species composition of fish predators and invertebrate prey apparently is being altered. The trophic patterns of fishes in the tidal creeks are now more similar to those of fishes in the main channel than was indicated from surveys in the 1970s (Lindquist 1998). Interestingly, prey richness in the 1970s was greatest at stations in the main channel and lowest in the tidal creeks (Barry et al. 1996). In the 1990s study, prey richness was similar between main channel and tidal creek sites (Lindquist 1998).

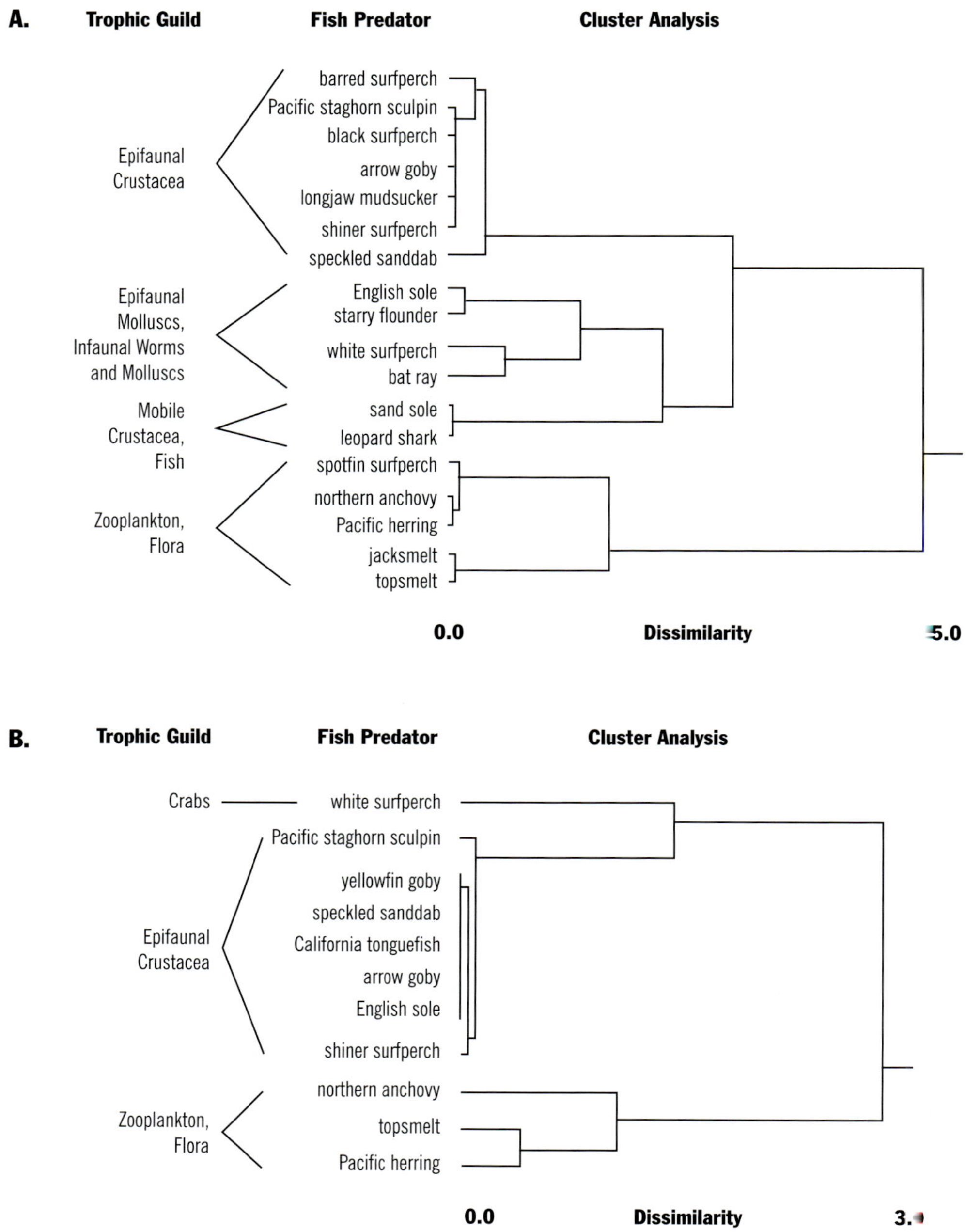

Figure 10.2. Cluster analysis of fishes in Elkhorn Slough, based on dietary information summed to trophic guilds. A. Fishes collected from 3 main channel, 2 tidal creek, and 2 nearshore ocean stations 1974-1980 (Barry et al. 1996). B. Fishes collected from 2 main channel and 2 tidal creek stations 1996-1997 (Lindquist 1998).

Sportfisheries

Sportfishing has been popular in Elkhorn Slough for many decades, but there is little historical information on the level of effort or impact on the fish populations. Two long-term studies of sportfish catches have provided some information about those species popular with anglers. These are shark derbies and creel surveys.

Shark Derbies

In the 1940s, the Pajaro Valley Rod and Gun Club (PVRGC) and the Castroville Rod and Gun Club (CRGC) initiated late spring–early summer angling derbies for elasmobranchs in Elkhorn Slough. In addition to their recreational value, these derbies were meant to control shark and ray populations that were believed to be reducing populations of more desirable finfish and shellfish in the slough. The last derbies (by either angling or archery) for elasmobranchs in Elkhorn Slough occurred in 1996.

Ichthyologists, most notably Earl S. Herald from the California Academy of Sciences, took an early interest in these derbies as a means of collecting data on estuarine elasmobranch populations. From 1951 to 1962, these researchers monitored the Elkhorn Slough derby catches each year recording species composition, total length, weight, sex, gut content, stage of sexual maturity, and fishing effort (Herald and Dempster 1952; Herald 1953; and Herald et al. 1960). In 1963 and 1964, the sponsoring clubs collected similar data from their own derbies. Collection of data from the elasmobranch derbies was resumed in 1971 by ichthyologists at MLML, California Department of Fish and Game, California Academy of Sciences, and San Francisco State University. Between 1980 and 1996, this was an annual event conducted by MLML personnel (King and Cailliet 1992). In 1988, MLML, the PVRGC, Monterey Bay Aquarium, and the Elkhorn Slough Foundation cooperatively initiated a tag-and-release program during the angling derbies, which continued until 1996.

Based on data collected from shark derbies (King and Cailliet 1992), and some presented by Talent (1985), we know that 7 species of large elasmobranchs are relatively common in Elkhorn Slough: bat ray, shovelnose guitarfish, leopard shark, gray smoothhound, brown smoothhound, thornback and round stingray. Two others, the spiny dogfish and Pacific electric ray, also have been caught in the slough.

Species composition and catch-per-unit-effort (CPUE) were evaluated from forty-eight elasmobranch derbies held in Elkhorn Slough during May, June, and July 1951–1990. The most noticeable change in species composition over these forty years was the nearly complete disappearance of the shovelnose guitarfish after 1972. From 1951 through 1960, catch ranged from 60 to 240 elasmobranchs per derby, with an average of about 100 fish per derby. Catch was lower in the more recent years, ranging from 60 to 170 fish, with an average of about 85. Although data were not available for all years, CPUE peaked in the mid to late 1950s (0.3–1.8 fish per fisherman), and declined since 1960 (0.3–0.4 fish per fisherman).

These researchers also presented trends in size structure and sex ratios for the most abundant species (bat ray, shovelnose guitarfish, and leopard shark) and noted that neither varied significantly over the years. One exception was that the smaller size classes of shovelnose guitarfish disappeared from the catch earlier (before the 1970s) than the larger fishes. Also, female bat rays and shovelnose guitarfish were larger than their male counterparts, and outnumbered males nearly 2:1, whereas female and male leopard sharks were nearly equal in size and sex ratio.

A young visitor to the Elkhorn Slough National Estuarine Research Reserve holds a small gray smoothhound. Photo credit: ESNERR.

Recreational Creel Surveys

From July 1974 to June 1976, Cailliet et al. (1977) conducted monthly creel censuses at popular sportfishing sites along the shores of Elkhorn Slough. Sampling sites were combined into those west of the Highway One bridge (including the Jetties, Skippers, and Bennett Slough) and those to the east (mainly Kirby Park; see fig. 10.1). The number of censuses at each of these two general locations ranged from 6 to 66 per month, and averaged 35.

At least 46 species of fishes were caught by sportfishermen in the marine waters west of the bridge; more than half of these were rare (less than 1% of the total; table 10.4). The ten most frequently reported species were Pacific staghorn sculpin, sand sole, walleye surfperch, white surfperch, shiner perch, white croaker, starry flounder, black surfperch, jacksmelt, and pile perch. Farther up the slough, catches were lower in abundance (only 37 specimens were surveyed), and only 8 species were reported. White surfperch, shiner perch, and rubberlip perch were reported to be the most abundant species.

In the 1980s and 1990s, creel censuses of similar seasonal coverage but much lower sample sizes were conducted by the National Marine Fisheries Service (NMFS), through the Marine Recreational Fishing Statistics Survey (MRFSS). NMFS used somewhat different survey techniques, but changes in sportfishery catches can be evaluated by comparing relative abundance (rank). The trends west of the bridge basically were similar. Only 25 species were caught, but eight of the top ten species were the same. The notable differences were lower rankings for sand sole (8th) and shiner perch (tied 14th) and higher rankings for silver surfperch (7th), barred surfperch (2nd), calico surfperch (tied 10th), and señorita (12th). The latter three species are more commonly found in southern California; increased numbers in the vicinity of Elkhorn Slough could reflect the coastwide warming trend in nearshore waters that has occurred since the late 1970s (Barry et al. 1995; Roemmich and McGowan 1995).

Based on small samples, 7 species were reported from upper slough sites. Five of these were the same as the 1970s. Pile perch (in 2nd rank) and brown rockfish (tied 6th) were new to the survey; shiner perch, starry flounder, and striped bass were absent.

During both time periods, catches east of the Highway One bridge were lower in abundance and diversity than catches in the west end of the slough. In the 1970s, the dominant species taken in the upper reaches of the slough were white surfperch, shiner surfperch, rubberlip perch, Pacific staghorn sculpin, striped surfperch, starry flounder, striped bass, and cabezon. Pacific staghorn sculpin dominated the catches in the 1980–1990s, followed by pile surfperch, striped surfperch, rubberlip perch, white surfperch, cabezon, and brown rockfish, the latter probably being primarily juveniles.

Management Issues and Research Recommendations

Elkhorn Slough provides habitat for many fishes, and serves as spawning and nursery grounds for a variety of ecologically, commercially, and recreationally important species. Juveniles of some commercially important fish species found in the slough include Pacific herring, English sole, and northern anchovy.

From the studies on fishes of Elkhorn Slough from the 1970s to the present, it is obvious that changes in habitat significantly influence the distribution, abundance, and trophic patterns of the slough's ichthyofauna. There are distinct relationships between the physical characteristics of these habitats, such as water depth, distance from the ocean, magnitude of tidal currents, temperature, and salinity, and the fish fauna that occupy them. As habitats are modified (e.g., broadening and deepening of tidal creeks due to tidal scour and erosion), the fish assemblages and their use of these habitats also change.

The contribution that estuaries and bays such as Elkhorn Slough make to the survival of coastal marine fishes is difficult to assess. The high relative abundance of marine-related (i.e., marine, marine-immigrant, and partial-resident life style categories) individuals and species entering Elkhorn Slough early in life or as spawning adults demonstrates the importance of this habitat to nearshore fish communities. Relatively large numbers of marine immigrants, especially in the highly productive shallow tidal creeks and upper areas of the slough, potentially can transport energy from nursery habitats to offshore waters. The scarcity of habitats like Elkhorn Slough along the central California coast implies that remaining areas are important to the success of nearshore marine fishes.

Here we suggest additional research topics that could help us to better understand the role of Elkhorn Slough and other estuaries in the life cycles of nearshore marine fishes; to measure the ongoing impacts of natural and human-induced changes on fish populations, their predators, and prey; and to interpret change in the fish assemblages through time.

Long-Term Monitoring of Fish and Ichthyoplankton

Surveys of the fishes occupying Elkhorn Slough and its habitats have been conducted over the past few decades. However, these were mostly opportunistic and related to environmental surveys required of industry or for land-use practices. It is now

Adult leopard shark collected for tag-and-release in the slough. Photo credit: ESNERR.

recognized that regular surveys of the fish fauna and associated environmental conditions could help us interpret fluctuations in these populations over time. With such data, a better assessment of fish responses to natural and human-induced environmental changes would be possible.

Ichthyoplankton surveys recently completed in association with the Moss Landing Power Plant modernization project (Tenera Environmental Services 2000) provide a third decade of information on the fish larvae dispersed throughout Elkhorn Slough over the seasons. Such longterm information will be useful in predicting the effects that entrainment of fish larvae into the power plant may have on the adult populations in the slough.

Use of Habitats by Fishes

Continued monitoring of the habitats in Elkhorn Slough, especially as it relates to erosional processes, is essential to predict change in fish assemblages. As we evaluate ways to control erosion, it will be important to document the types of available habitats and their distribution and relative proportion within the slough ecosystem.

Defining the patterns in habitat use by fishes will help us understand the functional relationships among habitats and the fish assemblages in the slough. Tag-recapture and tracking studies of various slough fishes could help us identify important feeding and nursery habitats. Indeed, it is known that several species of elasmobranchs pup in Elkhorn Slough in the spring, and the young sharks typically occupy tidal creeks for feeding and shelter (Barry and Cailliet 1981; Barry 1983; Barry et al. 1996). However, the physiographic and ecologic nature of these tidal creeks has been changing with continued erosion; the impact of such changes on the function of these habitats and associated fishes remains unknown.

Long-Term Predator and Prey Surveys

Because of the pronounced change in the prey assemblages consumed by fishes following two decades of increased tidal scour of Elkhorn Slough, it is important to monitor the ecological factors that could influence fish survival. Continued surveys of benthic infaunal and epifaunal organisms would enhance our ability to evaluate the importance of the slough as fish feeding grounds. Likewise, continued surveys of birds and mammals in the slough can help us determine potential effects that these predators may have on fishes and shared prey resources.

Contribution to Nearshore Fisheries in Monterey Bay and the Central California Coast

The prevalence of estuarine dependence by nearshore fishes, especially during their early life history, has often been suggested as one reason for the ecological importance of estuaries. However, few studies have adequately tested the hypothesis that some species of fishes are, indeed, estuarine dependent, or have identified the extent to which specific estuaries are used by coastal fishes during their early life history. Recently, scientists at the University of California, Santa Cruz (Pete Raimondi and Jennifer Brown, pers. comm.) have analyzed trace chemical elements in the *otoliths* of flatfishes to differentiate juveniles living in estuaries from those found

Table 10.4. Fish species taken in creel censuses from two general locations in Elkhorn Slough during the 1970s and the 1980s–1990s. Data are summarized as ranks (1 = most abundant; tr = trace numbers) due to differences in techniques from the 1970s (taken from Cailliet et al. 1977)) and the 1980s–1990s (summarized from the NMFS Marine Recreational Fishing Statistics Surveys [MRFSS] data base). Data are grouped into those sites west of the Highway One bridge (Jetties, Skippers, Bennett Slough) and those to the east (mainly Kirby Park).

		1970s		1980s–1990s	
Scientific Name	**Common Name**	**West**	**East**	**West**	**East**
Leptocottus armatus	Pacific staghorn sculpin	1	4.5	1	1
Psettichthys melanostictus	sand sole	2.5	-	8	-
Hyperprosopon argenteum	walleye surfperch	2.5	-	5	-
Phanerodon furcatus	white surfperch	4	1	9	5
Cymatogaster aggregata	shiner surfperch	5	2	14.5	-
Genyonemus lineatus	white croaker	6	-	6	-
Platichthys stellatus	starry flounder	7	6	9	-
Embiotoca jacksoni	black surfperch	8	4.5	10.5	3
Atherinopsis californiensis	jacksmelt	9	-	3	-
Rhacochilus vacca	pile surfperch	10	-	13	2
Sebastes paucispinis	bocaccio	11	-	-	-
Rhacochilus toxotes	rubberlip surfperch	12	3	17	4
Ophiodon elongatus	lingcod	13	-	24	-
Hexagrammos decagrammus	kelp greenling	14	-	-	-
Scorpaenichthys marmoratus	cabezon	15	8	17	6.5
Citharichthys stigmaeus	speckled sanddab	16	-	22.5	-
Embiotoca lateralis	striped surfperch	17	-	22.5	
Sebastes mystinus	blue rockfish	18	-	24	-
Morone saxatilis	striped bass	19	7	-	-
Hyperprosopon ellipticum	silver surfperch	20	-	7	-
Amphistichus rhodoterus	redtail surfperch	21	-	-	-
Amphistichus argenteus	barred surfperch	22	-	2	-
Sebastes auriculatus	brown rockfish	tr	-	14.5	6.5
Sebastes chrysomelas	black-and-yellow rockfish	tr	-	-	-
Amphistichus koelzi	calico surfperch	tr	-	10.5	-
Squalus acanthias	spiny dogfish	tr	-	-	-
Oxylebius pictus	painted greenling	tr	-	-	-
Porichthys notatus	plainfin midshipman	tr	-	-	-
Atherinops affinis	topsmelt	tr	-	-	-
Parophrys vetulus	English sole	tr	-	-	-
Neoclinus uninotatus	onespot fringehead	tr	-	-	-
Hypsurus caryi	rainbow surfperch	tr	-	-	-
Oncorhynchus tsawytscha	chinook salmon	tr	-	-	-
Hyperprosopon anale	spotfin surfperch	tr	-	-	-
Lyopsetta exilis	slender sole	tr	-	-	-
Ammodytes hexapterus	Pacific sand lance	tr	-	-	-
Trachurus symmetricus	jack mackerel	tr	-	-	-
Gibbonsia sp.	kelpfish	tr	-	-	-
Cebidichthys violaceus	monkeyface prickleback	tr	-	-	-
Engraulis mordax	northern anchovy	tr	-	-	-
Hypsopsetta guttulata	diamond turbot	tr	-	-	-
Oxyjulis californica	señorita	tr	-	12	-
Microgadus proximus	Pacific tomcod	tr	-	-	-
Myliobatis californica	bat ray	tr	-	-	-
Sebastes spp.	misc. rockfishes	tr	-	tr	-
Peprilus simillimus	Pacific pompano	tr	-	-	-
Triakis semifasciata	leopard shark	-	-	17	-
Anarrhichthys ocellatus	wolf-eel	-	-	22.5	-
TOTAL NUMBER OF SPECIMENS SURVEYED		**5,832**	**37**	**1,033**	**32**
TOTAL NUMBER OF SPECIES		**46**	**8**	**25**	**7**

offshore, and to potentially identify the actual estuary in which the young fishes grew up. Additional studies of this nature would identify and quantify essential estuarine habitats for fish species in central California.

Identification and Enumeration of Fishes from Archaeological Surveys

A great deal of information about the historical status of fishes can be obtained from archaeological samples. Several Native American archaeological sites around the Elkhorn Slough area have been sampled for fishes (Gordon 1985; Dietz, Hildebrandt, and Jones 1988; see chapter 6, "Archaeology and Prehistory"). Evidence to date indicates that these Native Americans harvested both freshwater and marine fishes, presumably at a time when the Salinas and Pajaro Rivers connected near the site where the Salinas River and Elkhorn Slough now converge (Gobalet 1990, 1993).

While data on species composition are readily available from bones, scales, and otoliths of fishes, less information about seasonal occupation of these sites both by Native Americans and by fishes, or about the reproductive status, size ranges, age composition, or mortality rates of the fish, is available. Modern techniques used to analyze growth zones and stable isotopes in such calcified structures as fish otoliths from these archaeological surveys could uncover details about the ecology of the fishes that occurred in this estuarine system thousands of years ago.

Acknowledgments

The authors would like to thank the Monterey Bay National Marine Sanctuary for providing funds for fish surveys; the Monterey Bay Aquarium for the use of photos and illustrations; Robert N. Lea of the California Department of Fish and Game, and Richard Parrish and Jan Mason of the National Marine Fisheries Service Pacific Fisheries Environmental Laboratory, for their helpful comments on early drafts of the manuscript; and R. Lea for revision and update of the fish species list. Thanks also to several anonymous reviewers of early drafts of the manuscript.

Appendix 10.1. Species of finfish (alphabetically ordered by Family) collected from Elkhorn Slough, California and adjacent areas (Moss Landing Harbor, Jetties and Bennett Slough).

Life style (M = marine, MI = marine immigrant, R = resident, PR = partial resident, and F = freshwater)
Life stage (A = adult, J = juvenile, L = larva, E = egg) '?' indicates that species or life stage were not verified.
NN indicates non-native species.
Common and scientific names follow Robins et al. (1991) and Robert N. Lea (pers. comm.).

Family	Species	Common Name	Life style	Life stage
Acipenseridae	*Acipenser medirostris*	green sturgeon	MI	A
Ammodytidae	*Ammodytes hexapterus*	Pacific sand lance	M	L, A
Anarhichadidae	*Anarrhichthys ocellatus*	wolf-eel	M	A
Atherinopsidae	*Atherinops affinis*	topsmelt	PR	L, J, A
	Atherinopsis californiensis	jacksmelt	PR	L, J, A
	Leuresthes tenuis	California grunion	M	A
Aulorhynchidae	*Aulorhynchus flavidus*	tubesnout	M	J, A
Bathylagidae	*Leuroglossus stilbius*	California smoothtongue	M	L
Batrachoididae	*Porichthys notatus*	plainfin midshipman	MI	J, A
Blenniidae	*Hypsoblennius gentilis*	bay blenny	M	J
Bothidae	*Citharichthys stigmaeus*	speckled sanddab	MI	E, L, J
	Paralichthys californicus	California halibut	MI	L, J, A
Carangidae	*Trachurus symmetricus*	jack mackerel	M	J, A
Carcharhinidae	*Mustelus californicus*	gray smoothhound	M	J, A

Family	Species	Common Name	Life style	Life stage
	Mustelus henlei	brown smoothhound	M	J, A
	Triakis semifasciata	leopard shark	PR	J, A
Clinidae	*Gibbonsia metzi*	striped kelpfish	M	L, A
	Heterostichus rostratus	giant kelpfish	M	J, A
	Neoclinus uninotatus	onespot fringehead	M	L, J, A
Clupeidae	*Alosa sapidissima* (NN)	American shad	F	J, A
	Clupea pallasi	Pacific herring	MI	L, J, A
	Dorosoma petenense	threadfin shad	F	A, ?
	Sardinops sagax	Pacific sardine	M	J, A
Cottidae	*Artedius harringtoni*	scalyhead sculpin	M	A, ?
	Clinocottus sp.	sculpin	M	L
	Cottus asper	prickly sculpin	F	?
	Leptocottus armatus	Pacific staghorn sculpin	R	L, J, A
	Scorpaenichthys marmoratus	cabezon	MI	J, A
Cyclopteridae	*Liparis* sp.[1]	snailfish	M	?
Embiotocidae	*Amphistichus argenteus*	barred surfperch	M	A
	Amphistichus koelzi	calico surfperch	M	A
	Amphistichus rhodoterus	redtail surfperch	M	A
	Cymatogaster aggregata	shiner surfperch	PR	J, A
	Embiotoca jacksoni	black surfperch	R	J, A
	Embiotoca lateralis	striped surfperch	M	A
	Hyperprosopon anale	spotfin surfperch	M	A
	Hyperprosopon argenteum	walleye surfperch	MI	J, A
	Hyperprosopon ellipticum	silver surfperch	M	A
	Hypsurus caryi	rainbow surfperch	M	A
	Micrometrus minimus	dwarf surfperch	MI	J, A
	Phanerodon furcatus	white surfperch	PR	J, A
	Rhacochilus toxotes	rubberlip surfperch	M	J, A
	Rhacochilus vacca	pile surfperch	MI	A
Engraulidae	*Engraulis mordax*	northern anchovy	MI	E, L, J, A
Gadidae	*Microgadus proximus*	Pacific tomcod	M	J, ?
Gasterosteidae	*Gasterosteus aculeatus*	threespine stickleback	F	J, A
Gobiidae	*Acanthogobius flavimanus* (NN)	yellowfin goby	R	J, A
	Clevelandia ios	arrow goby	R	L, J, A
	Eucyclogobius newberryi	tidewater goby	R	L, J, A
	Gillichthys mirabilis	longjaw mudsucker	R	L, J, A
	Lepidogobius lepidus	bay goby	R	L?, J?, A
	Rhinogobiops nicholsii	blackeye goby	M	L, J
	Tridentiger trigonocephalus ? (NN)	chameleon goby	M	?
Hexagrammidae	*Hexagrammos decagrammus*	kelp greenling	M	J, A
	Hexagrammos lagocephalus ?	rock greenling	M	?
	Ophiodon elongatus	lingcod	M	J, A
	Oxylebius pictus	painted greenling	M	J, A
Kyphosidae	*Girella nigricans*	opaleye	M	A
Labridae	*Oxyjulis californica*	señorita	M	L, A
Mugilidae	*Mugil cephalus*	striped mullet	MI	A
Myctophidae	*Stenobrachius leucopsarus*	northern lampfish	M	L
Myliobatidae	*Myliobatis californica*	bat ray	PR	J, A
Ophichthidae	*Ophichthus triserialis*	Pacific snake eel	M	A
Ophidiidae	*Chilara taylori*	spotted cusk-eel	M	J, A
Osmeridae	*Hypomesus pretiosus*	surf smelt	M	L, A
	Spirinchus starksi	night smelt	M	A

[1] *Liparis florae* would be expected.

Family	Species	Common Name	Life style	Life stage
Percichthyidae	*Morone saxatilis* (NN)	striped bass	F	J, A
Pleuronectidae	*Eopsetta exilis*	slender sole	M	L, A
	Hypsopsetta guttulata	diamond turbot	MI	J, A
	Platichthys stellatus	starry flounder	MI	L, J, A
	Parophrys vetulus	English sole	MI	J
	Pleuronichthys coenosus	C-O sole	M	E
	Pleuronichthys decurrens	curlfin turbot	M	E, J, A
	Pleuronichthys ritteri ?	spotted turbot	M	J, A
	Pleuronichthys verticalis	hornyhead turbot	M	E, L
	Psettichthys melanostictus	sand sole	M	L, J, A
Poeciliidae	*Gambusia affinis* (NN)	western mosquitofish	F	L, J, A
Rhinobatidae	*Platyrhinoidis triseriata*	thornback	M	A
	Rhinobatos productus	shovelnose guitarfish	M	A
Salmonidae	*Oncorhynchus mykiss*	rainbow trout (steelhead)	M	A
	Oncorhynchus tshawytscha	chinook salmon	M	A
Sciaenidae	*Seriphus politus*	queenfish	MI	J, A
	Genyonemus lineatus	white croaker	M	E, L, A
Scorpaenidae	*Sebastes atrovirens*	kelp rockfish	M	J
	Sebastes auriculatus	brown rockfish	M	J, A ?
	Sebastes carnatus	gopher rockfish	M	J
	Sebastes caurinus	copper rockfish	M	J
	Sebastes chrysomelas	black-and-yellow rockfish	M	J
	Sebastes dallii	calico rockfish	M	J
	Sebastes flavidus [2]	yellowtail rockfish	M	J
	Sebastes melanops	black rockfish	M	J
	Sebastes mystinus	blue rockfish	M	J
	Sebastes paucispinis	bocaccio	M	L, J
	Sebastes rastrelliger	grass rockfish	M	J, A
	Sebastes saxicola	stripetail rockfish	M	J
Soleidae	*Symphurus atricauda*	California tonguefish	MI	E, J, A
Squalidae	*Squalus acanthias*	spiny dogfish	M	A
Stichaeidae	*Cebidichthys violaceus*	monkeyface prickleback	M	L, A
Stromateidae	*Peprilus simillimus*	Pacific pompano	M	A
Syngnathidae	*Syngnathus californiensis*	kelp pipefish	M	A
	Syngnathus exilis	barcheek pipefish	M	A
	Syngnathus leptorhynchus	bay pipefish	R	L, J, A
Synodontidae	*Synodus lucioceps*	California lizardfish	M	J, A
Torpedinidae	*Torpedo californica*	Pacific electric ray	M	A
Urolophidae	*Urolophus halleri*	round stingray	M	J, A

[2] Difficult to distinguish from *Sebastes serranoides*.

References

ABA Consultants. 1989. Elkhorn Slough wetland management plan. Prepared for California State Coastal Conservancy and Monterey County Planning Department.

Allen, L. G. 1982. Seasonal abundance, composition, and productivity of the littoral fish assemblage in upper Newport Bay, California. *Fishery Bulletin U.S.* 80: 769–790.

Allen, L. G., and M. H. Horn. 1975. Abundance, diversity, and seasonality of fishes in Colorado Lagoon, Alamitos Bay, California. *Estuarine and Coastal Marine Science* 3: 371–380.

Ambrose, D. A. 1976. The distribution, abundance, and feeding ecology of four species of flatfish in the vicinity of Elkhorn Slough, California. M.A. thesis, Moss Landing Marine Laboratories. San Jose State University.

Antrim, B. S. 1981. Habitat and food resource utilization of three species of embiotocids in Elkhorn Slough, Monterey Bay, California. M.A. thesis, Moss Landing Marine Laboratories. California State University, Fresno.

Appiah, J. C. 1977. Species composition and abundance of fish populations in Bennett Slough, Monterey Bay, California. M.A. thesis, Moss Landing Marine Laboratories. California State University, Fresno.

Barry, J. P. 1983. Utilization of shallow marsh habitats by fishes in Elkhorn Slough, California. M.A. thesis, Moss Landing Marine Laboratories. San Jose State University.

Barry, J. P., and G. M. Cailliet. 1981. The utilization of shallow marsh habitats by commercially important fishes in Elkhorn Slough, California. *Cal-Neva Wildlife Transactions*, 38–47.

Barry, J. P., C. H. Baxter, R. D. Sagarin, and S. E. Gilman. 1995. Climate-related, long-term faunal changes in a California rocky intertidal community. *Science* 267: 672–675.

Barry, J. P., M. M. Yoklavich, G. M. Cailliet, D. A. Ambrose, and B. S. Antrim. 1996. Trophic ecology of the dominant fishes in Elkhorn Slough, California, 1974–1980. *Estuaries* 19: 115–138.

Blaber, S. J. M., D. T. Brewer, and J. P. Salini. 1989. Species composition and biomasses of fishes in different habitats of a tropical northern Australian estuary: Their occurrence in the adjoining sea and estuarine dependence. *Estuarine, Coastal, and Shelf Science* 29: 509–531.

Browning, B. M. 1972. *The Natural Resources of Elkhorn Slough: Their Present and Future Use.* California Department of Fish and Game, Coastal Wetlands Series 4.

Cailliet, G. M. 1992. Demography of the central California population of leopard shark (*Triakis semifasciata*). *Australian Journal of Marine and Freshwater Research* 43: 183–193.

Cailliet, G. M., B. Antrim, D. Ambrose, S. Pace, and M. Stevenson. 1977. Species composition, abundance, and ecological studies of fishes, larval fishes, and zooplankton in Elkhorn Slough. In J. Nybakken, G. Cailliet, and W. Broenkow, *Ecologic and Hydrographic Studies of Elkhorn Slough Moss Landing Harbor and Nearshore Coastal Waters, July 1974–June 1976*, pp. 216–386. Moss Landing, CA: Moss Landing Marine Laboratories Technical Publication.

Crampton, T. A. 1994. Long-term effects of Moss Landing Harbor on the wetlands of Elkhorn Slough. M.S. thesis, University of California, Santa Cruz.

Dietz, S. A., W. Hildebrandt, and T. Jones. 1988. *Archaeological Investigations at Elkhorn Slough: CA-MNT-229, a Middle Period Site on the Central California Coast.* Papers in Northern California Anthropology No. 3. Northern California Anthropological Research Group, Berkeley.

Eldridge, M. B., and Bryan, C. R. 1972. *Larval Fish Survey of Humboldt Bay, California.* National Oceanic and Atmospheric Administration Technical Report, National Marine Fisheries Service (Special Scientific Report, Fisheries, 665).

Elkhorn Slough Estuarine Sanctuary Advisory Committee. 1985. Elkhorn Slough National Estuarine Sanctuary Management Plan. California Department of Fish and Game, and Sanctuary Programs Division, NOAA.

Gobalet, K. W. 1990. Prehistoric status of freshwater fishes of the Pajaro-Salinas River system of California. *Copeia* 1990: 680–685.

———. 1993. Additional archaeological evidence for endemic fishes of California's Central Valley in the coastal Pajaro-Salinas basin. *Southwestern Naturalist* 38: 218–223.

Gordon, B. L. 1985. *Monterey Bay Area: Natural History and Cultural Imprints.* 2d ed. Pacific Grove, Calif.: Boxwood Press.

Gunderson, D. R., D. A. Armstrong, Y.-B. Shi, and R. A. McConnaughey. 1990. Patterns of estuarine use by juvenile English sole (*Parophrys vetulus*) and Dungeness crab (*Cancer magister*). *Estuaries* 13: 59–71.

Herald, E. S. 1953. The 1952 shark derbies at Elkhorn Slough and Coyote Point, San Francisco Bay. *California Fish and Game* 39: 237–243.

Herald, E. S., and R. P. Dempster. 1952. The 1951 shark derby at Elkhorn Slough, California. *California Fish and Game* 38: 133–134.

Herald, E. S., W. Schneebeli, N. Green, and K. Innes. 1960. Catch records for 17 shark derbies held at Elkhorn Slough, Monterey Bay, California. *California Fish and Game* 46: 59–67.

Hoff J. G., and R. M. Ibara. 1977. Factors affecting the seasonal abundance, composition, and diversity of fishes in a southeastern New England estuary. *Estuarine and Coastal Marine Science* 5: 665–678.

Horn, M. H., and Allen, L. G. 1985. Fish community ecology in southern California bays and estuaries. In A. Yanez-Arancibia, ed., *Fish Community Ecology in Estuaries and Coastal Lagoons: Towards an Ecosystem Integration*, pp. 169–190. Mexico City: Universidad Nacional Autonoma de Mexico.

Kao, J. S. 2000. Diet, daily ration and gastric evacuation of the leopard shark (*Triakis semifasiata*). M.S. thesis, Moss Landing Marine Laboratories. California State University, Hayward.

King, A. E., and G. M. Cailliet. 1992. The Elkhorn Slough shark derby: Past and present. *Chondros* 3: 1–4.

Krygier, E. E., and W. G. Pearcy. 1986. The role of estuarine and offshore nursery areas for young English sole, *Parophrys vetulus* Girard, off Oregon. *Fishery Bulletin U.S.* 84: 119–132.

Kukowski, G. E. 1972. *A Checklist of the Fishes of the Monterey Bay Area Including Elkhorn Slough, the San Lorenzo, Pajaro, and Salinas Rivers.* Moss Landing Marine Laboratories, Annual Report, Part 2. Technical Publication No. 72-2.

Kusher, D. I. 1987. Age and growth of the leopard shark, *Triakis semifasciata*, from central California. M.A. thesis, Moss Landing Marine Laboratories. San Francisco State University.

Kusher, D. I., S. E. Smith, and G. M. Cailliet. 1992. Validated age and growth of the leopard shark, *Triakis semifasciata*, with comments on reproduction. *Environmental Biology of Fishes* 35: 187–203.

Leithiser, R. M. 1981. Distribution and seasonal abundance of larval fishes in a pristine southern California salt marsh. *Rapports et procès-verbaux des réunions, Conseil permanent international pour l'exploration de la mer* 178: 174–175.

Lindquist, D. C. 1998. The effects of erosion on the trophic ecology of fishes in Elkhorn Slough, CA. M.S. thesis, Moss Landing Marine Laboratories. California State University, Hayward.

Loneragan, N. R., I. C. Potter, R. C. J. Lenanton, and N. Caputi. 1986. Spatial and seasonal differences in the fish fauna in the shallows of a large Australian estuary. *Marine Biology* 92: 575–586.

MacGinitie, G. E. 1935. Ecological aspects of a California marine estuary. *American Midland Naturalist* 16: 629–765.

Malzone, C., and R. Kvitek. 1994. Tidal scour, erosion, and habitat loss in Elkhorn Slough, California. A report of the Elkhorn Slough Foundation pursuant to National Oceanic and Atmospheric Administration Award #NA37OM0523.

Martin, L. K. 1982. Growth and reproduction of the bat ray, *Myliobatis californica* Gill, in California. M.A. thesis, Moss Landing Marine Laboratories. San Jose State University.

Martin, L. K. and G. M. Cailliet. 1988a. Aspects of reproduction of the bat ray, *Myliobatis californica*, in central California, USA. *Copeia* 1988: 754–762.

———. 1988b. Age and growth determination of the bat ray, *Myliobatis californica* Gill. *Copeia* 1988: 762–773.

McGowen, G. E. 1977. Ichthyoplankton populations in south San Diego Bay and related effects of an electricity generating station. M.S. thesis, San Diego State University.

Nordby, C. S. 1982. The comparative ecology of ichthyoplankton within Tijuana Estuary and in adjacent nearshore waters. M.S. thesis, San Diego State University.

Onuf, C. P., and M. L. Quammen. 1983. Fishes in a California coastal lagoon: Effects of major storms on distribution and abundance. *Marine Ecology Progress Series* 12: 1–14.

Oxman, D. S. 1995. Seasonal abundance, movements, and food habits of harbor seals (*Phoca vitulina*) in Elkhorn Slough, California. M.S. thesis, Moss Landing Marine Laboratories. California State University, Stanislaus.

Pacific Gas and Electric Co. 1983. Cooling water intake structures 316(B) demonstration. Report on file at Moss Landing Duke Power Plant, California.

Pearcy, W. G., and S. S. Myers. 1974. Larval fishes of Yaquina Bay, Oregon: A nursery ground for marine fishes? *Fishery Bulletin* 74: 201–213.

Prasad, R. R. 1948. The life history of *Clevelandia ios* (Jordan and Gilbert). Ph.D. diss., Stanford University.

Robins, C. R., R. M. Bailey, C. E. Bond, J. R. Brooker, E. A. Lachner, R. N. Lea, and W. B. Scott. 1991. *Common and Scientific Names of Fishes from the United States and Canada.* 5th ed. American Fisheries Society Special Publications No. 20.

Roemmich, D., and J. McGowan. 1995. Climatic warming and the decline of zooplankton in the California Current. *Science* 267: 1324–1326.

Ruagh, A. A. 1976. Feeding habits of silversides (family Atherinidae) in Elkhorn Slough, Monterey Bay, California. M.A. thesis, Moss Landing Marine Laboratories. California State University, Fresno.

San Filippo, Richard A. 1995. Diet, gastric evacuation, and estimates of daily ration of the gray smoothhound, *Mustelus californicus.* M.S. thesis, Moss Landing Marine Laboratories. San Jose State University.

Small, D. 1986. Fish colonization of a newly restored salt marsh in the Elkhorn Slough National Estuarine Sanctuary. Final report to the Office of Sanctuary Programs, National Oceanic and Atmospheric Administration, Washington, D.C.

Talent, L. G. 1976. Food habits of the leopard shark, *Triakis semifasciata*, in Elkhorn Slough, Monterey Bay, California. *California Fish and Game* 62: 286–298.

———. 1982. Food habits of the gray smoothhound (*Mustelus californicus*), the brown smoothhound (*Mustelus henlei*), the shovelnose guitarfish (*Rhinobatos productus*), and the bat ray (*Myliobatis californica*), in Elkhorn Slough, California. *California Fish and Game* 68: 224–234.

———. 1985. The occurrence, seasonal distribution, and reproductive condition of elasmobranch fishes in Elkhorn Slough, CA, USA. *California Fish and Game* 71: 210–219.

Tenera Environmental Services. 2000. Moss Landing Power Plant Modernization Project 316(b) Resource Assessment. Prepared for Duke Energy Moss Landing.

Toole, C. L. 1980. Intertidal recruitment and feeding in relation to optimal utilization of nursery areas by juvenile English sole (*Parophrys vetulus*). *Environmental Biology of Fishes* 5: 383–390.

U.S. Department of Commerce. 1979. Final environmental impact statement: Elkhorn Slough National Estuarine Sanctuary. Office of Coastal Zone Management, National Marine Fisheries Service, Washington, D.C.; and Department of Fish and Game, Resources Agency, Sacramento, California.

Wang, J. C. S. 1981. Taxonomy of the early life stages of fishes—Fishes of the Sacramento–San Joaquin estuary and Moss Landing Harbor–Elkhorn Slough, California. Report prepared for Ecological Analysts, Inc., Concord, California.

Weisel G. F., Jr. 1947. Breeding behavior and early development of the mudsucker, a gobiid fish of California. *Copeia* 2: 77–85.

Williams, P., and Associates, and Moffatt and Nichol Engineers. 1992. Elkhorn Slough tidal hydraulics erosion study. Report prepared for U.S. Army Corps of Engineers, San Francisco District, San Francisco.

Yoklavich, M. M. 1982a. Growth, food consumption, and conversion efficiency of juvenile English sole (*Parophrys vetulus*). In G. M. Cailliet and C. A. Simenstad, eds., *Fish Food Habit Studies*, pp. 97–105. Proceedings of the Third Pacific Workshop, University of Washington, Washington Sea Grant.

———. 1982b. Energy requirements of the English sole (*Parophrys vetulus*). M.A. thesis, Moss Landing Marine Laboratories. San Francisco State University.

Yoklavich, M. M., G. M. Cailliet, J. Barry, D. Ambrose, and B. Antrim. 1991. Temporal and spatial patterns in abundance and diversity of fish assemblages in Elkhorn Slough, California. *Estuaries* 14: 465–480.

Yoklavich, M. M., M. Stevenson, and G. M. Cailliet. 1992. Seasonal and spatial patterns of ichthyoplankton abundance in Elkhorn Slough, California. *Estuarine, Coastal, and Shelf Science* 34: 109–126.

Yudin, K. G. 1987. Age, growth, and aspects of the reproductive biology of two sharks, the gray smoothhound *Mustelus californicus* and the brown smoothhound *M. henlei*, from central California. M.A. thesis, Moss Landing Marine Laboratories. San Francisco State University.

Yudin, K. G., and G. M. Cailliet. 1990. Age and growth of the gray smoothhound, *Mustelus californicus*, and the brown smoothhound, *M. henlei*, sharks from central California. *Copeia* 1990: 191–204.

11

CHAPTER ELEVEN

Birds & Mammals

James T. Harvey, Sarah Connors

The diversity of habitats within the Elkhorn Slough watershed largely explains the vast number of bird and mammal species that use this region as year-round residents or seasonal visitors. One may encounter denizens of channel waters, coastal dunes and beaches, intertidal mudflats, salt marsh, oak woodlands, and grasslands. The moderate central coast climate allows for a diverse assemblage of species throughout the year.

Elkhorn Slough is recognized as a Globally Important Bird Area by the American Bird Conservancy. More than 265 bird species (73% of the California total) have been recorded in the Elkhorn Slough area (Roberson 1991; appendix 11.1). Most are seasonal visitors, but approximately 40 are year-round residents. Aquatic birds—shorebirds, seabirds, herons, and waterfowl—account for much of the slough's avian diversity. Roberson (1991) lists 137 aquatic species, including 57 that are common at some time each year. Many of these species are migrants. In fact, as one of the largest estuaries in California, Elkhorn Slough is a major stopover for birds migrating along the Pacific flyway. More than 20,000 sandpipers, plovers, and their relatives may be present at the peak of migration (Senner and Howe 1984; Ramer, Page, and Yoklavich 1991; Page et al. 1992). A number of these aquatic species nest in the Elkhorn Slough watershed: Great Egrets, Great Blue Herons, and Double-crested Cormorants nest in a grove of pines and eucalyptus in the Elkhorn Slough National Estuarine Research Reserve (ESNERR); Caspian Terns nest on man-made islands on the ESNERR; and threatened Snowy Plovers nest at the salt ponds and on beaches.

Fifty-nine species of mammals are believed to occur in the Elkhorn Slough watershed, five of which are marine (Schafer 1986; appendix 11.2). The sea otter (*Enhydra lutris*) returned to its historic range in the slough in the early 1980s, and the number using the slough has increased gradually since then. Harbor seal (*Phoca vitulina*) populations have also increased in recent years. California sea lion (*Zalophus californianus*), harbor porpoise (*Phocoena phocoena*), and juvenile gray whale (*Eschrichtius robustus*) are sighted infrequently in the lower reaches of the slough (Richman 1997).

In this chapter, we describe bird and mammal species and communities in Elkhorn Slough's primary aquatic and terrestrial habitats.* Although each habitat supports a characteristic assemblage of species, birds and mammals are highly mobile and often move to different habitats in response to seasonal changes in environmental conditions, prey availability, and habitat needs. We also discuss the effects of pollutants, nonnative red foxes, and other human activities on these faunas. Finally, we present recommendations for future study and management of these highly visible members of the Elkhorn Slough biota.

* This chapter covers the coastline and nearshore waters, including the harbor, from Jetty Road to Sandholdt Road, and slough and upland habitats within the borders of Highway 1, Salinas Road, Elkhorn Road, and Dolan Road.

History of Study

Naturalists have observed birds and mammals in Elkhorn Slough for over a century, but there have been few long-term studies or systematic surveys of these populations. The first bird surveys, conducted by George MacGinitie from 1926 to 1935, documented species composition and described the natural history of the slough's more common aquatic species (MacGinitie 1935). Half a century later, Bruce M. Browning (1972) produced a more comprehensive list of the slough's aquatic birds as part of his natural resource survey for the California Department of Fish and Game. A formal checklist of the birds in Elkhorn Slough and adjacent habitats followed (Ramer, Ramer, and Warriner 1978) and was updated by Don Roberson (1991). This checklist, based primarily on sightings by local birdwatching groups, is currently being revised. A thorough survey of birds nesting in the Elkhorn Slough watershed was conducted in the early 1990s (Roberson and Tenney 1993).

A typical scene at Elkhorn Slough depicting some of the many birds and mammals that inhabit the slough's channels and mudflats. Among the species that forage, breed, and rest in and around the slough are harbor seals and sea otters, and more than 265 bird species, including pelicans, terns, cormorants, grebes, and shorebirds. Illustration credit: Carleton Eyster, from photos by Hannah Nevins.

Detailed studies, focused primarily on aquatic species, have become much more common in recent decades. In the late 1970s, Ramer (1985) documented seasonal abundance, habitat use, and diet of migratory shorebirds in Elkhorn Slough. Between 1988 and 1993, shorebirds were censused one to four times each year as part of a larger effort by the Point Reyes Bird Observatory to detail critical habitats and population changes throughout central California (Page et al. 1992). From early 1998 through the spring of 2000, graduate student Sarah Connors replicated Ramer's original surveys to determine whether shorebird abundance and distribution in Elkhorn Slough have changed during the past twenty years (unpubl. data, fig. 11.1). Researchers at Moss Landing Marine Laboratories (MLML) have studied habitat use, diet, and other aspects of the biology of harbor seals, sea otters, and nesting seabirds (Caspian Tern, Forster's Tern, and Western Gull). Since 1984, the slough's population of threatened Snowy Plovers has received considerable attention in order to direct conservation strategies, specifically by monitoring nesting success and predation on the beaches of Monterey Bay and in the salt ponds.

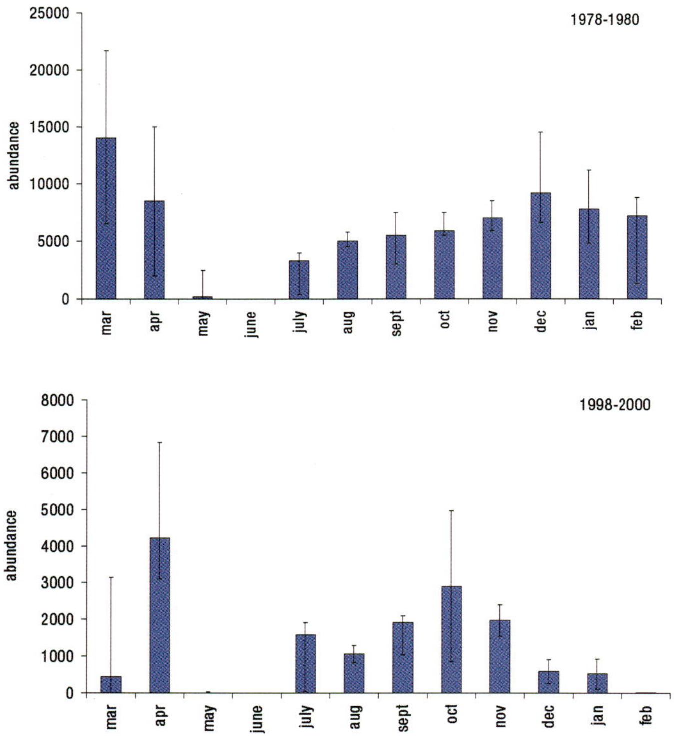

Figure 11.1. Abundance of Western Sandpipers at Elkhorn Slough from the mouth to Kirby Park in the late 1970s (Ramer 1985) and late 1990s (Connors, unpubl. data). The histogram shows median (bars), minimum, and maximum numbers per month. Note differences in y-axis values. Differences in abundance between studies may be due to changes in habitat availability in the slough and surrounding areas.

Terrestrial vertebrates in Elkhorn Slough have received much less attention than their aquatic counterparts. Land bird studies have included the breeding bird survey (Roberson and Tenney 1993) and an ongoing nest box project for cavity nesters, initiated on the ESNERR in 1992 (Thompson, unpubl. rpt.). Species composition, distribution, and habitat use of local terrestrial mammals are virtually unknown.

Aquatic Habitats

Channel, Harbor, and Coastal Waters

The deeper waters of the harbor and main slough channel attract seabirds and marine mammals from Monterey Bay. Most are visitors seeking food or shelter from rough ocean waters and predators. A few marine species enter Elkhorn Slough to bear young.

About 50 species of seabirds and 35 species of waterfowl have been observed in the slough and harbor or on adjacent bay waters and beaches. The more seagoing species typically remain near the mouth of the slough but sometimes follow the channel well inland. Others, including many of the ducks and geese, prefer calmer pond waters and only occasionally visit the main channel. About half of these species are seasonally common; the rest are seen infrequently, preferring either offshore waters or other latitudes. In fact, Moss Landing Harbor is well known as a spot to see species that are unusual in the region. Yellow-billed Loon, Red-necked Grebe, Harlequin Duck, Long-tailed Duck, and Black Skimmer are among the rare birds seen at the mouth of Elkhorn Slough.

Diving seabirds—heavy-bodied, underwater foragers such as cormorants, loons, grebes, and sea ducks—are common in the lower slough. Many probably visit the slough to rest or avoid stormy weather, but others take advantage of the rich prey resources. Most are seasonal visitors from breeding colonies on freshwater lakes farther north or inland. They are most numerous fall through spring, but some nonbreeding individuals may linger through summer. Species composition

Large flocks of shorebirds often can be seen during fall and winter, when they overwinter at the slough, and during migration in March and April. Photo credit: © Monterey Bay Aquarium Foundation.

varies from year to year, but there are some trends. Common Loons usually are more common than Pacific and Red-throated Loons. The larger Western and Clark's Grebes outnumber the smaller Horned and Eared Grebes. Surf and White-winged Scoters, the most common sea ducks, form sizable rafts (flocks of resting birds) near the slough mouth. Other diving ducks seen in the main channel include the Red-breasted Merganser, Common Goldeneye, Greater and Lesser Scaup, Bufflehead, and Ruddy Duck.

Numerous duck species, like this Cinnamon Teal, can be found in Elkhorn Slough's smaller channels and surrounding wetlands. Photo credit: © T. Rountree

In contrast to the aforementioned species, three cormorant species are year-round residents of Elkhorn Slough and Monterey Bay. Pelagic and Brandt's Cormorants, species that prefer open coast habitats, usually forage near the mouth of the slough and roost on rocks and pilings. Double-crested Cormorants feed and roost throughout the slough. Flocks rest and dry their wings on steep banks and mudflats. Actively feeding birds sometimes form lines along the channel edge to trap schools of fish.

Some seabirds, including the familiar Brown Pelicans and graceful terns, feed in the slough by aerial plunging; that is, they use gravity to plummet headfirst into the water to capture small fish below the surface. Brown Pelicans from nesting colonies farther south are common during summer and fall. Soaring on large wings, adult (white heads with gray bodies) and immature (brown with white bellies) pelicans cruise up and down the slough in search of prey. More than 1,000 pelicans also roost along the steep banks of the lower slough. Much more rare are American White Pelicans that occasionally visit the slough from inland colonies. Unlike their seagoing relatives, white pelicans do not plunge, but instead capture fish by scooping while swimming.

Plunging terns are a familiar sight in the main channel in all seasons except winter. Of the nine species recorded in Elkhorn Slough, three are seasonally common: the Caspian Tern and Forster's Tern, which nest in the slough area, and the Elegant Tern, which, like pelicans, visits from southerly breeding colonies during summer and fall. The larger Caspian Terns capture primarily shiner surfperch, smelt, Pacific butterfish, northern anchovy, Pacific herring, and staghorn sculpin in the main channel, whereas Forster's Terns usually catch threespine stickleback, northern anchovy, jacksmelt, and shiner surfperch and frequently feed in the smaller tidal sloughs (T. Harvey 1982; Parkin 1998). Although both terns feed on shiner surfperch and northern anchovy, the Caspian Tern primarily captures adult fish whereas the smaller Forster's Tern favors juveniles, illustrating resource partitioning between two species with similar foraging methods (Baltz, Morejohn, and Antrim 1979).

Gulls are common inhabitants of the slough, making their living by capturing near-surface prey, scavenging, or stealing from their neighbors. Fifteen species of gulls have been recorded (see appendix 11.1). Nine of these are fairly common during winter as visitors from inland and high-latitude colonies. Glaucous-winged, Thayer's, Herring, and Mew Gulls are usually restricted to coastal waters and beaches. California, Ring-billed, and Bonaparte's Gulls are more likely to venture up the slough. Heermann's Gulls are unique, nesting farther south and visiting the region in summer and fall, the same time as the pelicans they *kleptoparasitize*. The Western Gull, the only resident species, is common all year and often accompanies feeding otters, optimistically awaiting scraps of shellfish.

Only two species of marine mammals are consistently present in the channel waters of the slough: the harbor seal and the sea otter.

Harbor seals are year-round inhabitants of Elkhorn Slough. Hundreds of seals often haul out and rest on banks and mud flats in the lower reaches, especially around Seal Bend. Harbor seals occur solitarily or in small groups throughout the main channel. Although the harbor seal's diet is composed of many fishes commonly found in the slough, including Pacific staghorn sculpin, shiner surfperch, and topsmelt (Oxman 1995), many of the fishes consumed are larger than those which inhabit the slough; this evidence indicates that harbor seals feed primarily along the nearshore oceanic shelf

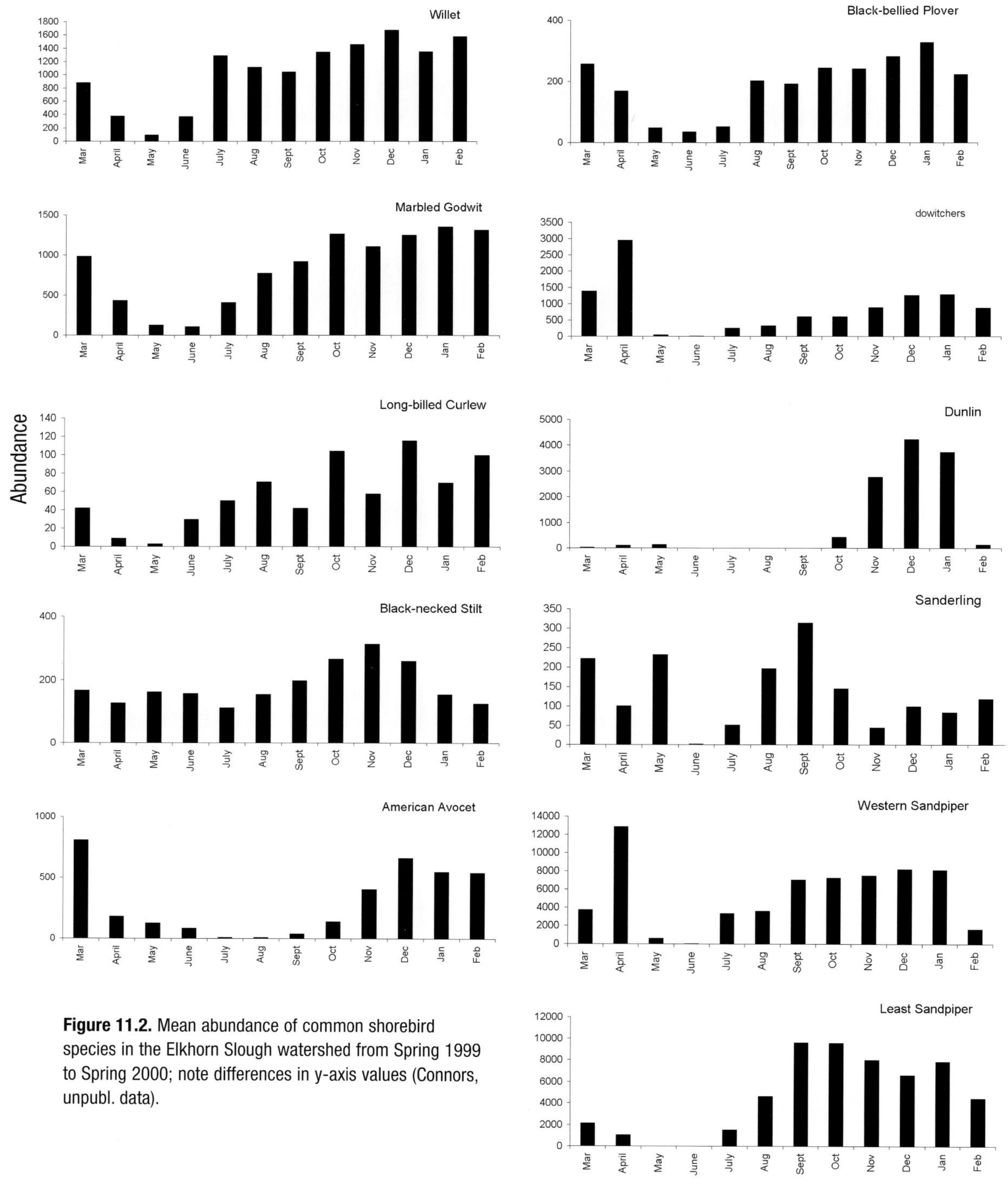

Figure 11.2. Mean abundance of common shorebird species in the Elkhorn Slough watershed from Spring 1999 to Spring 2000; note differences in y-axis values (Connors, unpubl. data).

outside the slough (Harvey, Helm, and Morejohn 1995). Thus, the primary importance of the slough for harbor seals may be as a place to rest, molt, and pup in safety.

Exact numbers of seals in the slough vary seasonally (Richman 1997) and with time of day (Oxman 1995). On a daily basis, harbor seals are most abundant in the daytime when individuals enter the slough to rest and remain until evening before departing for Monterey Bay to forage during the night (Oxman 1995; Eguchi 1998); seals begin to return to the slough in the morning, with numbers peaking in the afternoon (S. Oates, pers. comm.). Harbor seal abundance in the slough appears to be greatest from May through August, when pupping and molting take place. Since regular censuses

Harbor seals. Photo credit: Jim Harvey.

began in 1994, the highest single count was 339 seals in July 1997 (Richman 1997).

Harbor seals began to breed and give birth in Elkhorn Slough in 1989 (Osborn 1992), soon after the California Department of Fish and Game closed human access to the harbor seal haul-out sites near Seal Bend. Harbor seal pups are quite precocial, able to swim within hours of birth, so the actual number of pups born in the slough is not easily determined. The number of pups seen in the slough has increased steadily in the last ten years; at least 30 pups were present in 1999 (Harvey, unpubl. data). The reduction of human disturbance (e.g., fewer people walking along the dikes) has probably been the most important factor in the increased number of harbor seal births and pregnant females (Osborn 1985); females with young pups have recently established new haul-out sites in other parts of Elkhorn Slough (D. Greig, S. Oates, pers. comm.).

Sea otters are found throughout the slough but are observed most regularly foraging in the lower sections (Kvitek et al. 1988; Jolly 1997) or resting in rafts in the vicinity of Seal Bend. Otters form rafts of as many as 50 individuals, floating and resting close to each other, perhaps gaining an advantage in detecting potential predators. Individuals occasionally come ashore along the banks of the slough, a common practice in Alaska but unusual in California. Some individuals apparently leave the slough during midday, enter the harbor or ocean, and return later in the afternoon (Richman 1997; T. Darcey, pers. comm.). Most of the otters in the slough appear to be young males. In 1995, males accounted for 91% of the slough's otters, and 83% of these appeared to be juveniles (Feinholz 1998). This skewed sex ratio may explain why, since 1994, only two otter pups have been observed in the main channel (T. Kieckhaefer, pers. comm.). However, since 1997, 8 of 12 rehabilitated sea otters released in the slough by the Monterey Bay Aquarium were females (M. Staedler, pers. comm.); if these individuals remain in the region, the sex ratio may eventually become more balanced.

Sea otters returned to their historic range in Elkhorn Slough in the early 1980s, but their numbers remained low until the mid-1990s. The first systematic counts of the entire slough in July 1994 averaged only 5 otters per survey, but by July 1997 the average had increased dramatically to 43 individuals (Richman 1997). Reasons for the increase are unclear, but the relatively protected slough may provide a safe haven from rough weather and white sharks as well as plentiful prey, such as clams and crabs. Sea otters are currently counted weekly during wildlife tours guided by Yohn Gideon and by personnel from the Monterey Bay Aquarium. We hope that this effort will continue so that we have a continuous record of the number of sea otters using Elkhorn Slough.

In Elkhorn Slough, sea otters eat mostly clams, mussels, and other bivalves (Kvitek et al. 1988; Jolly 1997). Considering that each otter consumes prey at the rate of approximately 25% of its body weight per day (K. Mayer, pers. comm.), there has been concern that otters might be affecting bivalve and other invertebrate populations in the slough. Initially, the arrival of otters in the mid-1980s had no obvious effects (Kvitek et al. 1988), but by 1995, after ten years of otter occupancy and rising numbers, bivalve abundance, distribution, and average size had declined significantly (Jolly 1997).

Beaches

Although exposed to high winds, wave action, and human disturbance, the outer beaches provide seasonal feeding habitat for shorebirds and roosting sites for gulls and terns. Gulls roost on beaches throughout the year; during the nonbreeding season (late summer through winter) mixed-

Sea otter. Photo credit: Jim Harvey.

species roosting aggregations may total thousands of birds. Species composition varies seasonally. During fall, Heermann's Gulls, Elegant and other terns, and sometimes Brown Pelicans predominate. These species usually head south by December and are replaced by California, Glaucous-winged, and other more northern-nesting gull species.

The most noticeable shorebirds on coastal beaches are the Sanderlings. During the winter season, large flocks of these small, energetic birds scurry along the water's edge, staying just out of reach of incoming waves, actively probing in the wet sand for small crustaceans, primarily sandcrabs (*Emerita analoga*) and isopods (*Excirolana* spp.; Estelle 1991). Also common are Willets, Marbled Godwits, and Whimbrels, probers with sensory cells at the tip of their bills for detecting movement of prey, and Black-bellied and Snowy Plovers, large-eyed visual hunters that detect and capture invertebrate prey on the sand surface.

The upper beach (above the high tide line) provides important nesting habitat for the Snowy Plover, a tiny sand-colored shorebird whose West Coast population was federally listed as threatened in 1993. Snowy Plovers have historically nested on undisturbed beaches, habitat that is becoming scarce along the Pacific coast. The nest, a modest scrape in the sand often adorned with small shell fragments and situated near a tuft of vegetation or piece of driftwood, is virtually invisible to the untrained eye. The Point Reyes Bird Observatory has closely monitored the breeding population on Monterey Bay since 1984. Exposed ground nests are highly vulnerable to trampling and disturbance by humans and dogs as well as to predation by native and nonnative animals including red fox, feral cats, skunks, American Kestrels, Common Ravens, Northern Harriers, and Loggerhead Shrikes. Introduced red foxes (*Vulpes fulva*), which were initially brought to California for hunts and fur farms, are now resident throughout coastal central California. In recent decades, their populations have grown and reduced or eliminated many populations of ground-nesting birds.

Snowy Plover. Photo credit: Thomas Rountree.

Between 1985 and 1990, increased Snowy Plover nest loss on Monterey Bay was attributed to red fox predation. Consequently, in 1991, biologists began constructing wire fencing around nests in an effort to reduce predation. By 1993, most nests on beaches were protected by these exclosures. In 1994, the U.S. Fish and Wildlife Service initiated red fox removal in select regions of the Monterey Bay. Fewer exclosures are needed in areas where fox control has been effective and human disturbance is minimal. Although successful in keeping out mammalian predators, exclosures may be used as perches by avian predators such as shrikes, kestrels, and ravens, which prey on the precocial chicks as they leave the security of the exclosure within hours of hatching. The future of this delicate bird at Elkhorn Slough and throughout its Pacific coast range remains tenuous.

Mudflats

Mudflats are areas of little or no vegetation, exposed during low tides and typically bordered by salt marsh vegetation on one side and channel waters on the other. At Elkhorn Slough, they provide safe haul-out and roost sites for harbor seals and seabirds and are especially important as feeding areas for waders and shorebirds (Ramer 1985).

Elkhorn Slough, recently identified as a critically important shorebird site (one of forty-six officially recognized sites in seven countries) by the Western Hemisphere Shorebird Reserve Network (WHSRN), is an essential link among several migration stopovers along the Pacific flyway. Thirty-eight species of shorebirds—sandpipers, plovers, and their relatives—have been recorded in Elkhorn Slough (Senner and Howe 1984; Ramer, Page, and Yoklavich 1991; Page et al. 1992; see appendix 11.1). From 1977 to 1980, the most abundant shorebird species along the main channel (in descending order of abundance) were the Western Sandpiper, Dunlin, Least Sandpiper, Short-billed and Long-billed Dowitchers, Marbled Godwit, American Avocet, Willet, Black-bellied Plover, Long-

Marbled Godwit with a just-captured clam in its bill. Photo credit: Thomas Rountree.

billed Curlew, and Sanderling (Ramer 1985). During that period, the most abundant species, the Western Sandpiper, accounted for more than 75% of the shorebirds in the slough in all months except May and June. Recently, Western and Least Sandpipers generally made up 60–80% of all shorebirds in the slough watershed, with number of Least Sandpipers often exceeding that of Western Sandpipers (Connors, unpubl. data).

Species diversity and abundance vary seasonally. Most of the shorebird species are visitors from distant breeding colonies at higher latitudes or inland and thus are present only seasonally. Seasonal movement patterns and peaks in abundance vary considerably among species. Some species, such as Long-billed and Short-billed Dowitchers and Dunlin, are most numerous during migration as they stop to feed and rest on their way south to wintering areas or north to breeding areas. Others, including Willets, Marbled Godwits, Black-bellied Plovers, and Western Sandpipers, are winter residents in the slough, arriving in fall and staying until early spring when conditions on breeding grounds become suitable for nesting. Only four (Snowy Plover, Killdeer, Black-necked Stilt, and American Avocet) are considered permanent residents. Total shorebird abundance is greatest during spring and fall, sometimes exceeding 20,000 birds at the peak of migration, and lowest in late spring and early summer (fig. 11.2; Senner and Howe 1984; Ramer, Page, and Yoklavich 1991; Page et al. 1992).

Due to their active lifestyles and high metabolic rates, shorebirds require reliable, abundant food resources, particularly during migration (Schneider and Harrington 1981; Senner and Howe 1984). The Elkhorn Slough mudflats provide the most important feeding areas for shorebirds in this region (Ramer 1985). Mudflats harbor vast assemblages of invertebrates and often attract and sustain large shorebird populations. The arrival and departure of migrating shorebirds can be highly synchronized with the annual breeding cycle of the invertebrates (Harrington 1983; Myers et al. 1990), but this relationship has not been examined at Elkhorn Slough.

In Elkhorn Slough, shorebirds feed primarily on a wide variety of benthic invertebrates, including various copepods, clams, polychaetes, and crabs (Ramer 1985). The distribution of the many different prey species is partly a function of mudflat substrate. Mudflats are mostly sand near the slough entrance but become increasingly "muddier" (composed of finer-grained silt) farther up the main channel (Nybakken and Jong 1977). Therefore, different types of shorebirds can be seen as you venture up the slough. Ramer (1985) found that smaller shorebirds (e.g., Western Sandpiper, Least Sandpiper, and Dunlin) occurred in greater densities along the upper reaches of the slough, where they ate the smaller invertebrates associated with muddier habitats. Larger shorebirds (e.g., Marbled Godwit and Long-billed Curlew), which eat larger sand-dwelling prey, were more abundant near the mouth of the slough.

In spring and summer, Great Blue Herons and Great Egrets nest in a Monterey pine and eucalyptus grove at the Elkhorn Slough Natural Estuarine Research Reserve.
Photo credit: Paul Zaretsky.

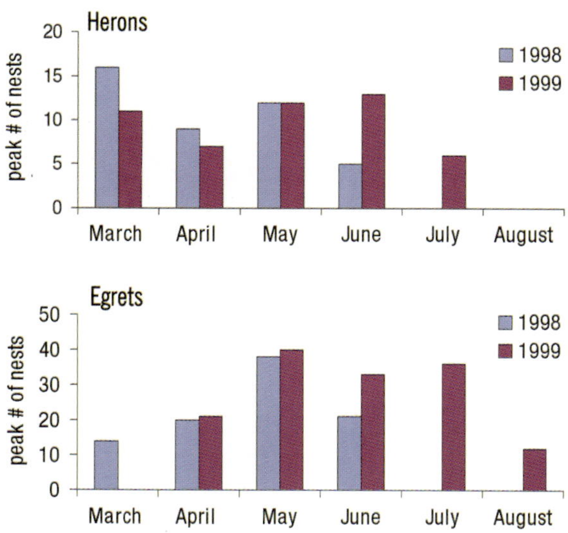

Figure 11.3. Peak numbers of Great Blue Heron and Great Egret nests at the ESNERR rookery in 1998 and 1999 (note differences in y-axis values).

Inter- and intraspecific competition for prey may be significant for shorebirds. Species or age classes can minimize competition by migrating at different times or along different routes, or by feeding in different places within a region. During spring migration, the peak of Least Sandpiper migration occurs before the peak of a morphologically similar species, the Western Sandpiper, thus reducing competition for food during this critical time (fig 11.2; Connors, unpubl. data). Competition among shorebird species that co-occur and feed together is reduced by differences in bill design and leg length, allowing different species to utilize different microhabitats (Green 1968); species with longer bills and longer legs, such as the Long-billed Curlew and the Marbled Godwit, can probe deeper in the substrate and feed in deeper water than smaller, shorter-billed species like the Least Sandpiper.

Shorebird distribution in Elkhorn Slough varies throughout the day in response to changes in tide level. In mudflats and other tidal habitats, food availability is intermittent. As tide level recedes and more mudflats are exposed, more prey become accessible. Shorebirds coordinate their daily movements in order to maximize foraging opportunities when resources are available at low tides. Available foraging space is believed to be the most significant factor in determining the size of migratory shorebird populations (Recher 1966). Tidally restricted mudflats adjacent to the main channel of the slough offer excellent opportunities for habitat enhancement. With proper water level management, these regions can provide alternate foraging and roosting areas for shorebirds during high tides when most other mudflat regions are flooded. These regions, such as the salt ponds and the marshes along Elkhorn Road, are important during migration when birds must maximize food intake (Connors, unpubl. data).

Herons, egrets, and other large wading birds also forage along the edges of Elkhorn Slough's mudflats. They do not feed by probing but instead use long legs and beaks to wade and catch small fish in the shallow waters along the slough channel (Byrnes 1997). Of the 9 such species recorded in the slough, only the Great Blue Heron, Great Egret, and Snowy Egret are common throughout the year. These aquatic birds also feed in ponds, marshes, and fields and roost and nest in trees.
In 1987, for unknown reasons, Great Egrets and Great Blue Herons established a nesting colony in the ESNERR. Pairs of both species built their large stick nests in a stand of Monterey pine and eucalyptus trees near the main channel of the slough.

In 1993, 27 pairs of Great Blue Herons and 61 pairs of Great Egrets nested in the reserve from March through June (Weed 1993a,b). In recent years, the efforts of dedicated docents have provided the reserve with valuable information on nesting activities in the rookery (fig. 11.3). In 1998, the greatest number of Great Blue Heron nests was recorded early in the season, with 16 nests counted in mid-March. Great Egret nesting activity appeared to peak in early May, with a total of 38 nests found. In 1999, heron nesting peaked in June with a high count of 13 nests, and egret nesting peaked in May with a count of 40 nests, with a large number of nests still active until the end of July. Double-crested Cormorants have also maintained a presence at the rookery since they were first observed nesting there in 1997 (A. Baldridge, pers. comm.). The number of cormorants appears to be increasing in the rookery, which may lead to competition for nest sites among the egrets, herons, and cormorants (ESNERR, unpubl. data). The entire colony may be short-lived, as guano from nesting birds appears to be killing the trees (M. Silberstein, pers. comm.).

Tidal [Salt and Brackish] Marshes

Salt marshes are tidal areas where saline-tolerant pickleweed (*Salicornia virginica*) predominates. Lower zones are flooded by tidal waters, although less often than mudflats. Upper zones usually stay dry except during the highest tides. Thus, salt marshes support wading birds but also provide breeding habitat for less aquatic birds and mammals.

Salt marshes are important to wading birds, particularly during higher tides when mudflats are unavailable. Waders and larger shorebirds, such as egrets, Great Blue Herons, Willets, Marbled Godwits, and Long-billed Curlews, roost and forage for polychaete worms, crabs, and fish. During exceptionally high tides when the uppermost sections of salt marsh are submerged these birds move to salt ponds, river mouths, and coastal beaches.

Since the Moss Landing Harbor was constructed in 1946, creating a direct channel between Elkhorn Slough and Monterey Bay, salt marsh vegetation has deteriorated (especially in the upper slough) and been replaced by mudflat (Crampton 1994; Lowe 1999). Conversion to mudflat has created additional foraging habitat for shorebirds, but the value may be outweighed by the loss of roosting areas for shorebirds and nesting habitat for marsh-dependent species. Little is known about these species at Elkhorn Slough. One such species, the elusive California Clapper Rail, previously sustained a breeding

An egret stabs the water in pursuit of prey. Photo credit: Thomas Rountree.

population in the slough's salt marshes. These shy waders relied on passageways through dense vegetation to come and go between nests built above tide level in the higher zones and feeding sites in the soft mud of the lower marsh. Clapper Rails were last seen in the slough in 1980 (Roberson 1993a). The exact cause of extirpation is unknown, but loss of salt marsh vegetation, contaminants, and predation by nonnative red foxes are likely factors (Thelander and Crabtree 1994).

Today, after years of tidal inflow and saltwater intrusion percolating through the substrate, the majority of the slough's marshes are salty (see chapter 4, "Hydrography"). In the nineteenth century, when the mouth of the Salinas River was located north of Moss Landing, the marshes in Elkhorn Slough were primarily fresh or brackish water. Although there are few historical records, these marshes undoubtedly supported quite different flora and fauna. In recent years, managers have begun to restore certain marshes to their natural seasonally variable condition in hope of reviving these lost ecosystems. For example, in 1995 tide gates at the head of the slough were repaired in order to prevent saltwater from the main channel from flowing into Porter Marsh, a pickleweed marsh that had once been a seasonal freshwater/brackish wetland and pasture. Now, this marsh is regaining its seasonal character. During winter storms, the marsh floods with rainwater and flow from Carneros and Watsonville Creeks; this freshwater completely covers the underlying salt marsh vegetation, and the area fills immediately with visiting ducks and geese that remain until the rainwater drains out into the slough. When spring arrives, a Northern Harrier pair builds a nest in the salt marsh and hunts the marsh and surrounding grasslands for rodents and small birds to feed their young. The harriers have fledged three young each year from 1997 to 1999. Pairs of White-tailed Kites, Red-shouldered Hawks, and Green Herons have begun to nest in adjacent stands of trees. Spring also brings Common Yellowthroats, Orange-crowned Warblers, Black-headed Grosbeaks, and other bird species to the lush riparian vegetation that borders the marsh to build their nests. In the fall, the dry pickleweed turns a brilliant reddish purple, and provides forage and cover for returning migrants.

Salt Ponds

Just north of the slough entrance is a series of ponds formed by levees and characterized by relatively low tidal flow. From 1916 until 1973 these were used as evaporation ponds for salt mining (Gordon 1996). Today, the area is managed for wildlife by the California Department of Fish and Game, and public access is limited.

Brown Pelican. Photo credit: Jim Harvey.

The salt ponds provide valuable habitat for birds. Even during the period of salt mining, the ponds supported numerous avian visitors. Slightly higher in elevation than the slough's mudflats, the salt ponds serve as an important refuge when high tides inundate the rest of Elkhorn Slough (Strong 1990). Currently, the ponds offer important feeding, roosting, and nesting habitat for a wide variety of shorebirds and seabirds.

When flooded, the salt ponds become inaccessible to mammalian predators and offer a safe roosting site for the endangered California Brown Pelican. Brown pelicans disperse to feed during the day but return each evening and congregate in large communal roosts in places that are safe from predation and disturbance (Jaques and Anderson 1988). On a summer or fall night, as many as 5,000 pelicans may roost on the salt pond levees or the dikes along the main slough channel (Briggs et al. 1983).

The salt ponds are especially important as a nesting site for the Snowy Plover. Management efforts such as limiting human access and controlling predators have contributed to increased productivity. Furthermore, since 1995, Point Reyes Bird Observatory biologists have carefully manipulated the amount of tidal water entering the salt ponds during the breeding season to create a mosaic of nesting and foraging habitat for the plovers. This successful management strategy has led to greater numbers of nests and higher reproductive success. For instance, during spring and summer 1997, 67 pairs of Snowy Plover nested in the salt ponds, representing about one-third of all Snowy Plover nests in the Monterey region. Especially high hatching rates (88%), greater than in any other region in Monterey Bay, produced 168 chicks and 82 fledglings (Page et al. 1997).

Salt pond management efforts have also benefited other nesting waders such as Black-necked Stilts, American Avocets, and Killdeer. However, the breeding biology of these species has not been closely studied in this area.

Over the years, the salt ponds have also supported three species of nesting seabirds—Caspian Tern, Forster's Tern, and Western Gull. Caspian Terns were first recorded nesting in the salt ponds in 1970 (Baldridge, Chandik, and DeSante 1970) and bred intermittently until 1978, by which time the colony had grown to 85 nesting pairs. Sometime after 1980, for unknown reasons, most of these birds relocated to dredged islands in the ESNERR (Parkin 1998). One of the earliest records of Forster's Terns breeding along the Pacific coast is from Elkhorn Slough. Some unknown number of pairs nested on small, marsh islands in 1932 (Ray, unpubl. data, in Harvey 1982). Sometime later, these terns established a breeding colony on the remnant dikes bordering the salt ponds; from 1978 to 1980, 90 to 300 breeding adults were present (Harvey 1982). In the mid 1990s, approximately 15-20 Forster's Tern nests were identified in the salt ponds. However, all the nests were abandoned before hatching, although eggs were still intact (D. George, pers. comm). Since then, there has been no documentation of breeding Forster's Terns in the salt ponds.

Caspian Tern parent and chick. Photo credit: Thomas Rountree.

Western Gulls nest along the salt pond levees and on various artificial structures, but the Elkhorn Slough colony has received little systematic study. The number of nesting pairs declined significantly between 1979 and 1989, from 61 to 5 (Roberson and Tenney 1993), after levee breaks in 1982 increased tidal action and flooding in some ponds.

Dredged Islands

In 1983, several small islands were constructed on the ESNERR using dredged bottom sediments in an effort to create habitat for birds. Safe from disturbance and predation, these islands quickly became a popular roosting site for Brown Pelicans and Double-crested Cormorants, and a nest site for a colony of Caspian Terns.

In the late 1980s, Caspian Terns, presumably from the salt pond colony, began to nest on the dredge islands. In 1992 there were 90 nesting pairs (Bailey 1993). The colony increased to 110 pairs and approximately 134 chicks in 1993; 200 pairs produced 189 chicks (150 of these fledged) in 1994. Unexpectedly, in 1995 colony production plummeted: approximately 160 tern pairs nested and laid eggs, but only 14 chicks hatched and only half of these fledged (Parkin 1998). Subsequent analysis of tern eggshells and unhatched chicks revealed high levels of the pesticide residue DDE (a metabolite of DDT; Parkin 1998) and indicated that the colony failure had been caused by exposure to pollutants. Runoff from agricultural fields following heavy rainfall during the winter of 1995 had probably contaminated slough waters with DDE, which the terns picked up through the food chain. Although the use of DDT in the U.S. was banned in 1972, residues of this persistent pesticide continue to contaminate soil in the watershed (see chapter 13, "Land Use and Contaminants").

Pollutants have not been the only threat to the Caspian Tern colony. In 1996 Caspian Terns started to nest, but an unidentified mammalian predator swam to the island and destroyed all the eggs. The terns abandoned the colony and attempted to renest at the Salinas River mouth; however, the few chicks that hatched at the new site were eaten by an avian predator, probably an owl (J. Parkin, pers. comm.). Following unsuccessful nesting efforts at the Salinas River mouth in 1997 and 1998, Caspian Terns returned to nest on the islands in Elkhorn Slough in 1999 and 2000, but again abandoned the site each year after a predator disturbed the colony. The terns bred successfully in 2001 due in large measure to researchers' efforts to deter predators. The future of the Elkhorn Slough Caspian Tern colony remains uncertain until predator access can be controlled.

Terrestrial Habitats

Birds and mammals inhabiting the terrestrial habitats of the Elkhorn Slough watershed have received very little attention from the research community. The only extensive studies—the breeding bird survey conducted from 1988 to 1992 by Roberson and Tenney (1993) and an ongoing nest box project for cavity nesters—are summarized below. We also present general descriptions of the fauna in each habitat, based primarily on the opportunistic observations of local naturalists and birding organizations.

Breeding Bird Survey and Nest Box Project

Between 1988 and 1992, a team of observers conducted intensive bird surveys throughout all of Monterey County to determine the breeding distribution of birds (Roberson and Tenney 1993). This report provides ESNERR with solid baseline information of birds that breed on the reserve.

In 1992, a project was initiated on the reserve by Andrew Thompson, a professor from Santa Clara University, to provide artificial nest sites (nest boxes) for cavity-nesting birds in oak woodlands. The project remains active today, managed by volunteers who faithfully maintain the nest boxes, determine species use and occupancy, band chicks, and record fledging success. There are 151 nest boxes located in nine different woodlots on the reserve. Since 1998, the nest boxes have been closely monitored by many dedicated volunteers at the slough. The Chestnut-backed Chickadee has been the most common occupant of the boxes during this time, followed by the Oak

Titmouse. During the 1998–2000 seasons, the peak egg-laying period occurred during the first two weeks of April for both of these species. Other species infrequently using the nest boxes include the Tree Swallow and Bewick's Wren. Continued monitoring of nest box use provides the reserve with baseline information that may prove useful in assessing the health of this habitat over time.

Oak Woodland–Savannah–Grassland

The predominant native terrestrial habitat in the Elkhorn Slough region is a mix of oak woodland (dense stands of trees separated by small grassy areas), savannah (grasslands with scattered trees), and grassland (trees absent). The oak canopy serves as a valuable foraging habitat for insectivorous foliage gleaners, including resident Oak Titmice, Chestnut-backed Chickadees, Hutton's Vireos, and Bushtits; summer visitors such as Orange-crowned Warblers; and winter visitors such as

Young White-Tailed Kites roosting above Porter Marsh, with American Crows in the background.
Photo credit: Carleton Eyster.

Ruby-crowned Kinglets and Townsend's and Yellow-rumped Warblers. Fall's release of acorns provides a generous seasonal supply of food for the Western Scrub-Jay, Acorn Woodpecker, and California Quail, as well as mammalian denizens such as the western gray squirrel. Beneath the trees, California Towhees, Spotted Towhees, and Dark-eyed Juncos persistently search or scratch in the leaf litter for seeds and invertebrates.

The more open savannahs and grasslands provide seasonally abundant food for grazers, browsers, and seed eaters, but little cover. For this reason, many species feed in grasslands but use other habitats for resting and breeding. Common mammals in this habitat include California ground squirrel (*Spermophilus beecheyi*), California meadow mouse (*Peromyscus* spp.), California vole (*Microtus californicus*), valley pocket gopher (*Thomomys bottae*), mule deer (*Odocoileus hemionus*), and the omnivorous coyote (*Canis latrans*). Western Meadowlarks and Savannah Sparrows typically nest in dense grasses, but cattle grazing has substantially reduced suitable nesting habitat for these species (Roberson and Tenney 1993). Predators such as owls, Golden Eagles, White-tailed Kites, Red-tailed Hawks, and American Kestrels often hunt in open fields but return to oaks for roosting or nesting. In winter, large mixed flocks of sparrows (usually including Golden-crowned Sparrow, White-crowned Sparrow, Savannah Sparrow, and Song Sparrow), finches, and blackbirds forage in open habitat.

Seasonal shifts in habitat use appear to be common, especially for small mammals. For example, California voles primarily inhabit grasslands during winter and spring when grasses are green and lush; however, during summer voles leave the dry grassland and move into the greener, more succulent salt marsh vegetation (E. Harding, pers. comm.). In this season, the salt marsh may provide more food and protection from avian predators. In years when California vole density is especially high, the western harvest mouse may be forced out of salt marshes and into other habitats, probably due to insufficient resources for both species (Harding 2000).

Chaparral

Chaparral or coastal scrub habitat consists of dense stands of coyote brush (*Baccharis pilularis*), manzanita (*Arctostaphylos* spp.), black sage (*Salvia mellifera*), ceanothus (*Ceanothus* spp.), and other shrubs. Both seedeaters and insectivores such as California Quail, goldfinches, Song Sparrow, towhees, Bewick's Wren, Wrentit, and Bushtit flourish in this dense, shrubby habitat. The brush rabbit (*Sylvilagus bachmani*) is abundant in chaparral, where it finds sufficient cover and an abundant source of food in herbaceous and woody vegetation (Chapman, Hockman, and Edwards 1982). Chaparral stands sometimes merge with oak woodlands. Mule deer commonly inhabit this oak woodland/chaparral ecotone, the oaks providing cover and food and chaparral providing abundant browse material (Mackie, Hamlin, and Pac 1982).

Riparian Woodland

Riparian woodlands lining the streams and rivers of the Elkhorn Slough area are composed of willows (*Salix* spp.), alders, maples (*Acer* spp.), and various water-loving understory species. These narrow corridors of lush streamside vegetation attract many bird and mammal species, especially during the dry summer and fall months. Birds that nest in this habitat include the White-tailed Kite, American Robin, Pacific-slope Flycatcher, Wilson's Warbler, Common Yellowthroat, Black-headed Grosbeak, Purple Finch, and Swainson's Thrush. Mammalian denizens include the raccoon (*Procyon lotor*), opossum (*Didelphis virginiana*), and dusky-footed woodrat (*Neotoma fuscipes*), all nocturnal species that require abundant water and cover (Burt and Grossenheider 1976; Gardner 1982; Kaufmann 1982).

Some residents of riparian woodlands rely on bordering habitat to meet their energy requirements. For example, White-tailed Kites often nest in large trees within a riparian corridor but hunt for rodents in nearby grasslands (Roberson 1993b). Long-tailed weasels inhabit the riparian woodland but may move to neighboring grasslands to feed on small rodents, shrews (*Sorex* spp.), and brush rabbits (Svendsen 1982). The gray fox (*Urocyon cinereoargenteus*) is commonly found in grassland/riparian woodland ecotones that provide open hunting grounds as well as sufficient cover (Samuel and Nelson 1982).

Freshwater Marshes and Ponds

Freshwater marshes and ponds are scarce in the Elkhorn Slough watershed and provide a rich environment for many species, some which are unique to this habitat alone in the slough. Typically surrounded by emergent cattails (*Typha latifolia*), bulrush (*Scirpus californicus*), and rush (*Juncus* spp.), freshwater ponds provide protected nesting habitat for terrestrial birds such as Marsh Wrens, Common Yellowthroats, and Red-winged Blackbirds. In summer, Barn, Cliff, Tree, and Violet-green Swallows hunt flying insects above the water surface; bats (hoary bat, *Lasiurus cinereus*; pallid bat, *Antrozous pallidus*; western pipistrel, *Pipistrellus hesperus*; California bat, *Myotis californicus*; and big brown bat, *Eptesicus fuscus*, are the most likely species) replace the swallows after sundown. Herons, egrets, and shy Soras and Virginia Rails forage along shorelines for fish, amphibians, and invertebrates. Waterfowl, primarily dabbling ducks such as Mallards, Northern Pintails, Gadwalls, Cinnamon and Green-winged Teals, and Northern Shovelers, often spend the winter on freshwater ponds, feeding on the green surface film of duckweed and other aquatic plants and animals. Most depart in spring, but a few Mallards and other ducks stay to raise broods. There appears to be suitable shoreline nesting habitat for American Coots, Pied-billed Grebes, and American Bitterns, but few nests have been found (Roberson and Tenney 1993).

Eucalyptus Forest

Some areas of the Elkhorn Slough watershed feature nearly pure stands of nonnative eucalyptus trees (*Eucalyptus* spp.). The species' allelopathic properties prevent most other trees and understory plant species from becoming established. Eucalyptus forests are thought to support fewer bird and mammal species than other wooded habitats in the watershed. During winter, large flocks of Yellow-rumped Warblers feed within the towering canopy. In the spring, Red-tailed Hawks often build their nests in the upper branches, which provide a commanding view of surrounding foraging areas.

Dunes

The coastal dunes system is a harsh environment, deficient in freshwater and continuously exposed to winds and salt spray. Still, the sparse assemblages of dune-binding vegetation support a few species of birds and mammals. In fact, some terrestrial mammals are well adapted to these conditions: the brush rabbit finds sufficient cover and food in herbaceous and woody vegetation; the western harvest mouse (*Reithrodontomys megalotis*) can drink saltwater to obtain fluids; the California ground squirrel and California mole (*Scapanus latimanus*) find a sandy substrate conducive to extensive burrowing. Several bird species are resident in dune habitat. White-crowned

Bobcat. Photo credit: ESNERR.

Sparrows commonly perch on bush lupines. Predators such as the Loggerhead Shrike and Say's Phoebe, though not numerous, are highly visible as they perch on posts or tall shrubs, searching for prey. In winter, Water Pipits search for ground-dwelling insects on the upper beach.

Agricultural and Residential Lands

Grazing pastures and row crops cover a large part of the Elkhorn Slough watershed. In spite of intense human activities, these lands support large populations of some species of birds and mammals. Enormous flocks of Red-winged, Tricolored, and Brewer's Blackbirds feed on grain near livestock and on invertebrates in newly tilled fields. Nonnative European Starlings and parasitic Brown-headed Cowbirds often join these flocks. Mixed flocks of sparrows and finches are also present. Swallows diligently search for insects over fields in summer.

Management Issues and Research Recommendations

Birds and mammals are ecologically important and highly visible members of the Elkhorn Slough watershed. In recent years, harbor seals, sea otters, Snowy Plovers, and several other species have been the focus of directed research and management efforts. But overall, the bird and mammal communities in Elkhorn Slough remain poorly studied. A thorough understanding of the status and biology of these species is essential if future restoration and conservation efforts are to be successful. Basic information about habitat requirements, population size, and community structure is missing for many species, especially terrestrial mammals. Without this baseline information, it is nearly impossible to detect changes and assess the effects of human activities in the slough. From both a scientific and management perspective, the opportunities and need for both basic and applied research on the bird and mammal species at Elkhorn Slough are enormous.

Basic Research

Basic ecological research, studying species and their roles within communities, is essential to understand the slough's complex ecosystems. Researchers have taken a close look at the biology of a few slough inhabitants. Continued research documenting changes in the patterns of these species is essential for an awareness of overall changes in the slough system, including: movement patterns, foraging ecology, reproduction and survival of harbor seals; sea otter abundance in the slough and effects of foraging on prey populations; relative abundance of shorebirds, herons, and egrets in regions throughout the watershed in relation to habitat changes; and breeding chronology and food habits of Caspian Terns and effects of pollutants on reproductive success.

A great deal of basic research topics still remain largely unexplored, including studies of the breeding biology of resident marine and terrestrial birds, effects of changing habitats on survival, and relative importance of prey species to birds and mammals in Elkhorn Slough.

Conservation and Management

Conservation-based research and informed management strategies are critical for future protection of the slough's birds and mammals. The most pressing concerns relate to anthropogenic factors, specifically habitat degradation, pollutants, and nonnative species. We recommend that the following subjects be studied in detail.

Impacts of Habitat Changes The aquatic habitats in Elkhorn Slough have undergone dramatic changes in the last century. Tidal scour, erosion, and sedimentation continue to affect mudflats and marshes throughout the slough (ABA 1989; Crampton 1994; Malzone and Kvitek 1994; see also chapter 4). Some of the key questions are: How have these changes in habitat affected shorebirds and waterfowl? Is one habitat more beneficial than another or is a balance of several habitats the best way to maximize diversity? Is it necessary or even possible to manage or restore freshwater marshes or other declining habitats? Will future management practices affect harbor seals or other species that appear to be thriving under current conditions?

We recommend that the amount and causes of scour be continually monitored, and that the effects of scour on marine birds (specifically shorebirds) and mammals be assessed via monitoring of habitat use. Monitoring of the Porter Marsh should continue to assess the value of restoring freshwater and brackish marshes to the Elkhorn Slough area.

Terrestrial habitats have also undergone dramatic changes. Throughout the Elkhorn Slough watershed, agricultural lands and housing developments have replaced native habitats.

Long-tailed weasel. Photo credit: John Sorenson.

Terrestrial bird and mammal communities have undoubtedly been altered, but these changes have not been documented. At this point, thorough assessment of populations in native and disturbed habitats is necessary to identify species at risk and plan appropriate management actions.

Efforts to continue and expand monitoring of all bird and mammal communities to assess the effects of increasing development in the Elkhorn Slough area should be implemented. As more lands are converted from agricultural or open space into housing, wildlands and migratory corridors decrease in size and number; the effects of such changes must be studied.

Effects of Pollutants Pollutants from agricultural activities and residential and commercial development pose a serious threat to birds and mammals in Elkhorn Slough. Still, very little is known about the effects of many pollutants on most birds and mammals and their prey in the slough. Toxins entering the food chain can adversely affect the health of individuals and lower reproductive rates (Parkin 1998). Although sea otter numbers have increased in Elkhorn Slough in recent years, the population is declining statewide. Do sea otters that regularly feed in the slough live as long and reproduce at the same rate as sea otters in the open ocean? Contaminant testing on sea otters that frequent the slough may shed some light on these questions. Long-term monitoring of contaminant levels in slough inhabitants would help us track the overall health of the ecosystem.

Control of Nonnative Species and Native Predators The negative impacts of introduced plant and animal species on an ecosystem can be severe, and often are irreversible. The effects of these introductions have received little attention in Elkhorn Slough, but could be widespread. Nonnative species may outcompete native species, resulting in a precarious, unbalanced system. As species at risk struggle to regain viable population sizes, additional pressures from nonnatives could potentially preclude recovery. Continued monitoring and selective removal of introduced species, such as the red fox, are necessary for the persistence of native species.

As human use along the Monterey Bay coast continues to increase, direct and indirect effects pervade our natural habitats. Particular native species, usually habitat "generalists" such as raccoons, crows, and ravens, tend to thrive amidst human habitation and become a threat to species at risk. Additionally, the negative impact of domestic species (e.g., dogs, cats, and rats) on native populations can be significant; the magnitude is generally linked to proximity of human communities to natural areas. With increased awareness of the sensitivity of our natural habitats and their inhabitants to our presence, we can positively influence the fate of Elkhorn Slough's unique environment.

Appendix 11.1. Birds of Elkhorn Slough

Seasons: Winter, November–February; Spring, (February) March–May; Summer, June–July; Fall, (July) August–October
Abundance: c: common, almost certain to be seen in suitable habitat; u: uncommon, present but not certain to be seen; o: occasional, seen only a few times during a season; r: rare, not present every year; x: extraordinary, 1 or 2 records

COMMON NAME	SCIENTIFIC NAME	Winter	Spring	Summer	Fall
<u>Loons — Family Gaviidae</u>					
Red-throated Loon	*Gavia stellata*	c	u	o	u
Pacific Loon	*Gavia pacifica*	o	o	–	o
Common Loon	*Gavia immer*	c	c	o	c
Yellow-billed Loon	*Gavia adamsii*	x	–	–	–
<u>Grebes — Family Podicipedidae</u>					
Horned Grebe	*Podiceps auritus*	u	o	o	u
Eared Grebe	*Podiceps nigricollis*	c	u	x	u
Pied-billed Grebe	*Podilymbus podiceps*	c	u	u	u
Red-necked Grebe	*Podiceps grisegena*	o	o	–	o
Clark's Grebe	*Aechmophorus clarkii*	u	u	u	u
Western Grebe	*Aechmophorus occidentalis*	c	c	u	c
<u>Pelicans — Family Pelecanidae</u>					
American White Pelican	*Pelecanus erythrorhynchos*	r	r	r	r
Brown Pelican	*Pelecanus occidentalis*	u	u	c	c
<u>Cormorants — Family Phalacrocoracidae</u>					
Double-crested Cormorant	*Phalacrocorax auritus*	c	c	o	c
Brandt's Cormorant	*Phalacrocorax penicillatus*	c	c	u	c
Pelagic Cormorant	*Phalacrocorax pelagicus*	u	u	u	u
<u>Herons, Bitterns — Family Ardeidae</u>					
American Bittern	*Botaurus lentiginosus*	r	r	o	r
Black-crowned Night-Heron	*Nycticorax nycticorax*	r	o	–	r
Green Heron	*Butorides virescens*	u	o	r	u
Reddish Egret	*Egretta rufescens*	–	–	–	x
Cattle Egret	*Bubulcus ibis*	o	r	–	r
Snowy Egret	*Egretta thula*	c	c	u	c
Great Egret	*Ardea alba*	c	c	c	c
Great Blue Heron	*Ardea herodias*	c	c	c	c
<u>Ibises — Family Threskiornithidae</u>					
White-faced Ibis	*Plegadis chihi*	r	–	r	r
<u>Ducks, Geese, Swans — Family Anatidae</u>					
Tundra Swan	*Cygnus columbianus*	r	–	–	–
Greater White-fronted Goose	*Anser albifrons*	o	–	–	[illegible]
Snow Goose	*Chen caerulescens*	r	–	–	[illegible]
Ross's Goose	*Chen rossii*	r	–	–	–
Emperor Goose	*Chen canagica*	x	–	–	–
Canada Goose	*Branta canadensis*	o	r	–	–
Brant	*Branta bernicla*	o	u	r	[illegible]
Wood Duck	*Aix sponsa*	o	–	–	o
Mallard	*Anas platyrhynchos*	c	c	c	c
Gadwall	*Anas strepera*	c	c	c	c

Scientific and common names follow the Check-list of North American Birds (7th ed., 1998) of the American Ornithologists' Union. Original species account compiled by B. Ramer, D. Ramer, and J. Warriner (1978); updated by D. Roberson (1991).

COMMON NAME	SCIENTIFIC NAME	Winter	Spring	Summer	Fall
Ducks, Geese, Swans — Family Anatidae (continued)					
Green-winged Teal	*Anas crecca*	c	u	–	c
American Wigeon	*Anas americana*	c	u	–	u
Eurasian Wigeon	*Anas penelope*	x	x	–	–
Northern Pintail	*Anas acuta*	c	c	u	c
Northern Shoveler	*Anas clypeata*	c	c	u	c
Blue-winged Teal	*Anas discors*	r	u	x	u
Cinnamon Teal	*Anas cyanoptera*	r	c	u	u
Canvasback	*Aythya valisineria*	c	u	–	u
Redhead	*Aythya americana*	r	r	–	r
Ring-necked Duck	*Aythya collaris*	u	r	–	r
Tufted Duck	*Aythya fuligula*	x	–	–	–
Greater Scaup	*Aythya marila*	c	u	o	u
Lesser Scaup	*Aythya affinis*	c	u	o	u
King Eider	*Somateria spectabilis*	x	x	x	–
Black Scoter	*Melanitta nigra*	o	r	–	r
White-winged Scoter	*Melanitta fusca*	c	u	o	u
Surf Scoter	*Melanitta perspicillata*	c	c	c	c
Harlequin Duck	*Histrionicus histrionicus*	o	x	–	–
Long-tailed Duck	*Clangula hyemalis*	o	r	x	–
Barrow's Goldeneye	*Bucephala islandica*	x	–	–	–
Common Goldeneye	*Bucephala clangula*	c	u	–	c
Bufflehead	*Bucephala albeola*	c	u	–	c
Common Merganser	*Mergus merganser*	x	x	x	x
Red-breasted Merganser	*Mergus serrator*	c	u	o	u
Hooded Merganser	*Lophodytes cucullatus*	r	–	–	–
Ruddy Duck	*Oxyura jamaicensis*	c	c	c	c
New World Vultures — Family Cathartidae					
Turkey Vulture	*Cathartes aura*	r	c	u	u
Hawks, Kites, Eagles — Family Accipitridae					
Osprey	*Pandion haliaetus*	o	o	x	o
White-tailed Kite	*Elanus leucurus*	r	u	u	u
Northern Harrier	*Circus cyaneus*	r	u	o	u
Golden Eagle	*Aquila chrysaetos*	r	u	u	u
Bald Eagle	*Haliaeetus leucocephalus*	x	–	–	–
Sharp-shinned Hawk	*Accipiter striatus*	u	u	–	u
Cooper's Hawk	*Accipiter cooperii*	u	u	o	u
Red-shouldered Hawk	*Buteo lineatus*	u	u	u	u
Red-tailed Hawk	*Buteo jamaicensis*	c	c	c	c
Rough-legged Hawk	*Buteo lagopus*	r	–	–	–
Ferruginous Hawk	*Buteo regalis*	r	–	–	–
American Kestrel	*Falco sparverius*	c	c	u	c
Merlin	*Falco columbarius*	c	r	–	r
Prairie Falcon	*Falco mexicanus*	r	r	x	r
Peregrine Falcon	*Falco peregrinus*	c	r	r	r
New World Quail — Family Odontophoridae					
California Quail	*Callipepla californica*	c	c	c	c
Wild Turkey — Family Phasianidae					
Wild Turkey	*Meleagris gallopavo*	r	r	r	r

COMMON NAME	SCIENTIFIC NAME	Winter	Spring	Summer	Fall
Rails, Coots — Family Rallidae					
Clapper Rail (not seen since 1980)	*Rallus longirostris*	?	?	?	?
Virginia Rail	*Rallus limicola*	u	o	–	o
Sora	*Porzana carolina*	u	o	–	o
Common Moorhen	*Gallinula chloropus*	r	r	–	r
American Coot	*Fulica americana*	c	c	c	c
Plovers — Family Charadriidae					
Black-bellied Plover	*Pluvialis squatarola*	c	c	u	c
Pacific Golden-Plover	*Pluvialis fulva*	x	x	–	r
American Golden-Plover	*Pluvialis dominica*	x	–	–	o
Snowy Plover	*Charadrius alexandrinus*	u	u	u	u
Semipalmated Plover	*Charadrius semipalmatus*	u	c	x	c
Mongolian Plover	*Charadrius mongolus*	–	–	–	x
Killdeer	*Charadrius vociferus*	c	c	c	c
Stilts, Avocets — Family Recurvirostridae					
American Avocet	*Recurvirostra americana*	c	c	c	c
Black-necked Stilt	*Himantopus mexicanus*	c	c	c	c
Sandpipers, Phalaropes — Family Scolopacidae					
Willet	*Catoptrophorus semipalmatus*	c	c	u	c
Greater Yellowlegs	*Tringa melanoleuca*	c	c	r	c
Lesser Yellowlegs	*Tringa flavipes*	o	o	–	u
Wandering Tattler	*Heteroscelus incanus*	x	x	–	x
Spotted Sandpiper	*Actitis macularia*	u	u	o	u
Whimbrel	*Numenius phaeopus*	u	c	r	c
Long-billed Curlew	*Numenius americanus*	c	c	o	c
Marbled Godwit	*Limosa fedoa*	c	c	u	c
Ruddy Turnstone	*Arenaria interpres*	c	c	r	c
Black Turnstone	*Arenaria melanocephala*	–	r	x	r
Surfbird	*Aphriza virgata*	–	x	–	x
Red Knot	*Calidris canutus*	o	u	x	u
Sanderling	*Calidris alba*	c	c	u	c
Dunlin	*Calidris alpina*	c	c	x	c
Semipalmated Sandpiper	*Calidris pusilla*	–	–	–	c
Western Sandpiper	*Calidris mauri*	c	c	x	c
Least Sandpiper	*Calidris minutilla*	c	c	r	c
Baird's Sandpiper	*Calidris bairdii*	–	–	–	u
Little Stint	*Calidris minuta*	–	–	–	x
Pectoral Sandpiper	*Calidris melanotos*	–	–	–	u
Sharp-tailed Sandpiper	*Calidris acuminata*	–	–	–	r
Buff-breasted Sandpiper	*Tryngites subruficollis*	–	–	–	x
Ruff	*Philomachus pugnax*	x	–	–	x
Short-billed Dowitcher	*Limnodromus griseus*	c	c	x	c
Long-billed Dowitcher	*Limnodromus scolopaceus*	c	c	–	c
Stilt Sandpiper	*Calidris himantopus*	–	–	–	
Common Snipe	*Gallinago gallinago*	u	u	–	[illegible]
Wilson's Phalarope	*Phalaropus tricolor*	x	u	o	[illegible]
Red-necked Phalarope	*Phalaropus lobatus*	x	u	r	[illegible]
Red Phalarope	*Phalaropus fulicaria*	r	r	x	[illegible]

COMMON NAME	SCIENTIFIC NAME	Winter	Spring	Summer	Fall
Jaegers, Gulls, Terns — Family Laridae					
Parasitic Jaeger	*Stercorarius parasiticus*	–	–	–	o
Heermann's Gull	*Larus heermanni*	u	o	c	c
Franklin's Gull	*Larus pipixcan*	x	x	x	r
Laughing Gull	*Larus atricilla*	x	–	–	–
Bonaparte's Gull	*Larus philadelphia*	c	c	o	c
Ring-billed Gull	*Larus delawarensis*	c	c	u	c
Mew Gull	*Larus canus*	u	o	–	o
California Gull	*Larus californicus*	c	c	u	c
Herring Gull	*Larus argentatus*	c	o	–	o
Glaucous Gull	*Larus hyperboreus*	r	x	–	–
Thayer's Gull	*Larus thayeri*	c	o	–	o
Western Gull	*Larus occidentalis*	c	c	c	c
Glaucous-winged Gull	*Larus glaucescens*	c	c	o	c
Black-legged Kittiwake	*Rissa tridactyla*	x	x	–	x
Sabine's Gull	*Xema sabini*	–	–	–	x
Swallow-tailed Gull	*Creagrus furcatus*	–	–	x	–
Elegant Tern	*Sterna elegans*	x	r	c	c
Royal Tern	*Sterna maxima*	–	–	–	x
Caspian Tern	*Sterna caspia*	x	c	c	c
Forster's Tern	*Sterna forsteri*	u	c	c	c
Common Tern	*Sterna hirundo*	–	u	r	u
Arctic Tern	*Sterna paradisaea*	–	o	–	o
Least Tern	*Sterna antillarum*	–	o	r	o
Black Tern	*Chlidonias niger*	–	o	–	o
Black Skimmer	*Rynchops niger*	–	x	–	x
Auks, Murres — Family Alcidae					
Common Murre	*Uria aalge*	r	r	r	r
Thick-billed Murre	*Uria lomvia*	x	–	–	–
Pigeon Guillemot	*Cepphus columba*	–	–	–	r
Craveri's Murrelet	*Synthliboramphus craveri*	–	–	–	x
Ancient Murrelet	*Synthliboramphus antiquus*	x	–	–	–
Cassin's Auklet	*Ptychoramphus aleuticus*	r	–	–	–
Rhinoceros Auklet	*Cerorhinca monocerata*	o	–	–	r
Pigeons, Doves — Family Columbidae					
Band-tailed Pigeon	*Columba fasciata*	u	c	u	u
Rock Dove	*Columba livia*	c	c	c	c
Mourning Dove	*Zenaida macroura*	c	c	c	c
White-winged Dove	*Zenaida asiatica*	–	–	–	x
Roadrunners — Family Cuculidae					
Greater Roadrunner	*Geococcyx californianus*	–	x	–	x
Owls — Families Tytonidae and Strigidae					
Barn Owl	*Tyto alba*	u	u	u	u
Short-eared Owl	*Asio flammeus*	r	r	r	r
Long-eared Owl	*Asio otus*	x	–	–	–
Great Horned Owl	*Bubo virginianus*	c	c	c	c
Burrowing Owl	*Athene cunicularia*	u	u	u	u
Swifts — Family Apodidae					
Vaux's Swift	*Chaetura vauxi*	–	o	–	o
White-throated Swift	*Aeronautes saxatalis*	–	x	–	–

COMMON NAME	SCIENTIFIC NAME	Winter	Spring	Summer	Fall
Hummingbirds — Family Trochilidae					
Anna's Hummingbird	*Calypte anna*	c	c	c	c
Rufous Hummingbird	*Selasphorus rufus*	–	o	–	o
Allen's Hummingbird	*Selasphorus sasin*	–	c	c	–
Kingfishers — Family Alcedinidae					
Belted Kingfisher	*Ceryle alcyon*	u	u	u	u
Woodpeckers — Family Picidae					
Acorn Woodpecker	*Melanerpes formicivorus*	u	u	u	u
Northern Flicker	*Colaptes auratus*	c	c	u	c
Red-breasted Sapsucker	*Sphyrapicus ruber*	o	o	–	o
Nuttall's Woodpecker	*Picoides nuttallii*	c	c	c	c
Downy Woodpecker	*Picoides pubescens*	u	u	u	u
Hairy Woodpecker	*Picoides villosus*	u	u	u	u
Tyrant Flycatchers — Family Tyrannidae					
Olive-sided Flycatcher	*Contopus cooperi*	–	u	u	–
Western Wood-Pewee	*Contopus sordidulus*	–	–	–	u
Pacific-slope Flycatcher	*Empidonax difficilis*	x	c	c	u
Black Phoebe	*Sayornis nigricans*	c	c	c	c
Say's Phoebe	*Sayornis saya*	u	r	–	u
Ash-throated Flycatcher	*Myiarchus cinerascens*	x	u	u	r
Western Kingbird	*Tyrannus verticalis*	–	r	x	r
Tropical Kingbird	*Tyrannus melancholicus*	x	–	–	r
Scissor-tailed Flycatcher	*Tyrannus forficatus*	–	x	–	–
Shrikes — Family Laniidae					
Loggerhead Shrike	*Lanius ludovicianus*	u	u	u	u
Vireos — Family Vireonidae					
Hutton's Vireo	*Vireo huttoni*	c	c	c	c
Red-eyed Vireo	*Vireo olivaceus*	–	–	–	x
Warbling Vireo	*Vireo gilvus*	–	c	u	c
Crows, Jays — Family Corvidae					
Steller's Jay	*Cyanocitta stelleri*	r	r	–	–
Western Scrub-Jay	*Aphelocoma californica*	c	c	c	c
American Crow	*Corvus brachyrhynchos*	c	c	c	c
Common Raven	*Corvus corax*	r	r	r	r
Larks — Family Alaudidae					
Horned Lark	*Eremophila alpestris*	u	r	–	u
Swallows — Family Hirundinidae					
Tree Swallow	*Tachycineta bicolor*	u	c	c	c
Violet-green Swallow	*Tachycineta thalassina*	r	c	c	c
Bank Swallow	*Riparia riparia*	x	r	r	r
Cliff Swallow	*Petrochelidon pyrrhonota*	–	c	c	c
Northern Rough-winged Swallow	*Stelgidopteryx serripennis*	x	o	r	r
Barn Swallow	*Hirundo rustica*	x	c	c	c
Babblers — Family Timaliidae					
Wrentit	*Chamaea fasciata*	c	c	c	c

COMMON NAME	SCIENTIFIC NAME	Winter	Spring	Summer	Fall
Chickadees, Titmice — Family Paridae					
Oak Titmouse	*Baeolophus inornatus*	c	c	c	c
Chestnut-backed Chickadee	*Poecile rufescens*	c	c	c	c
Bushtits — Family Aegithalidae					
Bushtit	*Psaltriparus minimus*	c	c	c	c
Creepers — Family Certhiidae					
Brown Creeper	*Certhia americana*	u	u	u	u
Nuthatches — Family Sittidae					
Red-breasted Nuthatch	*Sitta canadensis*	o	–	–	o
Pygmy Nuthatch	*Sitta pygmaea*	u	u	u	u
Wrens — Family Troglodytidae					
House Wren	*Troglodytes aedon*	x	u	–	u
Winter Wren	*Troglodytes troglodytes*	u	r	–	r
Bewick's Wren	*Thyromanes bewickii*	c	c	c	c
Marsh Wren	*Cistothorus palustris*	c	c	c	c
Kinglets — Family Regulidae					
Ruby-crowned Kinglet	*Regulus calendula*	c	c	–	c
Thrushes — Family Turdidae					
Western Bluebird	*Sialia mexicana*	u	r	–	–
Townsend's Solitaire	*Myadestes townsendi*	x	–	–	–
Swainson's Thrush	*Catharus ustulatus*	–	c	u	r
Hermit Thrush	*Catharus guttatus*	u	u	–	u
Varied Thrush	*Ixoreus naevius*	u	r	–	r
American Robin	*Turdus migratorius*	c	c	u	c
Mockingbirds, Thrashers — Family Mimidae					
Northern Mockingbird	*Mimus polyglottos*	u	u	u	u
California Thrasher	*Toxostoma redivivum*	o	o	o	o
Starlings — Family Sturnidae					
European Starling	*Sturnus vulgaris*	c	c	c	c
Wagtails, Pipits — Family Motacillidae					
White Wagtail	*Motacilla alba*	x	–	–	–
American Pipit	*Anthus rubescens*	c	u	–	c
Red-throated Pipit	*Anthus cervinus*	–	–	–	x
Waxwings — Family Bombycillidae					
Cedar Waxwing	*Bombycilla cedrorum*	c	c	–	c
Wood-Warblers — Family Parulidae					
Orange-crowned Warbler	*Vermivora celata*	u	c	c	c
Nashville Warbler	*Vermivora ruficapilla*	x	r	–	r
Yellow-rumped Warbler	*Dendroica coronata*	c	c	–	c
Black-throated Gray Warbler	*Dendroica nigrescens*	–	o	–	o
Townsend's Warbler	*Dendroica townsendi*	c	u	–	u
Yellow Warbler	*Dendroica petechia*	–	u	o	u
MacGillivray's Warbler	*Oporornis tolmiei*	–	o	–	o
Wilson's Warbler	*Wilsonia pusilla*	–	c	c	c
Northern Waterthrush	*Seiurus noveboracensis*	x	–	–	–
Common Yellowthroat	*Geothlypis trichas*	u	u	u	u
American Redstart	*Setophaga ruticilla*	x	–	–	–

COMMON NAME	SCIENTIFIC NAME	Winter	Spring	Summer	Fall
Tanagers — Family Thraupidae					
Western Tanager	*Piranga ludoviciana*	–	u	–	u
Emberizids (Towhees, Sparrows) — Family Emberizidae					
California Towhee	*Pipilo crissalis*	c	c	c	c
Spotted Towhee	*Pipilo maculatus*	c	c	c	c
Brewer's Sparrow	*Spizella breweri*	–	x	–	–
Grasshopper Sparrow	*Ammodramus savannarum*	r	–	–	r
Saltmarsh Sharp-tailed Sparrow	*Ammodramus caudacutus*	–	–	–	x
Fox Sparrow	*Passerella iliaca*	u	u	–	u
Lark Bunting	*Calamospiza melanocorys*	–	–	–	x
Savannah Sparrow	*Passerculus sandwichensis*	c	c	c	c
Lincoln's Sparrow	*Melospiza lincolnii*	u	u	–	u
Song Sparrow	*Melospiza melodia*	c	c	c	c
Swamp Sparrow	*Melospiza georgiana*	o	–	–	o
White-throated Sparrow	*Zonotrichia albicollis*	x	–	–	–
White-crowned Sparrow	*Zonotrichia leucophrys*	c	c	c	c
Golden-crowned Sparrow	*Zonotrichia atricapilla*	c	c	–	c
Dark-eyed Junco	*Junco hyemalis*	c	c	u	c
Smith's Longspur	*Calcarius pictus*	–	–	–	x
Cardinals (Grosbeaks) — Family Cardinalidae					
Black-headed Grosbeak	*Pheucticus melanocephalus*	–	u	u	u
Blackbirds — Family Icteridae					
Bobolink	*Dolichonyx oryzivorus*	–	–	–	r
Western Meadowlark	*Sturnella neglecta*	c	c	c	c
Yellow-headed Blackbird	*Xanthocephalus xanthocephalus*	x	o	–	o
Red-winged Blackbird	*Agelaius phoeniceus*	c	c	c	c
Tricolored Blackbird	*Agelaius tricolor*	c	u	o	c
Rusty Blackbird	*Euphagus carolinus*	x	–	–	–
Brewer's Blackbird	*Euphagus cyanocephalus*	c	c	c	c
Brown-headed Cowbird	*Molothrus ater*	u	u	u	u
Bullock's Oriole	*Icterus bullockii*	r	u	–	u
Finches — Family Fringillidae					
Purple Finch	*Carpodacus purpureus*	c	c	c	c
House Finch	*Carpodacus mexicanus*	c	c	c	c
Red Crossbill	*Loxia curvirostra*	r	–	–	–
Pine Siskin	*Carduelis pinus*	c	c	u	c
American Goldfinch	*Carduelis tristis*	c	c	c	c
Lesser Goldfinch	*Carduelis psaltria*	c	c	c	c
Lawrence's Goldfinch	*Carduelis lawrencei*	–	r	–	–
Old World Sparrows — Family Passeridae					
House Sparrow	*Passer domesticus*	c	c	c	c

Appendix 11.2. Mammals of the Elkhorn Slough Area

Marsupials — Order Marsupialia	
Virginia opossum	*Didelphis virginiana*

Shrews and Moles — Order Insectivora	
Trowbridge's shrew	*Sorex trowbridgii*
vagrant shrew	*Sorex vagrans*
ornate shrew	*Sorex ornatus*
shrew-mole	*Neurotrichus gibbsii*
broad-footed mole	*Scapanus latimanus*

Bats — Order Chiroptera	
little brown bat	*Myotis lucifugus*
Yuma bat	*Myotis yumanensis*
long-eared bat	*Myotis evotis*
fringed bat	*Myotis thysanodes*
long-legged bat	*Myotis volans*
California bat	*Myotis californicus*
western small-footed bat	*Myotis ciliolabrum*
western pipistrel	*Pipistrellus hesperus*
big brown bat	*Eptesicus fuscus*
western red bat	*Lasiurus blossevillii*
hoary bat	*Lasiurus cinereus*
pallid bat	*Antrozous pallidus*
guano bat	*Tadarida brasiliensis*

Carnivores — Order Carnivora	
raccoon	*Procyon lotor*
ringtail	*Bassariscus astutus*
long-tailed weasel	*Mustela frenata*
sea otter	*Enhydra lutris*
badger	*Taxidea taxus*
spotted skunk	*Spilogale putorius*
striped skunk	*Mephitis mephitis*
coyote	*Canis latrans*
red fox	*Vulpes vulpes*
gray fox	*Urocyon cinereoargenteus*
mountain lion	*Felis concolor*
bobcat	*Lynx rufus*

Sea Lions and Seals — Order Pinnipedia	
California sea lion	*Zalophus californianus*
harbor seal	*Phoca vitulina*

Rodents — Order Rodentia	
California ground squirrel	*Spermophilus beecheyi*
western gray squirrel	*Sciurus griseus*
eastern gray squirrel	*Sciurus carolinensis*
fox squirrel	*Sciurus niger*
Botta's pocket gopher	*Thomomys bottae*
California pocket mouse	*Chaetodipus californicus*
Heermann's kangaroo rat	*Dipodomys heermanni*
narrow-faced kangaroo rat	*Dipodomys venustus*
beaver	*Castor canadensis*
western harvest mouse	*Reithrodontomys megalotis*
parasitic mouse	*Peromyscus californicus*
deer mouse	*Peromyscus maniculatus*
piñon mouse	*Peromyscus truei*
dusky-footed woodrat	*Neotoma fuscipes*
California meadow vole	*Microtus californicus*
muskrat	*Ondatra zibethica*
house mouse	*Mus musculus*
Norway rat	*Rattus norvegicus*
black rat	*Rattus rattus*

Rabbits — Order Lagomorpha	
black-tailed jackrabbit	*Lepus californicus*
Audubon's cottontail	*Sylvilagus audubonii*
brush rabbit	*Sylvilagus bachmani*

Ungulates — Order Artiodactyla	
black-tailed deer	*Odocoileus hemionus*

Whales — Order Cetacea	
gray whale	*Eschrichtius robustus*
harbor porpoise	*Phocoena phocoena*

List compiled by E. Schafer (1986) for the Elkhorn Slough National Estuarine Research Reserve. Little information is available regarding occurrence and abundance of mammal species in this region and therefore is not presented in this appendix. Known information is included in the text. This list represents both known and potential presence at the slough.

References

ABA Consultants. 1989. *Elkhorn Slough Wetland Management Plan*. Prepared for the California State Coastal Conservancy and Monterey County Planning Department. Capitola, CA: ABA Consultants.

Bailey, S. F. 1993. Caspian Tern, *Sterna caspia*. In D. Roberson and C. Tenney, eds., *Atlas of the Breeding Birds of Monterey County*, pp. 138–139. Carmel, CA: Monterey Peninsula Audubon Society.

Baldridge, A., T. Chandik, and D. DeSante. 1970. Middle Pacific Coast Region. *American Birds* 24 (5): 713–715.

Baltz, D. M., G. V. Morejohn, and B. S. Antrim. 1979. Size selective predation and food habits of two California terns. *Western Birds* 10: 17–24.

Briggs, K. T., W. B. Tyler, D. B. Lewis, P. R. Kelly, and D. A. Croll. 1983. Brown pelicans in central and northern California. *Journal of Field Ornithology* 54 (4): 353–373.

Browning, B. M. 1972. *The Natural Resources of Elkhorn Slough: Their Present and Future Use*. California Department of Fish Game, Coastal Wetland Series No. 4.

Burt, W. H., and R. P. Grossenheider. 1976. *A Field Guide to the Mammals of America North of Mexico*. 3d ed. New York: Houghton Mifflin.

Byrnes, P. E. 1997. Habitat use, behavior, and morphology of herons and egrets in Elkhorn Slough, California. Master's thesis, Moss Landing Marine Laboratories. San Francisco State University.

Chapman, J. A., J. G. Hockman, and W. R. Edwards. 1982. Cottontails (*Sylvilagus floridanus* and allies). In J. A. Chapman and G. A. Feldhamer, eds., *Wild Mammals of North America*, pp. 83–123. Baltimore: Johns Hopkins University Press.

Crampton, A. C. 1994. Long-term effects of Moss Landing Harbor on the wetlands of Elkhorn Slough. M.S. thesis, University of California, Santa Cruz.

Eguchi, T. 1998. Morphology of the Pacific harbor seal (*Phoca vitulina richardsi*) using Elkhorn Slough, California, and their movements, diving behavior in the Monterey Bay area. Master's thesis, Moss Landing Marine Laboratories. Fresno State University.

Estelle, V. B. 1991. Spring foraging of Sanderling (*Calidris alba*) in response to aggregations of *Emerita analoga* megalopae. Master's thesis, Moss Landing Marine Laboratories. San Jose State University.

Feinholz, D. M. 1998. Abundance, distribution, and behavior of the southern sea otter (*Enhydra lutris nereis*) in a California estuary. *Aquatic Mammals* 24: 105–116.

Gardner, A. L. 1982. Virginia opossum (*Didelphis virginiana*). In J. A. Chapman and G. A. Feldhamer, eds., *Wild Mammals of North America*, pp. 3–36. Baltimore: Johns Hopkins University Press.

Gordon, B. L. 1996. *Monterey Bay Area: Natural History and Cultural Imprints*. 3d ed. Pacific Grove, CA: Boxwood Press.

Green, J. 1968. *The Biology of Estuarine Animals*. Seattle: University of Washington Publications

Harding, E. K. 2000. Landscape heterogeneity and its importance for community dynamics and conservation of a marsh-grassland system. Ph.D. diss., University of California, Santa Cruz.

Harrington, B. 1983. The migration of the Red Knot. *Oceanus* 26 44–48.

Harvey, J. T., R. C. Helm, and G. V. Morejohn. 1995. Food habits of harbor seals inhabiting Elkhorn Slough, California. *California Fish Game* 81: 1–9.

Harvey, T. E. 1982. The ecology of three larids breeding in the Elkhorn Slough salt ponds, California. Master's thesis, Moss Landing Marine Laboratories. San Francisco State University.

Jaques, D. L., and D. W. Anderson. 1988. *Brown Pelican Use of the Moss Landing Wildlife Management Area*. California Department of Fish and Game, Wildlife Management Division, Nongame Bird and Mammal Section Report

Jolly, J. M. 1997. Foraging ecology of the sea otter, *Enhydra lutris*, in a soft sediment community. Master's thesis University of California, Santa Cruz.

Kaufmann, J. H. 1982. Raccoon and allies (*Procyon lotor* and Allies). In J. A. Chapman and G. A. Feldhamer, eds., *Wild Mammals of North America*, pp. 567–585. Baltimore: Johns Hopkins University Press.

Kvitek, R. G., A. K. Fukayama, B. S. Anderson, and B. K. Grimm. 1988. Sea otter foraging on deep-burrowing bivalves in a California coastal lagoon. *Marine Biology* 98: 157–167.

Lowe, P. B. 1999. Marsh loss in Elkhorn Slough, California: Patterns, mechanisms, and impact on shorebirds. Master's thesis, Moss Landing Marine Laboratories. San Jose State University.

MacGinitie, G. E. 1935. Ecological aspects of a California marine estuary. *American Midland Naturalist* 16: 629–765.

Mackie, R. J., K. L. Hamlin, and D. F. Pac. 1982. Mule deer (*Odocoileus virginianus*). In J. A. Chapman and G. A. Feldhamer, eds., *Wild Mammals of North America*, pp. 862–877. Baltimore: Johns Hopkins University Press.

Malzone, C., and R. Kvitek. 1994. Tidal scour, erosion, and habitat loss in Elkhorn Slough, California. A report of the Elkhorn Slough Foundation pursuant to National Oceanic and Atmospheric Administration Award #NA37OM0523.

Moser, G. D. 1996. Trace elements and organochlorine compounds in the harbor seal *Phoca vitulina richardsi*, along the Pacific coast. Master's thesis, Moss Landing Marine Laboratories. San Jose State University.

Myers, J. P., M. Sallaberry, E. Ortiz, G. Castro, L. M. Gordon, J. L. Maron, C. T. Schick, E. Talibo, P. Antas, and T. Below. 1990. Migration routes of New World Sanderling (*Calidris alba*). *Auk* 107: 172–180.

Nybakken, J., and C. Jong. 1977. Benthic invertebrate baseline studies of the Moss Landing—Elkhorn Slough environment. In J. Nybakken, G. Calliet, and W. Broenkow, eds., *Ecologic and Hydrologic Studies of Elkhorn Slough, Moss Landing Harbor, and Nearshore Coastal Waters, July 1974–June 1976*, pp. 1–215. Moss Landing, CA: Moss Landing Marine Laboratories Technical Publication.

Osborn, L. S. 1985. Population dynamics, behavior, and the effect of disturbance on haulout patterns of the harbor seal, *Phoca vitulina richardsi*. Senior thesis, University of California, Santa Cruz.

———. 1992. A report of preliminary findings: Harbor seal studies in Elkhorn Slough, 1989–1991. California Department of Fish and Game, Monterey.

Oxman, D. S. 1995. Seasonal abundance, movements, and food habits of harbor seals (*Phoca vitulina richardsi*) in Elkhorn Slough, CA. Master's thesis, Moss Landing Marine Laboratories. California State University, Stanislaus.

Page, G. W., W. D. Shuford, J. E. Kjelmyr, and L. E. Stenzel. 1992. Shorebird numbers in wetlands of the Pacific flyway: A summary of counts from April 1988 to January 1992. Report, Point Reyes Bird Observatory, Stinson Beach, Calif.

Page, G. W., J. S. Warriner, J. C. Warriner, and L. E. Stenzel. 1997. Nesting success of snowy plovers on Monterey Bay in 1997. Report, Point Reyes Bird Observatory, Stinson Beach, Calif.

Parkin, J. 1998. Ecology of breeding Caspian Terns (*Sterna caspia*) in Elkhorn Slough, California. Master's thesis, Moss Landing Marine Laboratories. San Jose State University.

Ramer, B. A. 1985. Seasonal abundance, habitat use, and diet of selected shorebirds in Elkhorn Slough. Master's thesis, Moss Landing Marine Laboratories. California State University, Hayward.

Ramer, B. A., G. W. Page, and M. M. Yoklavich. 1991. Seasonal abundance, habitat use, and diet of shorebirds in Elkhorn Slough, California. *Western Birds* 22: 157–174.

Ramer, B., D. Ramer, and J. Warriner. 1978. *Checklist of the Birds of Elkhorn Slough*. N.p.; produced by the Elkhorn Slough Foundation.

Recher, H. F. 1966. Some aspects of the ecology of migrant shorebirds. *Ecology* 47: 393–407.

Richman, S. E. 1997. Seasonal abundance and distribution of marine mammals and birds in Elkhorn Slough, Moss Landing, California, from July 1994 to July 1997. Independent study submitted to Elkhorn Slough Foundation.

Roberson, D. 1985. *Monterey Birds: Status and Distribution of Birds in Monterey Co., California*. Carmel, CA: Monterey Peninsula Audubon Society.

———. 1991. Update of the checklist of the birds of Elkhorn Slough. Report for Elkhorn Slough National Research Reserve.

———. 1993a. Clapper Rail, *Rallus longirostris*. In D. Roberson and C. Tenney, eds., *Atlas of the Breeding Birds of Monterey County*, pp. 413–414. Carmel, CA: Monterey Peninsula Audubon Society.

———. 1993b. White-tailed Kite, *Elanus (caerulus) leucurus*. In D. Roberson and C. Tenney, eds., *Atlas of the Breeding Birds of Monterey County*, pp. 88–89. Carmel, CA: Monterey Peninsula Audubon Society.

Roberson, D., and C. Tenney, eds. 1993. *Atlas of the Breeding Birds of Monterey County*. Carmel, CA: Monterey Peninsula Audubon Society.

Samuel, D. E., and B. B. Nelson. 1982. Foxes (*Vulpes vulpes* and allies). In J. A. Chapman and G. A. Feldhamer, eds., *Wild Mammals of North America*, pp. 475–490. Baltimore: Johns Hopkins University Press.

Schafer, E. 1986. Checklist of the amphibians, reptiles, and mammals of Elkhorn Slough National Estuarine Sanctuary and vicinity. Report for Elkhorn Slough National Research Reserve.

Schneider, D. C., and B. A. Harrington. 1981. Timing of shorebird migration in relation to prey depletion. *Auk* 98: 801–811.

Senner, S. E., and M. A. Howe. 1984. Conservation of nearctic shorebirds. In J. Burger and B. L. Olla, eds., *Behavior of Marine Animals*, vol. 5: *Shorebirds: Breeding Behavior and Populations*, pp. 379–421. New York: Plenum Press.

Strong, C. 1990. Bird use of the Moss Landing Wildlife Management area, April 1988 to April 1989. Final report to Santa Clara Valley Audubon Society.

Svendsen, G. E. 1982. Weasels (*Mustela* species). In J. A. Chapman and G. A. Feldhamer, eds., *Wild Mammals of North America*, pp. 613–628. Baltimore: Johns Hopkins University Press.

Thelander, C., and M. Crabtree. 1994. *Life on the Edge, A Guide to California's Endangered Natural Resources: Wildlife*. Berkeley, CA: Heyday Books/Biosystems Books.

Weed, B. J. 1993a. Great Blue Heron, *Ardea herodias*. In D. Roberson and C. Tenney, eds., *Atlas of the Breeding Birds of Monterey County*, pp. 60–61. Carmel, CA: Monterey Peninsula Audubon Society.

———. 1993b. Great Egret, *Casmerodius albus*. In D. Roberson and C. Tenney, eds., *Atlas of the Breeding Birds of Monterey County*, p. 62. Carmel, CA: Monterey Peninsula Audubon Society.

Additional Sources

Davis, J., and A. Baldridge. 1980. *The Bird Year: A Book for Birders, with Special Reference to the Monterey Bay Area*. Pacific Grove, CA: Boxwood Press.

Ehrlich, P. R., D .S. Dobkin, and D. Wheye. 1988. *The Birder's Handbook*. New York: Simon and Schuster.

Ingles, L. G. 1965. *Mammals of the Pacific States*. Stanford: Stanford University Press.

Peterson, R. T. 1990. *A Field Guide to Western Birds*. 3d ed. New York: Houghton Mifflin.

Silberstein, M., and E. Campbell. 1989. *Elkhorn Slough*. Monterey, CA: Monterey Bay Aquarium.

12

CHAPTER TWELVE

Biogeochemical Cycling

Jane Caffrey

Historically, nutrient cycling and ecosystem function in Elkhorn Slough received little attention from local scientists and managers. In recent years, however, concerns about water quality and nonpoint source pollution have driven new research and monitoring programs. The results of these studies have helped us begin to understand nutrient cycling in the slough and how nutrient levels are affecting Elkhorn Slough's ecosystem.

Nutrient cycling is critical to the proper functioning of any ecosystem. All the biotic and abiotic components of an ecosystem are interconnected through the exchange of energy and the chemical elements, or nutrients, that are the building blocks of the member organisms. Each nutrient is transferred through a unique biogeochemical cycle. Considered in total, the combined energy and nutrient flow is termed ecosystem function. Nutrient balance is essential for ecosystem health. Lack of nutrients limits plant growth and primary production, and thus consumer populations. Overabundance of nutrients can cause changes in community diversity and trigger eutrophication, or excessive primary production. Natural ecosystems conserve nutrients well, whereas crop fields and disturbed areas are sieves for nutrients (Brewer 1988).

In this chapter, I discuss the carbon cycle, especially decomposition and net ecosystem metabolism (the balance between production and respiration); and macronutrients (nitrogen, phosphorous, and silica)—the primary sources, past and present levels, and specific aspects of local nitrogen and phosphorous cycles. My focus here is primarily on nutrient cycling in the estuary itself, because little research has been done in terrestrial and groundwater areas of the Elkhorn Slough watershed. Nutrient levels in the slough may be influenced by external sources of nutrient input such as atmospheric processes, water exchange with Monterey Bay, and terrestrial inputs associated with human activities. The chapter closes with suggestions for both basic and applied research that will give us a more complete picture of nutrient sources, as well as links between various habitats and the groundwater system, to better understand nutrient cycling in the slough.

Effects of Land Use Change

Prior to the 1820s, Elkhorn Slough was surrounded by native woodlands, grasslands, and marshes. These terrestrial habitats undoubtedly made sporadic contributions of organic carbon and nutrients to the estuary following fires and floods. Large flocks of migratory wading birds and waterfowl might also have made significant nutrient contributions to the slough in the spring and fall. Although the narrow estuarine channel was open to Monterey Bay only during the rainy season, marine upwelling may have been a source of nutrients and phytoplankton for the slough as well.

Since that time, human activities in the Elkhorn Slough watershed—such as grazing, agriculture, development, and channel dredging—have dramatically altered the landscape (see chapter 7, "History of Land Use"). Changes in the estuary's function and nutrient cycling are unstudied but may be equally dramatic. Historical reconstructions of nutrient budgets in several East Coast estuarine systems suggest that urbanization has led to increased nutrient loading (Peierls et al. 1991; Nixon 1995; Howarth et al. 1996; Valiela et al. 1997). For example, nitrogen inputs to Narragansett Bay, Rhode Island, are thought to have increased tenfold between 1870 and 1920 as populations grew and cities and towns developed sewage treatment systems (Nixon 1995). Nitrogen and phosphorus inputs to the Patuxent River estuary increased by four to ten times between 1960 and 1990 (Boynton et al. 1995). Although similar changes in West Coast estuaries have likely occurred, no systematic nutrient budgets have been developed.

While urbanization has not occurred to any great extent within the Elkhorn Slough watershed, land use here has changed dramatically. Increases in intensity of agriculture and residential development have been especially significant over the last thirty years. Use of fertilizers, pesticides, herbicides, and soil erosion have undoubtedly affected nutrient levels in the estuary. (The effects of contaminants and xenobiotic compounds on the slough ecosystem are discussed in chapter 13, "Land Use and Contaminants.")

In addition, substantial hydrologic changes within the watershed have undoubtedly influenced nutrient flow. The dredging of Moss Landing Harbor and, particularly, the opening of the harbor mouth initiated a fundamental change in slough circulation (see chapters 2 and 4, "Geology" and "Hydrography"). The direct connection of the slough and Monterey Bay led to increased tidal currents and changes in sediment deposition patterns. Railroad construction in the 1870s isolated marshes and altered the flow of water between marshes and channels. Diking and draining of wetlands and their subsequent conversion to pasture resulted in a loss of marsh habitat and a loss of wetland functions such as filtering and buffering (Mitsch and Gosselink 1986). Intact marshes trap and filter sediments and transform nutrients. In the absence of healthy marshes, animal wastes, fertilizers, and other sediments can run off upland fields and directly into Elkhorn Slough.

Carbon, Organic Matter, and Decomposition

Carbon is the building block for all biotic communities. All ecosystems are fueled by the energy stored in the chemical bonds of carbon compounds. Carbon cycling mostly involves gas exchange during photosynthesis (use of CO_2 to produce carbon compounds) and respiration (the release of CO_2 during breakdown of carbon compounds). (Carbon—or primary—production is discussed in chapter 8, "Primary Producers.") Carbon is also recycled through the process of decomposition, where nonliving carbon-based structures (bodies, leaves, feces, etc.) are broken down by biotic or abiotic forces into particulate or dissolved organic matter. This organic matter is an especially important source of energy for aquatic food chains. Decomposition of organic matter takes place in the water column and in bottom sediments. In shallow systems like the slough, sediment decomposition is probably more important (Hargrave 1973; Graf 1992). Decomposition of dead organic matter is carried out primarily by heterotrophic bacteria that gain energy by converting organic carbon into carbon dioxide by means of oxygen, nitrate, iron, sulfate, or some other electron acceptor. Thus measurements of bacterial production and oxygen consumption can provide indirect information about organic matter levels and decomposition cycles.

Sediment oxygen consumption, a measure of bacterial activity, was compared at various subtidal sites in Elkhorn Slough in 1994–1995 (table 12.1). Rates were compared at sites adjacent to agricultural land, grazing land, and the more natural lands of the Elkhorn Slough National Estuarine Research Reserve (ESNERR) (fig. 12.1; Caffrey 1996). Values varied widely (range: 19.4–133.4 mmol m^{-2} d^{-1}) among seasons and from site to site, but in general, sediment oxygen consumption was highest near agricultural lands, intermediate in grazed areas, and lowest within the reserve. There were few clear seasonal trends, but rates in natural sites were highest in January and lowest in May. Overall, sediment oxygen consumption in Elkhorn Slough was two to four times higher than the average annual rate in South San Francisco Bay (Caffrey et al. 1996) and Tomales Bay (Dollar et al. 1991). These high sediment oxygen consumption rates suggest that Elkhorn Slough may be quite rich in organic matter compared to other estuarine systems. Fertilizer runoff from agricultural fields undoubtedly contributes to high rates of primary production in these areas, and much of this organic matter is broken down in the sediments.

Direct measurements of bacterial production and abundance were recently made in the intertidal microbial mats of Elkhorn Slough (Colet 1997; Hogan and Ward 1998). Mean bacterial production measured using the thymidine incorporation method was 0.09 mg C m^{-2} hr^{-1}, which is low compared to sandy North Sea sediments (van Duyl, van Raaphorst, and Kop 1993) or tropical seagrass beds (Moriarty et al. 1984). When this technique is applied in sediments, it does not measure the activity of the whole bacterial community, particularly the activity of some anaerobes such as sulfate reducers (Findlay, Meyer, and Edwards 1984; Pollard and Moriarty 1984). In many estuaries, bacterial decomposers primarily use sulfate reduction to break down organic matter (Moriarty et al. 1984; Sørensen 1987; Capone and Kiene 1988). Sulfur cycling has not been investigated at Elkhorn Slough, but the presence of black, anaerobic sediments, smell of hydrogen sulfide, and appearance of *Beggiatoa* and purple and green photosynthetic sulfur bacteria suggest that sulfate reduction is an important process here. Bacterial abundance ranged from 0.4 to 1.3 x 10^9 cells g^{-1} wet weight with greater numbers near the sediment-water interface. These values are comparable to estimates from other estuarine sediments (Capone and Kiene 1988; Jørgensen and Revsbech 1989; van Duyl, van Raaphorst, and Kop 1993). The abundances of bacteria have been measured in many different aquatic systems, and the number of bacteria appears to be relatively constant. This is in contrast to measurements of bacterial activity that generally have a much wider variation in rates in different environments.

Sediment surface at low tide with Japanese mud snails interspersed with patches of sulfur-oxidizing bacteria, Beggiatoa. *Photo credit: Jane Caffrey.*

Net Ecosystem Metabolism

Net ecosystem metabolism is the balance between primary production and respiration (Odum 1969; Kemp et al. 1997; Smith and Hollibaugh 1997). If production is greater than respiration, then the system is considered to be autotrophic; systems where respiration is greater than production are heterotrophic. Net ecosystem metabolism has been characterized for only a few estuarine systems and appears to be controlled by the ratio of inorganic nitrogen to organic carbon loading (Kemp et al. 1997). Autotrophic estuaries such as Chesapeake Bay and Narragansett Bay are dominated by inorganic inputs, while heterotrophic estuaries are dominated by organic inputs (Kemp et al. 1997; Smith and Hollibaugh 1997). Net production and respiration in Elkhorn Slough were estimated in March and August 1972 (Smith 1974) from diurnal dissolved oxygen curves (Odum 1956). Based on these estimates, net ecosystem metabolism was slightly positive, 0.7 mmol O_2 m^{-2} d^{-1}, but given the uncertainties and assumptions involved in the calculations probably not significantly different than zero. Net ecosystem metabolism from nearby Tomales and South San Francisco Bays has shown that both of these systems are net heterotrophic on an annual basis (Smith and Hollibaugh 1997; Caffrey, Cloern, and Grenz 1998).

Excessive production or input of organic matter can lead to eutrophication and severe degradation of estuarine systems (Nixon 1995). Excessive nutrient inputs have caused problems such as nuisance algal blooms and oxygen depletion in many estuaries. This phenomenon does not seem to be widespread within Elkhorn Slough. Since June 1995, dissolved oxygen concentrations have been measured continuously at two sites within the slough: South Marsh, a well-flushed site similar hydrographically to the main channel; and Azevedo Upper Pond, which has restricted circulation and receives agricultural runoff. In summer, dissolved oxygen concentrations ranged from 60 to 120% saturation at the South Marsh site; in contrast, at the Azevedo site, dissolved oxygen concentrations varied from near 0 to >250% saturation (fig. 12.2). Even during winter, when lower metabolic rates and rapid exchange between the atmosphere and water column are expected, dissolved concentrations were more variable at Azevedo than South Marsh. Measurements of pH showed a similar diel (24-hour) pattern, indicating that carbon fixation and respiration were occurring. This diel pattern is characteristic of natural waters and represents the balance between primary production and

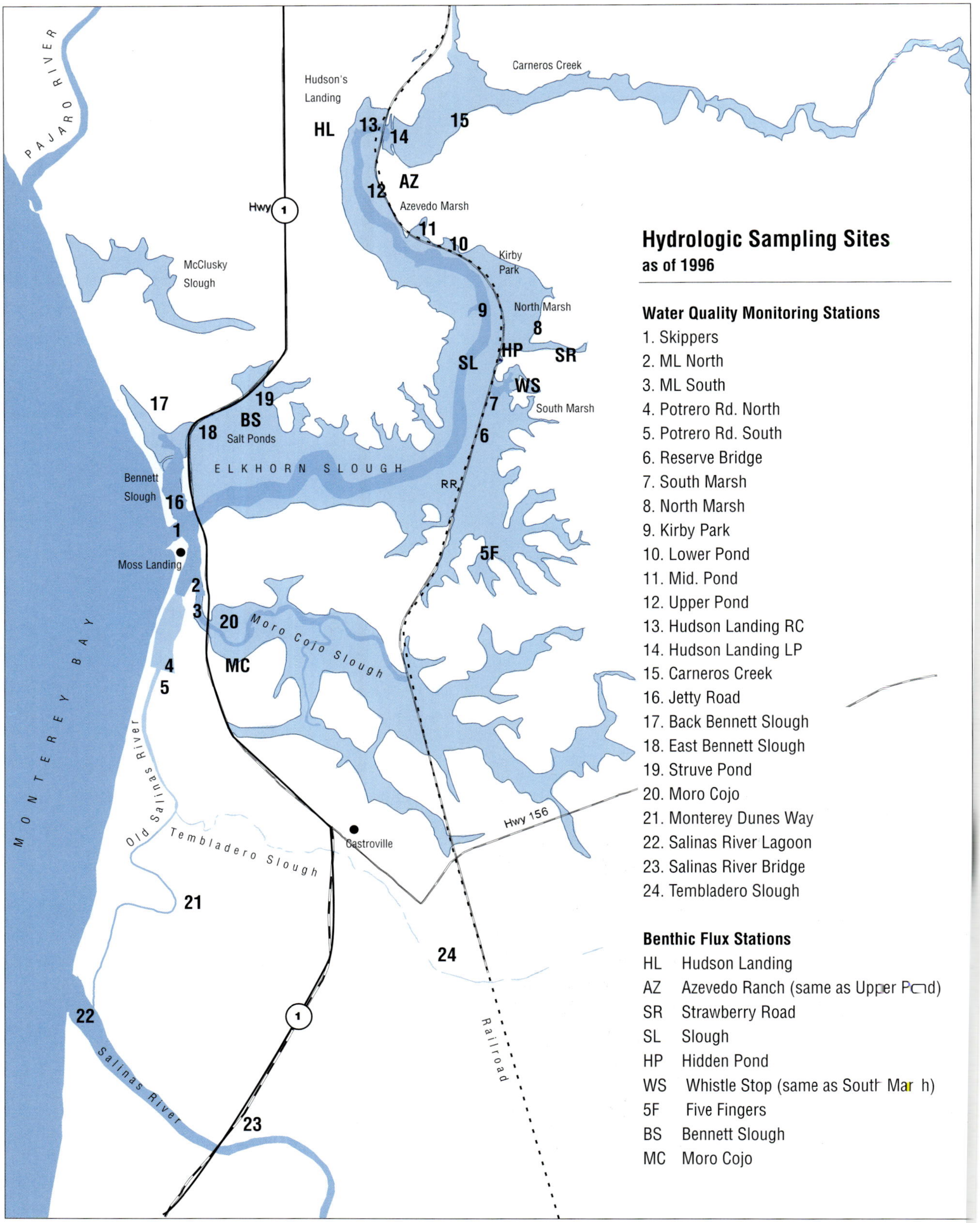

Figure 12.1. Location of sampling stations for water monitoring and benthic fluxes.

Table 12.1. Rates of sediment oxygen consumption, ammonium mineralization, ammonium flux, nitrate flux, nitrite flux, intact core nitrification, potential nitrification, and denitrification. All rates in mmol m^{-2} d^{-1} except for potential nitrification which is in nmol cm^{-3} h^{-1}. (Mean + S.E., n = 3)

	Sediment Oxygen Consumption mean (se)	Ammonium mineralization mean (se)	Ammonium Flux mean (se)	Nitrate Flux mean (se)	Nitrite Flux mean (se)	Dissolved inorganic P flux mean (se)	Potential Nitrification mean (se)	Nitrification mean (se)	Denitrification mean (se)
AGRICULTURAL									
AZEVEDO RANCH									
Sept.	73 (15)	9.1 (2.5)	4.8 (1.1)	−1.1 (0.3)	−0.1 (0.1)	−1.0 (0.3)	0.0 (0.1)	1.1 (1.1)	2.2 (0.8)
Jan.	133 (9)	4.4 (2.8)	37.2 (6.2)	−9.6 (9.7)	2.5 (0.8)	0.5 (0.4)	0.0 (0.4)	0.0 (11.2)	9.6 (0.0)
May	105 (8)	13.7 (1.8)	0.5 (0.3)	0.0 (0.0)	0.0 (0.0)	1.7 (0.1)	10.6 (8.1)	0.0 (0.3)	0.0 (1.23)
BENNETT SLOUGH									
Sept.	68 ([illegible]9)	6.2 (2.8)	−0.2 (0.1)	0.1 (0.1)	0.0 (0.0)	−0.1 (0.5)	1.3 (0.8)	0.2 (0.3)	0.0 (0.3)
Jan.	17 ([illegible]2)	5.4 (5.2)	11.7 (6.1)	−74.9 (32.8)	0.5 (0.4)	0.0 (0.0)	0.0 (0.4)	0.0 (6.4)	65.7 (37.1)
May	99 (2)	7.2 (2.5)	0.1 (0.1)	0.3 (0.0)	0.0 (0.0)	0.0 (0.0)	3.7 (1.2)	0.0 (0.1)	0.0 (0.10)
STRAWBERRY ROAD									
Sept.	0 (4)	6.1 (3.3)	5.1 (0.3)	−0.3 (0.1)	0.0 (0.0)	0.5 (0.3)	0.2 (0.0)	0.0 (1.5)	0.2 (1.4)
Jan.	69 (17)	2.8 (1.9)	35.7 (14.9)	−38.4 (46.0)	−0.3 (0.2)	−0.1 (0.1)	0.8 (0.6)	0.0 (11.5)	11.3 (43.8)
May	82 (2)	4.7 (1.0)	0.3 (0.1)	0.2 (0.1)	−0.2 (0.0)	−2.9 (0.2)	0.4 (0.0)	1.0 (0.7)	1.8 (1.5)
GRAZED									
5 FINGERS									
Sept.	24 (4)	8.6 (1.6)	1.0 (0.5)	1.1 (0.1)	−0.2 (0.1)	−0.8 (0.3)	74.3 (4.4)	0.6 (0.9)	0.0 (1.0)
Jan.	50 (1[illegible])	2.0 (2.5)	2.4 (0.4)	−0.7 (8.3)	0.0 (0.0)	−0.2 (0.1)	[illegible]3.4 (6.1)	0.0 (0.6)	0.7 (8.6)
May	51 ([illegible]5)	2.4 (0.3)	6.5 (0.8)	0.4 (1.0)	0.0 (0.0)	0.0 (0.0)	[illegible]06.1 (10.4)	0.0 (0.6)	0.0 (0.8)
HUDSON'S LANDING									
Sept.	122 (42)	3.5 (1.2)	−0.5 (0.2)	−1.0 (0.4)	0.0 (0.0)	0.0 (0.0)	0.0 (0.0)	0.2 (0.6)	1.1 (1.0)
Jan.	52 (10)	6.4 (5.0)	0.0 (0.0)	9.9 (11.4)	0.0 (0.0)	0.0 (0.0)	0.0 (5.6)	9.8 (4.5)	0.0 (8.7)
May	52 ([illegible])	7.9 (1.2)	3.2 (0.9)	−0.1 (0.2)	−0.2 (0.2)	0.0 (0.0)	6.3 (1.0)	0.0 (0.8)	0.1 (4.8)
MORO COJO									
Sept.	66 ([illegible]3)	1.4 (0.7)	0.7 (0.7)	0.1 (0.2)	−0.0 (0.0)	0.5 (0.5)	0.0 (0.0)	0.0 (0.7)	0.0 (0.6)
Jan.	30 ([illegible]0)	4.7 (3.5)	8.8 (7.2)	1.5 (1.3)	6.6 (1.8)	−2.3 (1.9)	19.4 (7.8)	0.0 (10.9)	0.0 (10.9)
May	29 (6)	3.3 (0.7)	0.7 (0.3)	0.0 (0.0)	0.0 (0.0)	0.8 (0.7)	33.1 (8.9)	0.1 (0.3)	0.0 (0.3)
RESERVE									
HIDDEN POND									
Sept.	34 (10)	4.5 (2.3)	4.6 (0.5)	−0.1 (0.0)	0.0 (0.0)	2.2 (0.9)	2[illegible].1 (2.1)	0.0 (0.1)	0.1 (0.1)
Jan.	57 (1[illegible])	2.1 (1.9)	3.5 (0.2)	6.8 (3.0)	0.0 (0.0)	0.0 (0.0)	36.7 (1.5)	3.9 (3.8)	0.0 (0.0)
May	25 (5)	11.4 (2.5)	5.4 (1.0)	0.1 (0.1)	0.0 (0.0)	0.0 (0.0)	23.6 (2.2)	0.0 (1.8)	0.4 (3.7)
SLOUGH									
Sept.	19 (1[illegible])	5.4 (1.2)	4.2 (0.4)	0.2 (0.0)	0.1 (0.0)	−0.1 (0.1)	4.4 (3.1)	0.0 (0.7)	0.0 (0.7)
Jan.	45 (1[illegible])	3.7 (2.7)	−0.1 (0.1)	3.0 (2.4)	0.2 (0.1)	0.7 (0.1)	33.7 (5.5)	3.3 (1.5)	0.3 (2.8)
May	20 (8)	6.9 (1.6)	10.6 (2.1)	−0.4 (0.2)	0.0 (0.0)	0.3 (0.0)	110.1 (28.5)	1.5 (0.3)	1.5 (0.3)
WHISTLE STOP									
Sept.	5[illegible] (6)	6.4 (1.9)	2.1 (0.3)	0.5 (0.2)	0.0 (0.0)	0.1 (0.3)	55.7 (4.4)	1.5 (0.3)	1.1 (0.1)
Jan.	8[illegible] (2[illegible])	2.9 (1.7)	−0.8 (0.3)	20. 7 (8.6)	0.0 (0.0)	0.0 (0.0)	22.8 (9.1)	1.2 (0.6)	0.0 (8.0)
May	20 (6)	3.0 (0.9)	1.0 (0.4)	0.8 (0.1)	0.2 (0.0)	−0.5 (0.4)	65.6 (1.7)	0.0 (0.5)	0. 0 (0.5)

Table 12.2. Nitrogen inputs to Elkhorn Slough watershed.

		Area or no. of dwelling units (DU) DU or km^2 *	Estimated application rate kg N/yr +	Estimated NO_3- input to groundwater kg N/yr *	%
Urban residential	DU	4,631		65,100	12%
Suburban residential	DU	4,295		60,400	11%
Rural residential	DU	392		7,300	1%
Commercial/office	km^2	0.8		0	0%
Industrial	km^2	0.3		0	0%
Truck crops	km^2	29.7	568,000	229,000	42%
Berries	km^2	17.0	343,000	128,000	24%
Field crops	km^2	2.0	37,700	8,380	2%
Flowers/nursery/mushrooms	km^2	4.2	81,000	39,300	7%
Fruits/nuts	km^2	1.9	36,000	4,460	1%
Pasture	km^2	1.3		0	0%
Vineyard	km^2	0.1		133	0%
Grazing/open	km^2	110.7		0	0%
Other agriculture	km^2	2.1		0	0%
Recreation open space	km^2	0.5		0	0%
Conservation open space	km^2	2.1		0	0%
Public facilities	km^2	1.1		0	0%
Utilities/common	km^2	0.2		0	0%
Total acreage	**km^2**	**173**	**1,066,870**	**542,073**	

* From Table 18 of the North County Hydrogeologic Report (Monterey County Water Resource Agency 1995).
+ Estimated application rate for California from USDA.

respiration. Respiration occurs continuously over a twenty-four-hour period, while primary production occurs only during daylight hours.

Nutrients and Nutrient Cycles

In what follows, the focus is primarily on nutrient cycling in the estuary because little research has been done in terrestrial and groundwater areas of the Elkhorn Slough watershed. In particular, attention is given to nitrogen, phosphorous, and silica, the macronutrients considered essential for growth of primary producers (Ryther and Dunstan 1971; Conley, Schelske, and Stoermer 1993; Conley and Malone 1992). Trace substances, known to be important in controlling production in some habitats (e.g., iron in the open ocean; Martin et al. 1994), remain little studied in this or other estuaries.

Nutrient Sources

Ecosystems recycle nutrients internally, but they also exchange (gain or lose) nutrients with outside sources. In estuaries and other coastal ecosystems, freshwater runoff is the primary source of nutrient input (Nixon 1981; Nixon and Pilson 1984), but groundwater (Capone and Bautista 1985; Valiela et al. 1990) and atmospheric sources—atmospheric nitrogen deposition (Paerl 1995, 1997; Howarth et al. 1996) and nitrogen fixation (Howarth, Marino, and Cole 1988; Howarth et al. 1988)—may also be important. Although a comprehensive nutrient budget has not been developed for the Elkhorn Slough watershed, all three major sources are thought to be involved.

Runoff Inputs Freshwater runoff into the slough is highest during the rainy season, usually October to May (see chapter 4 for a discussion of freshwater flow). Runoff varies considerably from year to year. For example, average flow in Carneros Creek, the Elkhorn Slough watershed's main source of freshwater, has ranged from 120 m^3 d^{-1} (in 1988, an extreme drought year) to 34,000 m^3 d^{-1} (in 1993; Monterey County Water Resources Agency, unpublished data). Assuming that nutrient levels are constant, higher freshwater flow would lead to higher nutrient loading to the slough.

During the 2001 water year (October 1, 2000 to September 30, 2001), researchers from the Center for Agroecology and Sustainable Food Systems at UC Santa Cruz collected water samples and estimated annual nitrogen loads in the Pajaro River, Corralitos Creek, Watsonville Slough, and Carneros Creek. Nitrate concentrations at Dunbarton Road (the

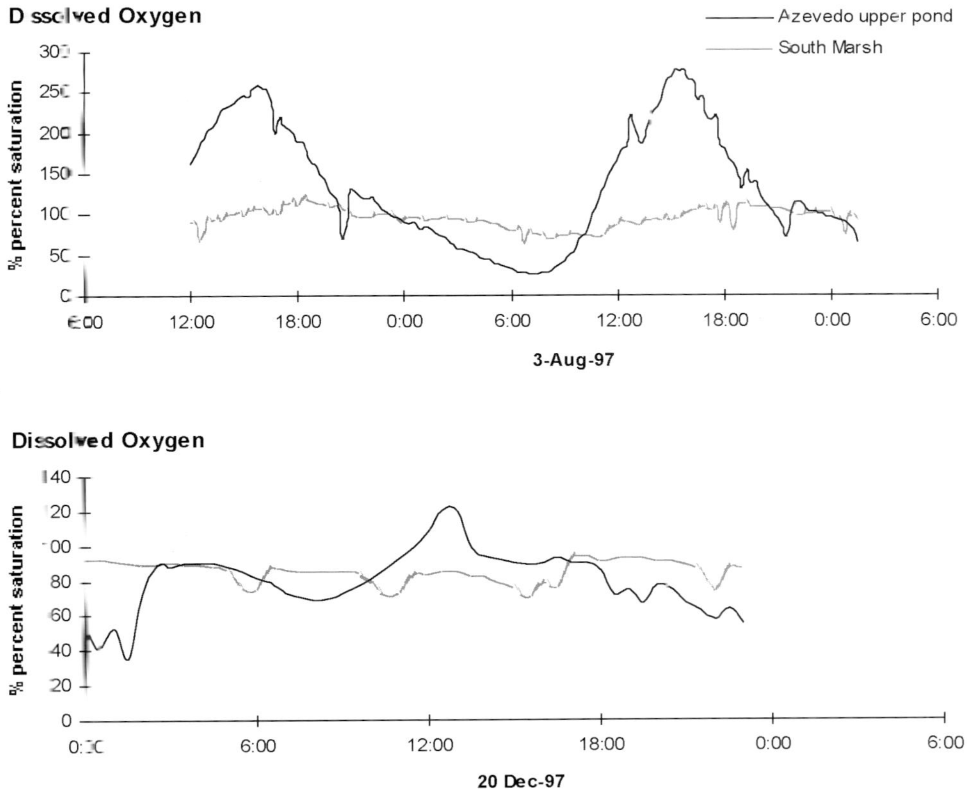

Figure 12.2. Diurnal oxygen concentrations (percent saturation) at South Marsh and Azevedo Upper Pond monitoring sites in August and December 1997.

upstream sampling site on Carneros Creek) were always lower than those collected at San Miguel Canyon Road (the downstream site), where nitrate-N concentrations varied from 0.7 to 17.1 mg N l^{-1}. Although nitrate loss from grazing land and oak woodlands above Dunbarton Road was small, the combination of agricultural activities and low-density housing lower in the watershed contributed to elevated nitrate concentrations downstream in Carneros Creek (Los Huertos, Gentry, and Shennan 2001).

The UCSC researchers estimated that Carneros Creek carried an annual load of 2.97 tons of nitrogen during the 2001 season. Of the nitrogen measured, 23% was in an organic form, reflecting a relatively high sediment load compared with the Pajaro River and Corralitos Creek. After each storm, nitrate concentration increased in sequential ISCO samples collected at the San Miguel Canyon Road site (Los Huertos, Gentry, and Shennan 2001). Nitrate concentrations elsewhere in the slough are also highest during the rainy season, suggesting that runoff is a significant source of nitrate and perhaps other nutrients.

Groundwater Input Groundwater nutrient levels in the Elkhorn Slough watershed are regularly monitored by the Monterey County Water Resources Agency, and some wells in this region have nitrate concentrations above the state drinking water standards of 40 ppm (560 mM). Nutrients occur naturally in groundwater, sometimes in high concentrations (Strathouse et al. 1980; Viets and Hagemen 1971), but elevated levels are often the result of human activities. In Monterey County, the Water Resources Agency estimates that 76% of the nitrates in groundwater are from agricultural fertilizers and the rest are from residential development (MCWRA 1995, table 12.2). The residential input is probably associated with septic systems (Capone and Bautista 1985; Valiela et al. 1990), which are widely used within the Elkhorn Slough watershed (MCWRA 1995).

No matter what the groundwater nutrient concentrations are, groundwater probably contributes little to nutrient levels in Elkhorn Slough, primarily because the flow of groundwater into the slough has declined significantly as agriculture has increased. Fifty years ago, artesian wells, natural seeps, and springs were found throughout the Elkhorn Slough watershed. By 1994, however, irrigation and other human-related water use had led to a serious overdraft (where natural recharge can not keep up with use) of local aquifers (MCWRA 1995). The overuse of groundwater has also caused another significant

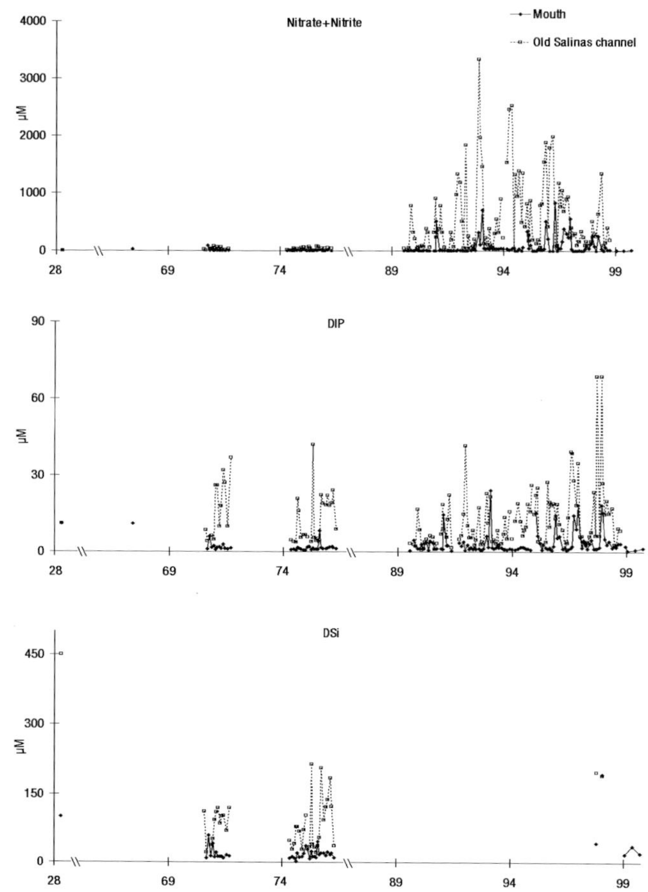

Figure 12.3. Nitrate, phosphate, silica concentrations (µM) at the mouth of Elkhorn Slough and the old Salinas River channel from 1928 to present.

problem: saltwater intrusion. Over the last twenty to thirty years, ocean water has steadily intruded farther and farther inland, raising chloride concentrations in wells near Monterey Bay and Elkhorn Slough (MCWRA 1995).

Atmospheric Inputs Atmospheric sources of nutrients include wet deposition (dissolved nutrients in rain, fog, or other precipitation), dryfall (settling of airborne dust or particles), and fixation of di-nitrogen gas by specialized bacteria. Atmospheric nutrient contributions are unknown for Elkhorn Slough, but some have been measured at Point Reyes National Seashore, 193 kilometers father north. There, nitrogen inputs are 31 µmol m^{-2} d^{-1} for rain plus dryfall and 82 µmol m^{-2} d^{-1} from fog condensate (Holton, Barbour, and Martens 1991). These values are comparable to estimates for Tomales Bay (Smith and Hollibaugh 1997) and Lake Tahoe (Jassby et al. 1994) but generally at the lower end of the range of atmospheric inputs measured worldwide (Paerl 1995). The highest cited values (up to 200 µmol m^{-2} d^{-1}) are in North Sea and Baltic regions with extensive livestock operations. Grazing and feedlots can produce high nutrient inputs, particularly of ammonia. Manure contains high concentrations of ammonia and urea, which readily decomposes and forms ammonia. Ammonia volatilizes into the atmosphere as animal waste decomposes. The effects of livestock operations on atmospheric nutrients in the Elkhorn Slough watershed are unknown at this time.

Furthermore, although atmospheric inputs can be the dominant nitrogen source for some watersheds, they may not contribute as much to the estuary itself because some nitrogen is sequestered in plants or removed by denitrification (Valiela et al. 1997). Atmospheric nitrogen (di-nitrogen gas) is not available to most plants, but specialized nitrogen-fixing bacteria can convert it to a usable organic form. Although nitrogen fixation has not been measured in Elkhorn Slough, research in other estuaries suggests that fixation rates in sediments and the water column are generally low (Howarth, Marino, and Cole 1988; Howarth et al. 1988). Fixation may be more important in freshwater and chaparral habitats; however, since these habitats cover a very small fraction of the Elkhorn Slough watershed, nitrogen fixation is probably not an important source here.

Nutrient Levels: Past and Present

Nutrient levels in Elkhorn Slough have increased significantly during the past sixty years. Nitrate levels, especially, have reached astounding concentrations in some areas of the slough, presumably reflecting the regional shift to intensive agriculture. Estuarine nutrient concentrations were first measured near the mouth of Elkhorn Slough in July 1928. At that time, the nitrate level was 0.5 µM, comparable to concentrations in Monterey Bay and at the Salinas River bridge (fig. 12.3; Bigelow and Leslie 1932; MacGinitie 1935). Dissolved inorganic phosphate concentration was 11 µM, the same as the Salinas River and four times higher than the bay. Dissolved silica was about 100 µM, also four times higher than levels in the bay but 4.5 times lower than levels in the Salinas River. The low concentrations in Monterey Bay are typical of the coastal ocean. In contrast, nutrient concentrations in rivers are often high in phosphorus and silica due to release from soil and rock weathering.

By the time slough nutrient levels were measured again, in October 1967 by the Bureau of Sanitary Engineering, nitrate and phosphate concentrations had increased substantially

Table 12.3. Range of dissolved nutrient concentrations (nitrate and nitrite, ammonium, phosphate, silica) in Elkhorn Slough and Lower Salinas River and other estuarine and coastal environments.

	Nitrate/Nitrite µM-	Ammonium µM	Phosphate µM	Silica µM	References
Elkhorn Slough	0-2000	0-590	0-50	0-100	Smith 1974; Nybakken, Cailliet, and Broenkow 1977; Caffrey et al. 1997
Salinas River	0-5800	0-6500	0-80	0-270	Smith 1974; Nybakken, Cailliet, and Broenkow 1977; Caffrey et al. 1997
Monterey Bay	0-30		0.4-0.6	4-10	Shea and Broenkow 1982
Tomales Bay	0-30	0-8	0.5-4	5-60	Smith et al. 1987; Smith and Hollibaugh 1997
North San Francisco Bay	5-30	1-10	2-5	30-300	Conomos et al. 1979; Schemel and Hager 1986
South San Francisco Bay	5-100	0-30	1-25	0-140	Hager and Schemel 1996
Narragansett Bay, R.I.	0.1-30	0.1-35	0.2-3	0-40	Kremer and Nixon 1978; Nixon and Pilson 1984
Chesapeake Bay					
- riverine	30-120	1-17	0.1-1	16-98	Fisher et al. 1988
- shelf	<1-3	1-5	0.1-1	1-17	Fisher at al. 1988
Delaware Bay					
- riverine	100-140	2-50	1.5-4.3	15-110	Fisher at al. 1988
- shelf	4-7	1-12	0.2-1.7	1-15	Fisher at al. 1988
Hudson River					
- riverine	25-40	2-55	1.2-4	8	Fisher at al 1988
- shelf	3	6	1.2-4	5	Fisher at al. 1988
Neuse River, N.C.	0.02-44	0.03-40	0.03-8		Paerl et al. 1995
Septic tank effluent	300-1100				Clark 1974
Secondary Sewage Treatment Plant	340-3400		800		Clark 1974

(fig. 12.3; Foster et al. 1967). Dissolved silica was not measured at this time. Nitrate concentrations in Elkhorn Slough's main channel were up to 30 µM, while Moss Landing Harbor averaged 90 µM and Tembladero Slough ranged from 1,000 to 3,000 µM. Phosphate concentrations showed a similar pattern, with 8–17 µM in the slough, 19–42 µM in Moss Landing Harbor, and 150–300 µM in Tembladero Slough. Increased agriculture and residential development in the watershed is probably responsible for much of these increases. High concentrations in Tembladero Slough were probably a result of discharge from the Castroville Sewage Treatment Plant. Studies in the 1970s reported similar values. In the old Salinas River channel, winter peaks reached 95 µM nitrate, 46 µM phosphate, and 200 µM silica (fig. 12.3). Peak concentrations in the upper reaches of the slough were somewhat lower (40, 5, and 80 µM, respectively; see Smith 1974; Nybakken, Cailliet, and Broenkow 1977).

Smith and Nybakken also demonstrated that there were significant seasonal, daily, hourly, and tide-level variations in slough nutrient concentrations. Smith (1974) found significant diel variation in dissolved inorganic nutrient concentrations in the lower slough. For example, hourly phosphate levels ranged from 0.5 to 4.0 µM; concentrations were lowest during the day, probably associated with plant uptake during photosynthesis. Statistical analyses indicated that ammonium concentrations varied in conjunction with the semidaily tides, suggesting that ammonium concentrations in the slough were being diluted with Monterey Bay water during high tides (Nybakken, Cailliet, and Broenkow 1977). Nutrient

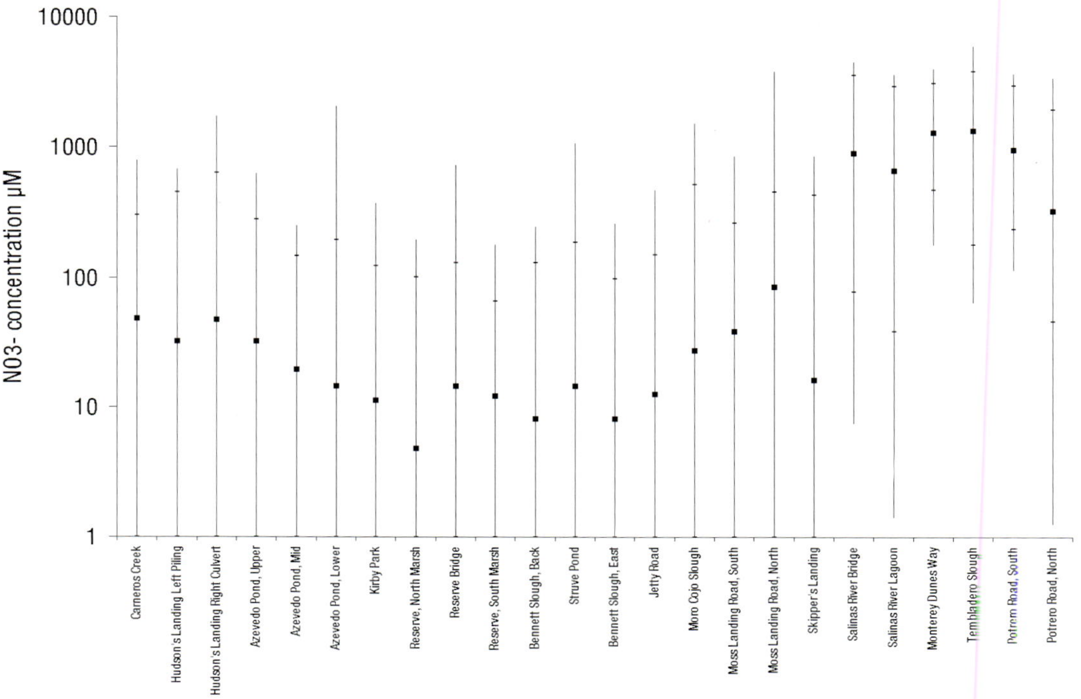

Figure 12.4. Box plot of nitrate concentrations (µM) within Elkhorn Slough and the lower Salinas River between 1989 and 1999. Square indicates median concentration, horizontal dash represents 5th and 95th percentiles, vertical line indicates range of concentrations.

concentrations were also found to be higher during the winter rainy season than in other months (fig. 12.3; Smith 1974; Nybakken, Cailliet, and Broenkow 1977).

Since 1989, nutrient concentrations have been measured monthly at twenty-four stations (fig. 12.1) throughout the Elkhorn Slough watershed as part of a systematic water quality monitoring program. These measurements show that nutrient concentrations, particularly nitrate, have continued to increase (fig. 12.3 and 12.4). For example, at the Carneros Creek station in the uppermost part of the estuary, median nitrate concentration during the nine-year study was 48 µM and the maximum was 790 µM (Caffrey et al. 1997). Nitrate concentrations in Tembladero Slough and the lower Salinas River were even higher, averaging more than 1,000 µM.

Differences in nutrient levels at different sites reflect different inputs and losses. For example, the high levels in the lower Salinas River and old Salinas River channel are probably a result of high nitrogen inputs and relatively low denitrification rates, while in the well-flushed reaches of Elkhorn Slough dilution with low-nutrient Monterey Bay water, uptake by primary producers, and bacterial removal by denitrification result in lower concentrations. The monitoring data also show that nitrate is the dominant form of inorganic nitrogen in much of the slough (nitrate is a more oxidized form). Although primary producers use both nitrate and ammonium, different species have different affinities for these nutrients, so species composition could be affected as the relative availability of the nutrients changes. Nitrate levels are approximately one hundred times higher than ammonium levels at the head of the estuary, although concentrations are comparable in the main channel.

Dissolved phosphorus concentrations have also increased in the Elkhorn Slough watershed, and now range as high as 135 µM in sites receiving agricultural runoff. Peak dissolved phosphorus in the old Salinas River channel appears to have been cut in half since the 1970s, perhaps as result of the diversion of the Castroville Sewage Treatment Plant. Approximately 90% of the inorganic phosphate in the slough is in the dissolved form (Caffrey et al. 1997) despite the high sediment inputs from soil erosion. Although dissolved silica has not been measured routinely in Elkhorn Slough since the 1970s, recent measurements suggest that concentrations have remained the same (fig. 12.3; Caffrey, unpublished data).

Nitrate concentrations in the Elkhorn Slough watershed since the early 1990s are extraordinarily high compared to other estuaries (table 12.3). Levels in the lower Salinas River and old Salinas River channel average almost 1,000 µM and sometimes exceed 5,000 µM, the highest measured for any estuary to date. These high concentrations are probably the result of the intensive agricultural production and drainage of agricultural ditches. Peak phosphate and ammonium concentrations in the slough and the Salinas River are much higher than in other estuaries, and even median phosphate and ammonium concentrations (9 and 4.5 µM respectively) are high relative to many other estuaries (Caffrey et al. 1997).

Nutrient levels in Elkhorn Slough are best understood in contrast to two other nearby, well-studied California estuaries: Tomales Bay and San Francisco Bay. Present-day nutrient concentrations in Tomales Bay and its feeder streams are comparable to levels found in Elkhorn Slough in the 1970s (table 12.3; Smith and Hollibaugh 1997). The lands around Tomales Bay are predominantly used for grazing, as was previously the case for Elkhorn Slough. San Francisco Bay, a highly urbanized estuary, shows slightly higher nutrient levels than Tomales Bay, but still much lower than Elkhorn Slough (table 12.3). Nutrient concentrations in South San Francisco Bay are dominated by inputs from wastewater treatment plants (Nichols et al. 1986), although concentrations can be greatly drawn down during the annual spring phytoplankton bloom (Cloern 1996). Nutrient concentrations in the effluent from wastewater treatment plant are two to six times higher than the average concentrations in the South Bay (Hager and Schemel 1996). North San Francisco Bay, fed by sewage treatment plants and agricultural runoff from the Central Valley, has slightly lower nutrient levels (Conomos et al. 1979; Schemel and Hager 1986).

Nitrogen Cycling

Nitrogen, an essential nutrient for plants, animals, and microorganisms, occurs in a number of organic and inorganic forms, each with unique chemical characteristics. In its organic form, nitrogen combines with oxygen and carbon to build proteins. Atmospheric nitrogen includes di-nitrogen (N_2), the harmless gas that makes up 79% of the earth's atmosphere; nitric oxide (NO) and nitrous oxide (N_2O), rarer gases thought to reduce air quality, deplete stratospheric ozone, and contribute to the greenhouse effect; and ammonia (NH_3) gas, a by-product of respiration. These gaseous forms are not readily available as nutrients for bacteria, algae, fungi, and plants, which instead assimilate water-soluble compounds such as ammonium (NH_4^+), nitrite (NO_2^-), and nitrate (NO_3^-). Plants and microbes take up ammonium and nitrate to make proteins during growth. Plants and microbes are also eaten by animals, which assimilate and concentrate organic nitrogen. When these organisms die or senesce, they become detritus or dead organic matter. This organic matter is then broken down and reconverted into ammonium by bacteria (decomposers). Other bacteria may then convert ammonium to nitrite and nitrate through the process called nitrification. Some nitrogen-fixing bacteria can convert molecular (atmospheric) nitrogen (N_2) into organic nitrogen, which is broken down into ammonium.

In natural, undisturbed ecosystems, nitrogen is typically conserved; that is, inputs and outputs remain equal over time. Although ecosystems are sometimes able to assimilate additional nitrogen inputs (e.g., fertilizers), scientists have begun to recognize that ecosystems can become nitrogen saturated. In these cases, the excess nitrogen is quickly lost to the atmosphere, surface water, or groundwater supplies. The result can be rapid growth of nonnative plants, contamination of ground water, etc. Given its generally high concentrations of nitrogen, Elkhorn Slough could support very active nitrogen cycling.

Aquatic and Estuarine Nitrogen Cycling As discussed previously, nitrogen enters aquatic systems via runoff, atmospheric deposition, groundwater input, and nitrogen fixation. Bacteria transform this nitrogen into more and less usable forms (fig. 12.5). Forms such as nitrate or ammonium are readily available to primary producers from diatoms to pickleweed. Some forms are more easily recycled, while other forms are more easily lost from the system. Clearly, though, the bacterial transformations determine the levels and availability of nitrogen within the estuary.

Nitrogen cycling in the waters of Elkhorn Slough has not been studied, but an in-depth study of nitrogen cycling in slough sediments was conducted in 1994–1995 (Caffrey 1996). The rates of the major nitrogen transformation processes were compared in September, January, and May at nine sites adjacent to three habitat types: agricultural land, grazing land, and the more natural land of the ESNERR (fig. 12.1, table 12.1). Measurements included ammonium remineralization (a measure of the ammonium created when bacterial decomposers break down organic matter), nitrification, and denitrification rates in the sediments and the flux (or exchange) of ammonium, nitrite, and nitrate across the sediment and water column interface.

During this study, remineralization rates ranged from 1.4 to 14 mmol m^{-2} d^{-1}, with no clear differences between habitats (table 12.1). High rates (>10 mmol m^{-2} d^{-1}) were recorded at two

Figure 12.5. Diagram of nitrogen cycle.

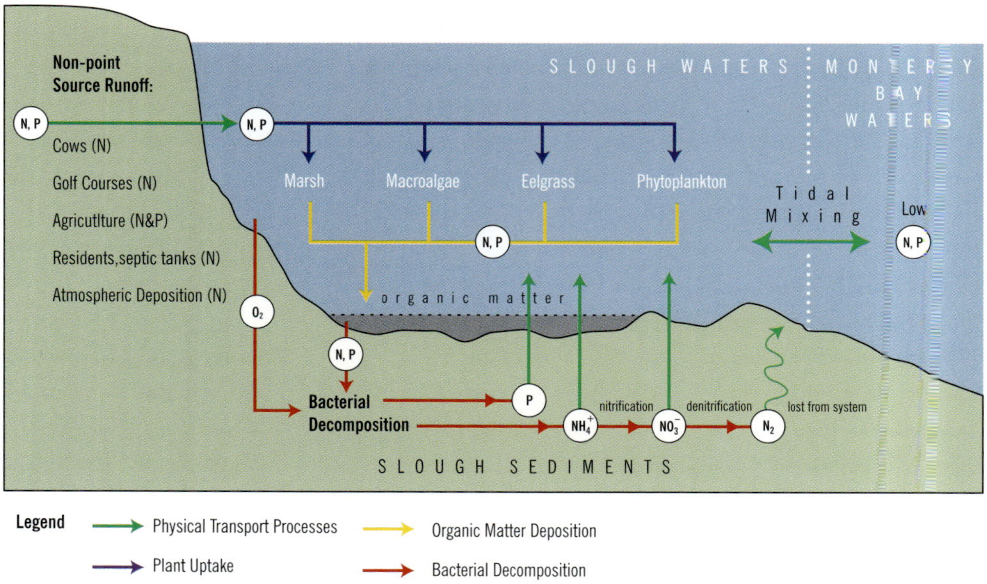

sites, Azevedo Ranch (agriculture) and Hidden Pond (reserve), in May. Rates were generally lowest (<7 mmol m^{-2} d^{-1}) throughout the slough in January. These rates are comparable to rates measured in San Francisco Bay and other coastal sediments (Caffrey 1995). Remineralization rates were generally higher in the top 4 centimeters than deeper in the sediments. This pattern is consistent with that found in San Francisco Bay and other systems, suggesting that organic matter deposited at the sediment surface is rapidly broken down. Extractable ammonium concentrations in sediments (*porewater* ammonium plus that bound to sediment particles) from the agricultural sites were extremely high, averaging nearly double the concentrations at the grazed sites and triple the concentrations at the sites on the reserve.

Although sediments can lose ammonium through nitrification or uptake by benthic algae (Sundbäck et al. 1991; Cerco and Seitzinger 1997), ammonium is usually lost to the adjacent waters through diffusion. Fluxes or exchanges of ammonium between sediments and the water column occur whenever there is a sufficient concentration difference between the two zones. In this study, the highest flux rates were recorded at the agricultural sites in January, when 12 to 37 mmol m^{-2} d^{-1} of sediment ammonium were lost; flux at these sites was much lower in September and near zero in May (table 12.1). Reserve and grazed sites generally had low to moderate (<11 mmol m^{-2} d^{-1}) ammonium loss from sediments. Rates were generally highest in reserve sites in May and lowest in grazed sites in September. Low ammonium uptake by sediments was measured at one grazed and one agricultural site in September and two natural sites in January. Overall, ammonium fluxes in Elkhorn Slough were greater than fluxes in either San Francisco Bay (Caffrey et al. 1996) or Tomales Bay (Dollar et al. 1991). Bacterial processes associated with decomposition such as sediment oxygen consumption and ammonium remineralization are high throughout the slough, particularly in areas adjacent to agricultural fields. This breakdown of organic matter in sediments leads to an accumulation of ammonium with the sediments and ultimately to ammonium fluxes out of the sediments. The high rates of these processes in sediments near agricultural fields suggests that they have a greater supply of organic matter than in the ESNERR or well-flushed main channel.

While ecosystems usually recycle nutrients efficiently, fixed (usable) nitrogen can be lost from the system. Elkhorn Slough ecosystems may lose nitrogen by burial in sediments, advection to Monterey Bay, and the bacterial processes of nitrification and denitrification. In nitrification, specialized (nitrifying) bacteria oxidize ammonium to make nitrite and then oxidize nitrite to create nitrate. These chemoautrophic bacteria use carbon dioxide and ammonium to produce their own energy (Kaplan 1983; Henriksen and Kemp 1988). They also require molecular oxygen, so nitrification is restricted to the narrow aerobic zone near the sediment-water interface (Kaplan 1983; Henriksen and Kemp 1988). Nitrification by itself does not necessarily result in a net loss of nitrogen from the system. Nitrite and nitrate are highly soluble and susceptible to leaching, but they are often conserved by rapid uptake into plants. It is the second step, denitrification, that actually removes fixed nitrogen from the system. Denitrifying bacteria transform nitrate into di-nitrogen gas (N_2), which cannot be used by most producers. Denitrifying bacteria are heterotrophic, requiring a source of organic carbon as well as nitrate, and only function in anaerobic environments, so denitrification is limited to sediments in most estuaries (Koike and Sørensen 1988; Seitzinger 1990). Many studies have shown there is a strong coupling between nitrification and denitrification in estuarine sediments (Jenkins and Kemp 1984), which means that most of the nitrate produced during nitrification is rapidly denitrified and lost to the system. Thus, denitrification, which can remove as much as 50% of the nitrogen input (Seitzinger 1988), is a key process controlling nitrogen availability.

In this study, nitrification rates were measured using two different techniques. Intact core nitrification, measured in intact cores at ambient temperatures, is the amount of activity actually occurring at the time of sampling (Sloth, Nielsen, and Blackburn 1990; Caffrey and Miller 1995). Potential nitrification, measured with sediment slurries under optimal conditions, represents the maximum potential activity of the nitrifying bacteria present at the site (Henriksen and Kemp 1988). Intact core nitrification rates were low (0–1.6 mmol m^{-2} d^{-1}) at most sites in all seasons (table 12.1). Higher rates were recorded only in January at Hudson's Landing (9.8 mmol m^{-2} d^{-1}), a grazed site, and at two ESNERR sites (3–4 mmol m^{-2} d^{-1}). In San Francisco Bay, intact core nitrification rates ranged between 0.3 and 3 mmol m^{-2} d^{-1} (Caffrey and Miller 1995). Potential nitrification rates were considerably higher (2.4–26 mmol m^{-2} d^{-1}) at the natural and grazed sites and were similar to rates measured in San Francisco Bay (Caffrey, unpublished data) and Tomales Bay (Joye 1993). Rates were near zero at all agricultural sites in most months. In general, potential nitrification rates were highest in May throughout the study area. Inhibition of nitrification was probably occurring in some of the Elkhorn Slough sediments, particularly at the agricultural sites. Oxygen is an essential requirement for nitrification, so low oxygen concentrations in porewaters may have contributed to low rates. In addition, sulfide can inhibit nitrification (Joye and Hollibaugh 1995), and that may have contributed to the low rates.

Fluxes of nitrate and nitrite across the sediment water interface were also measured (table 12.1). In January, significant nitrate fluxes (3.0–21 mmol m^{-2} d^{-1}) out of the sediment were measured at all the ESNERR sites, while nitrate fluxes of 9.6–75 mmol m^{-2} d^{-1} were into the sediment at all three agricultural sites that had very high nitrate concentrations in the overlying water. Nitrate fluxes were near zero at all sites in September and May. Nitrite fluxes were low (<0.5 mmol m^{-2} d^{-1}) at most sites throughout the study. Significant nitrite fluxes (0.5–6.6 mmol m^{-2} d^{-1}) out of the sediment occurred only in January at one grazed site (Moro Cojo) and two agricultural sites (Azevedo Ranch and Bennett Slough). Nitrite concentrations and fluxes are generally very low because nitrite is rapidly converted to nitrate by nitrifying bacteria.

Denitrification rates were estimated based on nitrification rates and nitrate fluxes. The major assumptions in this calculation are that nitrate concentration in porewater is constant over the incubation period and that nitrate reduction to ammonium is not significant. Denitrification was generally low (<2.2 mmol m^{-2} d^{-1}) at all sites in May and September. In January, rates exceeded 10 mmol m^{-2} d^{-1} at two agricultural sites and were zero at the ESNERR and grazed sites (table 12.1). These high denitrification rates at the agricultural sites were a result of high nitrate concentrations in the water column, which led to high nitrate fluxes into the sediments.

The 1994–1995 study discovered (1) high rates of sediment oxygen consumption, ammonium remineralization, and ammonium flux (into the water column), indicating high rates of decomposition of nitrogen-rich sediment organic matter, and (2) generally low rates of nitrification and denitrification, indicating that relatively little of the ammonium regenerated in the sediments was being removed by denitrification. In particular, there does not appear to be very close coupling of nitrification with denitrification. This pattern has been observed in other eutrophic estuaries, particularly Chesapeake Bay (Kemp et al. 1990), and suggests that the nitrogen removal mechanisms in the slough are not very efficient. A variety of factors can contribute to low nitrification rates or inhibition of nitrification. For example, porewater oxygen concentrations may be insufficient to support nitrifying bacteria. Or nitrifiers may be inhibited by sulfide (Joye and Hollibaugh 1995), organochlorine and carbamate pesticides (Martens and Bremner 1997), and other compounds likely to be present in Elkhorn Slough.

Terrestrial Nitrogen Cycling The Mediterranean climate (wet winter, dry summer) drives terrestrial nitrogen dynamics in the Elkhorn Slough watershed. As the soils dry after the rainy season, usually about April and May, annual grasses senesce and soil microbes die off, creating nitrogen-rich organic matter. The remaining soil microbes break down this organic matter, releasing inorganic nitrogen, primarily in the form of ammonium. Some of the ammonium is oxidized to nitrate. Without rain, nitrification slows but there is no leaching, so both nitrate and ammonium concentrations accumulate during the summer drought. In fact, available inorganic nitrogen can increase tenfold during the summer months (Los Huertos 1999). However, without soil moisture, few organisms are able to use these nutrients. When the rains return and soil moisture increases, plants and soil microbes begin to assimilate the available nitrogen, but if early storms bring heavy precipitation, the nitrate can be leached out before the organisms use it. Natural leaching levels in the Elkhorn Slough watershed are unknown, but there is evidence that the annual grasses have soil water nitrate concentrations above 710 μM (10 ppm-N), which is the safe drinking water standard.

Core from upper Azevedo pond showing black, reduced sediments up to the sediment-water interface with Beggiatoa. *Photo credit: Jane Caffrey.*

Because nitrogen is the primary limitation to plant growth, nitrogen fertilizers are used widely to increase agricultural production. As a result, nitrogen saturation is common in agroecosystems, even where the consequences of nitrogen excess threaten human and ecosystem health. In fact, nitrogen saturation is probably required to maintain economically viable agroecosystems in California (Los Huertos, pers. comm.).

Approximately 16% of the Elkhorn Slough watershed is used to grow strawberries, raspberries, cut flowers, and other vegetable crops. Nitrogen budgets have not been calculated for agriculture lands of the central California coast, but fertilizers are thought to add 50–200 kg of nitrogen/hectare per year (D. Mountjoy, pers. comm.). Assimilation rates vary with fertilizer application methods, crop variety, and weather conditions, but it is reasonable to assume that crops use about 50% of this added nitrogen. The remaining 25–100 kg/hectare of nitrogen that is not assimilated may be incorporated into soil organic matter, lost as gases, or washed into groundwater or surface water.

Numerous studies have shown that the timing of fertilizer application can significantly affect water quality (USDA 1994). Applications just prior to major rainstorms can lead to significant nitrogen loss due to runoff. For example, in newly fertilized strawberry fields in the Elkhorn Slough watershed, nitrate concentration in surface runoff during an early winter storm was over 1300 μM (18 ppm-N) (Werner et al. 1997). Subsequent storms produced runoff with very low nitrate concentrations, suggesting that the available nitrate had already been washed from the fields.

Elevated nitrogen concentrations and possible nitrogen pollution in Elkhorn Slough are a major management concern. Dairy, grazing, and agricultural activities probably account for a significant proportion of the nitrates in the slough, but their exact contributions are unknown. Crop regimes and agricultural practices have changed dramatically over the last two decades, so current fertilizer use may not be the only cause of elevated nitrogen levels (Keeney 1982). Researchers cannot evaluate possible causes of nitrogen pollution until they determine the slough's nutrient transport routes, transport rates, and transformation processes. Furthermore, since water quality monitoring is costly, it is difficult to obtain enough data to determine how management practices affect nitrogen levels in Elkhorn Slough.

Vegetative buffer strips (VBSs, also known as vegetated filter strips, biogeochemical barriers, protective zones, shelterbelts, and hyporheic zones; Vought et al. 1995) are zones of natural vegetation, situated between cropland and open waterways, that can remove excess nutrients and sediments from agricultural runoff (Karr and Schlosser 1978; Peterjohn and Correll 1984; Lowrance et al. 1984). Grass buffer strips (Dillha et al. 1989; Muscutt et al. 1993) and riparian forests

(Lowrance 1992; Fail, Haines, and Todd 1987; Chescheir et al. 1991; Cooper and Gilliam, 1987) have received the most attention to date. VBSs appear to reduce nitrogen loading through plant assimilation and denitrification, although the relative importance of the two processes isn't yet clear. For example, in eastern riparian forests, denitrification may be spatially limited to soil surfaces with sufficient organic carbon and temporally restricted to periods of high water and low oxygen (Lowrance 1992); nonetheless, it can still account for 50% of the nitrate loss (Hanson, Groffman, and Gold 1994). Furthermore, mature riparian forests may not function as nutrient filters, because they are expected to have little or no net annual uptake of nutrients (Omernik, Abernathy, and Male 1981) Such forests might have to be harvested periodically if they are to serve as nutrient sinks (Lowrance et al. 1984).

The relative importance of denitrification and plant assimilation in the Mediterranean climate of the central coast of California is unknown. Because they can reduce nitrogen and other nutrients in agricultural runoff, VBSs appear to be a promising tool for conservation and restoration, but further testing is needed. For example, nutrient concentrations are sometimes higher in the effluent leaving buffer strips than in the incoming runoff. This suggests that nutrients may be trapped by buffer strips but then released later (Dillha et al. 1989).

Work by UC Santa Cruz graduate student Marc Los Huertos appears to confirm this observation. From 1996 to 1999, Los Huertos studied the efficacy of VBSs planted on the Azevedo Ranch at reducing nutrients entering Elkhorn Slough from adjacent cropped fields. He found that although the VBSs and soil microbes took up nitrogen during the growing season (winter and spring), nitrate and ammonium forms of nitrogen were released into the soil when plant material and soil microbes died in the summer and fall. High concentrations of soil nitrate can then be carried to the slough in the fall after the first rainfall events via subsurface flow. Los Huertos concluded that nitrogen retention by vegetation buffers in California's climate may not be as successful as grass buffers in other parts of the country and that their use should be combined with other strategies to control nitrogen losses from agricultural fields (Los Huertos 1999).

Phosphorus Cycling

Phosphorus is another essential macronutrient that can limit primary production. Approximately 90% of the phosphate in the slough is dissolved inorganic phosphate (Caffrey et al. 1997) Phosphorus cycling is strongly controlled by physical processes such as dissolution of particulate phosphorus and diffusion across the sediment-water interface. Regeneration or decomposition of organic matter and interactions with iron cycling can also affect phosphorus levels. In several estuaries, dissolved inorganic phosphate fluxes across the sediment-water interface have been correlated with mineralization rates, sediment oxygen consumption (Reay, Gallagher, and Simmons 1995), or bottom water oxygen concentrations (Kemp and Boynton 1984; Koop et al. 1990). In 1994–1995, dissolved inorganic phosphate fluxes were low (<3 mmol m^{-2} d^{-1}) throughout Elkhorn Slough (table 12.1). Phosphate fluxes into and out of sediments were observed in all months. Dissolved inorganic phosphate flux did not correlate with sediment oxygen consumption, ammonium flux, or dissolved inorganic phosphate concentration in the overlying water. Thus, phosphorus release from sediments in Elkhorn Slough does not seem to be directly related to regeneration of organic matter or the physical process of diffusion across the sediment water interface.

The role of phosphorus in lakes has been well documented (Schindler 1981; Vollenweider 1976; Hecky and Kilham 1988), but its importance (relative to that of nitrogen) in controlling production in estuarine and marine environments is still debated (Nixon and Pilson 1984; Smith 1984; Hecky and Kilham 1988; Caraco, Cole, and Likens 1990; Conley 1998). Usually, the ratio of dissolved inorganic nitrogen (DIN) to dissolved inorganic phosphorus (DIP) in the water column is compared with the 16N:1P ratio in phytoplankton (the so-called Redfield ratio). If the DIN:DIP ratio in a system is less than 16, nitrogen is probably limiting; if the ratio is greater than 16, phosphorus is limiting. DIN:DIP ratios measured in Elkhorn Slough in the 1970s were well below 16, suggesting that phosphorus was not limiting (Table 12.4; Smith 1974; Nybakken, Cailliet, and Broenkow 1977). Recent (1990s) studies, however, reported increased ratios at comparable study sites and much higher ratios elsewhere in the slough. In these more extensive studies, DIN:DIP ratios ranged from 0.7 to 366 (table 12.4). The highest ratios (>200) were found in the lower Salinas River and old Salinas River channel, while the lowest ratios (<7) were in areas where circulation with the main channel of Elkhorn Slough is restricted. These geographic differences among ratios probably result from local differences in nitrogen input and cycling rather than changes in phosphorus levels.

Table 12.4. Dissolved inorganic nitrogen to dissolved inorganic phosphorus (DIN:DIP) ratios in Elkhorn Slough water column.

	DIN:DIP RATIO	
	1970s	**1990s**
Skippers Restaurant	3.6	13.1
Old Salinas River channel	4.4	46.0
Kirby Park	5.0	8.5
Carneros	—	16.7
Hudson's Landing RC	—	8.1
Hudson's Landing LP	—	7.2
Lower Pond	—	0.7
Middle Pond	—	5.1
Upper Pond	—	17.2
Reserve Bridge	—	9.5
South Marsh	—	9.2
North Marsh	—	26.9
Moss Landing North	—	32.0
Moss Landing South	—	15.8
Moro Cojo Slough	—	6.5
Jetty Road	—	7.9
East Bennett	—	3.7
Struve Pond	—	1.2
Back Bennett	—	5.9
South Potrero	—	72.0
Tembladero Slough	—	81.0
Monterey Dunes Way	—	160.0
Salinas River Lagoon	—	207.0
Salinas River Bridge		366.0

Sources: Data for the 1970s are from Smith 1974 and Nybakken, Cailliet, and Broenkow 1977; 1990s data are from Caffrey et al. 1997.

Management Issues and Research Recommendations

Change within Elkhorn Slough and its watershed occurs from natural variations such as the semidiurnal tides, seasonal rainfall patterns, and interannual disturbances such as fires and earthquakes. However, human activities have had a major impact on the function of terrestrial and estuarine ecosystems. Although nutrient cycling in the estuary has been examined, little is known about such areas as nutrient and other biogeochemical cycles in the varied terrestrial ecosystems in the Elkhorn Slough watershed, or the nutrient link between Elkhorn Slough and Monterey Bay. Below are suggestions for additional areas of applied research that would provide information for management decisions, as well as basic research to expand our understanding of productivity and nutrient cycling in the watershed and slough.

Agricultural Runoff and Groundwater Issues

While some studies on the transport and fate of agricultural runoff through vegetative buffer strips have been done (Los Huertos 1999; Rein 1999), more work is needed. Seasonal rainfall, soil types, and slopes in Elkhorn Slough suggest that runoff characteristics in this area may be very different than on the East Coast, where most research has been conducted. Restoration in conservation easements, particularly in riparian corridors, is increasing in Elkhorn Slough, and research on nutrient retention and transport through these areas is required.

In addition, one of the most serious issues facing managers in the Elkhorn Slough watershed is groundwater availability and quality. Research is needed into exchanges between surface water and groundwater.

Nitrate's Impact on Ecosystem Function

In the past sixty years, nutrient levels—particularly nitrates—have risen dramatically, due in large part to the growth of commercial agriculture in the watershed. One of the greatest management challenges is trying to determine the effects of these changes on ecosystem functions and what, if anything, should be done to reduce them. The effects of nonpoint-source pollution from agricultural runoff on estuarine biota and ecosystem processes are of particular concern. This chapter has outlined the way that adjacent land use affects nutrient concentrations and different bacterial transformations. Areas that receive runoff from agricultural fields generally have higher nutrient concentrations and rates of organic matter degradation but lower rates of oxygen-sensitive processes like nitrification.

Continued water quality monitoring in Elkhorn Slough is critical, particularly in subwatersheds where restoration is occurring or new management practices are being tested. Although this chapter has discussed the importance of a variety of nutrient sources to Elkhorn Slough, the magnitudes of some of these sources are unknown. Is the Salinas River and Old Salinas River channel an important source of nutrients to Elkhorn Slough proper, or does the drainage exit only through the harbor into Monterey Bay? Is groundwater or the atmosphere a significant source of nutrient inputs to Elkhorn Slough? In addition, reconstruction of historical nutrient inputs and budgets for several estuaries has provided useful information about baseline conditions before increases in anthropogenic inputs. Such a reconstruction could be useful for Elkhorn Slough as restoration progresses.

Water quality monitoring programs are needed to assess effects of restoration programs and new land management practices. Photo credit: Jane Caffrey.

Productivity and Nutrient Cycling

Applied research in the area of productivity and nutrient cycling can provide useful information for managers, with many issues ripe for analysis. However, essential basic research on estuarine productivity and terrestrial ecosystems still needs to be done. We know very little about organic matter and nutrient cycling or nutrient contributions of terrestrial ecosystems, particularly the native chaparral, grassland, and oak woodland communities within the watershed. Questions to address include the nutrient links between uplands and marshes, and the way that changes in land use in the uplands affect nutrient and sediment supplies to the marshes and sloughs. Most work on organic matter and nutrient cycling in salt marshes has focused on East Coast *Spartina alterniflora* marshes with relatively little work done in pickleweed marshes like those in Elkhorn Slough.

The slough is an organic-rich system, but nothing is known about the dynamics of dissolved organic compounds—specifically, organic carbon and nitrogen—in this system. Similarly, trace element cycling is an open book, particularly the role of iron in organic decomposition and the relationship between iron and sulfur cycling. In most estuaries, sulfate-reducing bacteria decompose a majority of the organic matter in sediments. The fate of the reduced sulfur compounds, whether they become buried in sediments or diffuse and are reoxidized back to sulfate, depends on the availability of iron.

Another major area of basic research is the relationship between Elkhorn Slough and Monterey Bay. Studies are needed to address whether the export of nutrients, trace substances, and organic matter from the slough have an effect on bay processes.

References

Bigelow, H. B., and M. Leslie. 1930. Reconnaissance of the waters and plankton of Monterey Bay, July 1928. *Bulletin of the Museum of Comparative Zoology at Harvard College* 52 (5): 430-581.

Boynton, W. R., J. H. Garber, R. Summers, and W. M. Kemp. 1995. Inputs, transformations, and transport of nitrogen and phosphorus in Chesapeake Bay and selected tributaries. *Estuaries* 18: 285-314.

Brewer, R. 1988. *The Science of Ecology*. Philadelphia: W. B. Saunders.

Caffrey, J. M. 1995. Spatial and seasonal patterns in sediment ammonium concentrations and regeneration rates in San Francisco Bay, CA. *Estuaries* 18: 219-233.

———. 1996. Effect of land-use practices on nutrient dynamics in subtidal estuarine sediments in Elkhorn Slough, CA. Final report to National Oceanic and Atmospheric Administration, Sanctuaries and Reserves Division. Grant #NA47OR0198.

Caffrey, J. M., J. E. Cloern, and C. Grenz. 1998. Changes in production and respiration during a spring phytoplankton bloom in San Francisco Bay, California: Implications for net ecosystem metabolism. *Marine Ecology Progress Series* 172: 1-12.

Caffrey, J. M., D. E. Hammond, J. S. Kuwabara, L. G. Miller, and R. R. Twilley. 1996. Benthic processes in south San Francisco Bay: The role of organic inputs and bioturbation. In J. T. Hollibaugh, ed., *San Francisco Bay: The Ecosystem*, pp. 425-444. San Francisco: American Association for the Advancement of Science Pacific Division.

Caffrey, J. M., and L. G. Miller. 1995. A comparison of two nitrification inhibitors used to measure nitrification rates in estuarine sediment. *FEMS Microbiology Ecology* 17: 213-220.

Caffrey, J. M., S. Shaw, M. Silberstein, A. De Vogelaere, and M. White. 1997. Water quality monitoring in Elkhorn Slough: A summary of results, 1988-1996. Elkhorn Slough Foundation and Elkhorn Slough National Estuarine Research Reserve Report.

Capone, D. G., and M. G. Bautista. 1985. A groundwater source of nitrate in nearshore sediments. *Nature* 313: 214-216.

Capone, D. G., and R. P. Kiene. 1988. Comparison of microbial dynamics in marine and freshwater sediments: Contrasts in anaerobic carbon catabolism. *Limnology and Oceanography* 33: 725-749.

Caraco, N., J. Cole, and G. E. Likens. 1990. A comparison of phosphorus immobilization in sediments of freshwater and coastal marine systems. *Biogeochemistry* 9: 277-290.

Cerco, C. F., and S. P. Seitzinger 1997. Measured and modeled effects of benthic algae on eutrophication in Indian River-Rehoboth Bay, Delaware. *Estuaries* 20: 231-248.

Chescheir, G., J. W. Gilliam, R. Skaggs, and R. Broadhead. 1991. Nutrient and sediment removal in forested wetlands receiving pumped agricultural drainage water. *Wetlands* 11: 89-103.

Clark, J. 1974. *Coastal Ecosystems: Ecological Considerations for Management of the Coastal Zone*. Washington, D.C.: Conservation Foundation.

Cloern, J. E. 1996. Phytoplankton bloom dynamics in coastal ecosystems: A review with some general lessons from sustained investigations of San Francisco Bay, California. *Reviews of Geophysics* 34: 127-168.

Conley, D. J. 1998. Biogeochemical nutrient cycles and nutrient management strategies. *Developments in Hydrobiology: Man and River Systems. Hydrobiologia* 410: 87-96..

Conley, D. J., and T. C. Malone 1992. Annual cycle of dissolved silicate in Chesapeake Bay: Implications for the production and fate of phytoplankton biomass. *Marine Ecology Progress Series* 81: 121-128.

Conley, D. J., C. L. Schelske, and E. F. Stoermer. 1993. Modification of the biogeochemical cycle of silica with eutrophication. *Marine Ecology Progress Series* 101: 179-192.

Conomos, T. J., R. E. Smith, D. H. Peterson, S. W. Hager, and L. E. Schemel. 1979. Processes affecting seasonal distributions of water properties in the San Francisco Bay estuarine system. In T. J. Conomos, ed., *San Francisco Bay: The Urbanized Estuary*, pp. 115-142. San Francisco: American Association for the Advancement of Science Pacific Division.

Cooper, J., and J. Gilliam. 1987. Phosphorus redistribution from cultivated fields into riparian areas. *Soil Science Society of America Journal* 51: 1600-1604.

Dillha T. A., R. B. Reneau, S. Mostaghimi, and D. Lee. 1989. Vegetative filter strips for agricultural nonpoint source pollution control. *Transactions of the American Society of Agricultural Engineers* 32: 513-519.

Dollar, S. J., S. V. Smith, S. M. Vink, S. Obrebski, and J. T. Hollibaugh. 1991. Annual cycle of benthic nutrient fluxes in Tomales Bay, California, and contribution of the benthos to total ecosystem metabolism. *Marine Ecology Progress Series* 79: 115-125.

Fail, J. L. J., B. L. Haines, and R. L. Todd. 1987. Riparian forests communities and their role in nutrient conservation in an agricultural watershed. *American Journal of Alternative Agriculture* 2: 114-121.

Findlay, S. E. G., J. L. Meyer, and R. T. Edwards. 1984. Measuring bacterial production via rate of incorporation of [^{3}H]thymidine into DNA. *Journal of Microbiological Methods* 2: 57-72.

Fisher, T. R., L. W. Harding Jr., D. W. Stanley, and L. G. Ward. 1988. Phytoplankton, nutrients, and turbidity in the Chesapeake, Delaware, and Hudson estuaries. *Estuarine Coastal and Shelf Science* 27: 61-93.

Foster, H. B., H. J. Ongerth, M. E. Stewart, and R. J. Hee. 1967. Sanitary engineering investigation of quality of water and shellfish in the estuary system at Moss Landing, Monterey County, California. California Department of Public Health, Bureau of Sanitary Engineering.

Golet, D. S. 1997. Fine-scale vertical distribution of denitrification potential, denitrifying bacteria, and benzoate utilization in intertidal microbial mat communities. M.S. thesis, University of California, Santa Cruz.

Graf, G. 1992. Benthic-pelagic coupling: A benthic view. *Oceanography Marine Biology Annual Review* 30: 149-190.

Hager, S. W., and L. E Schemel. 1996. Dissolved inorganic nitrogen, phosphorus, and silicon in South San Francisco Bay. I. Major factors affecting distribution. In J. T. Hollibaugh, ed., *San Francisco Bay. The Ecosystem*, pp. 217-236. San Francisco: American Association for the Advancement of Science Pacific Division.

Hanson, G. C., P. M. Groffman, and A. J. Gold. 1994. Denitrification in riparian wetlands receiving high and low groundwater nitrate inputs. *Journal of Environmental Quality* 23: 917-922

Hargrave, B. T. 1973. Coupling carbon flow through some pelagic and benthic communities. *Journal of the Fisheries Research Board of Canada* 30: 1317-1326.

Hecky, R. E., and P. Kilham. 1988. Nutrient limitation of phytoplankton in freshwater and marine environments: A review of recent evidence on the effects of enrichment. *Limnology and Oceanography* 33: 796-822.

Henriksen, K., and W. M Kemp. 1988. Nitrification in estuarine and coastal marine sediments: Methods, patterns, and regulating factors. In T. H. Blackburn and J. Sørensen, eds., *Nitrogen Cycling in Coastal Marine Environments*, pp. 207-250. New York: John Wiley.

Hogan, M. E., and B. B. Ward 1998. Acclimation of a marine microbial sediment community to simulated in situ exposure of 2,4-dichlorophenoxyacetic acid. *Microbial Ecology* 35: 72-82.

Holton, B., Jr., M. G. Barbour, and S. N. Martens. 1991. Some aspects of the nitrogen cycle in a California strand ecosystem. *Madroño* 38: 170-184.

Howarth, R. W., G. Billen, D. Swaney, A. Townsend, N. Jaworski, K. Lajtha, and J.A. Downing. 1996. Regional nitrogen budgets and riverine N and P fluxes for the drainages to the North Atlantic Ocean: Natural and human influences. In R. W. Howarth, ed., *Nitrogen Cycling in the North Atlantic Ocean and Its Watersheds*, pp. 75-140. Dordrecht: Kluwer Academic Publishers.

Howarth, R. W., R. Marino, and J. J. Cole. 1988. Nitrogen fixation in freshwater, estuarine, and marine ecosystems. 2. Biogeochemical controls. *Limnology and Oceanography* 33: 688-701.

Howarth, R. W., R. Marino, J. Lane, and J. J. Cole. 1988. Nitrogen fixation in freshwater, estuarine, and marine ecosystems. 1. Rates and importance. *Limnology and Oceanography* 33: 669-687.

Jassby, A. D., J. E. Reuter, R. P. Axler, C. R. Goldman, and S. H. Hackley. 1994. Atmospheric deposition of nitrogen and phosphorus in the annual nutrient load of Lake Tahoe (California-Nevada). *Water Resources Research* 30: 2207-2216.

Jenkins, M. C., and W. M. Kemp. 1984. The coupling of nitrification and denitrification in two estuarine sediments. *Limnology and Oceanography* 29: 609-619.

Jørgensen, B. B., and N. P. Revsbech. 1989. Oxygen uptake, bacterial distribution, and carbon-nitrogen-sulfur cycling in sediments from the Baltic Sea-North Sea transition. *Ophelia* 31: 29-49.

Joye, S. B. 1993. Spatial and temporal patterns of nitrogen fixation and denitrification in the intertidal and subtidal sediments of a Mediterranean-type estuary: Tomales Bay, California. Ph.D. diss., University of North Carolina at Chapel Hill.

Joye, S. B., and J. T. Hollibaugh. 1995. Influence of sulfide inhibition of nitrification on nitrogen regeneration in sediments. *Science* 270: 623-625.

Kaplan, H. 1983. Nitrification. In E. J. Carpenter and D. G. Capone, eds., *Nitrogen in the Marine Environment*, pp. 139-190. New York: Academic Press.

Karr, J. R., and I. J. Schlosser. 1978. Water resources and the land-water interface. *Science* 201: 229-234.

Keeney, D. R. 1982. Nitrogen management for maximum efficiency and minimum pollution. In F. J. Stevenson, ed., *Nitrogen in Agricultural Soils*, pp. 605-649. Madison, WI.: American Society of Agronomy.

Kemp, W. M., and W. R. Boynton. 1984. Spatial and temporal coupling of nutrient inputs to estuarine primary production: The role of particulate transport and decomposition. *Bulletin of Marine Science* 35: 522-535.

Kemp, W. M., P. Sampou, J. Caffrey, M. Mayer, K. Henriksen, and W. R. Boynton. 1990. Ammonium recycling versus denitrification in Chesapeake Bay sediments. *Limnology and Oceanography* 35: 1545-1563.

Kemp, W. M., E. M. Smith, M. Marvin-DiPasquale, and W.R. Boynton. 1997. Organic carbon balance and net ecosystem metabolism in Chesapeake Bay. *Marine Ecology Progress Series* 150: 229-248.

Koike, I., and J. Sørensen. 1988. Nitrate reduction and denitrification in marine sediments. In T. H. Blackburn and J. Sørensen, eds., *Nitrogen Cycling in Coastal Marine Environments*, pp. 251-273. New York: John Wiley.

Koop, K., W. R. Boynton, F. Wulff and R. Carman. 1990. Sediment-water oxygen and nutrient exchanges along a depth gradient in the Baltic Sea. *Marine Ecology Progress Series* 63: 65-77.

Kremer, J. N., and S. W. Nixon. 1978. *A Coastal Marine Ecoystem Simulation and Analysis*. Berlin: Springer-Verlag.

Los Huertos, M. W. 1999. Nitrogen dynamics in vegetative buffer strips receiving nitrogen runoff in Elkhorn Slough Watershed, California. Ph.D. diss.. University of California, Santa Cruz.

Los Huertos, M., L. Gentry, and C. Shennan. 2001. Land use and stream nitrogen concentrations in agricultural watersheds along the central coast of California. In *Optimizing Nitrogen Management in Food and Energy Production and Environmen-tal Protection: Proceedings of the 2nd International Nitrogen Conference on Science and Policy. The Scientific World 1* (www.thescientificworld.com).

Lowrance, R. 1992. Groundwater nitrate and denitrification in a coastal plain riparian forest. *Journal of Environmental Quality* 21: 401-405

Lowrance, R., R. Todd, J. Fail, D. Hedrickson, R. Leonard and L. Asmussen. 1984. Riparian forests as nutrient sinks in agricultural watersheds. *BioScience* 34: 374-377.

MacGinitie, G. 1935. Ecological aspects of a California marine estuary. *American Midland Naturalist* 16: 629-765.

Martens, D. A., and J. M. Bremner. 1997. Inhibitory effects of fungicides on hydrolysis of urea and nitrification of urea nitrogen in soil. *Pesticide Science* 49: 344-352.

Martin, J. H., K. H. Coale, K. S. Johnson, et al. 1994. Testing the iron hypothesis in ecosystems of the equatorial Pacific Ocean. *Nature* 371: 123-129.

Mitsch, W. J., and J. G. Gosselink. 1986. *Wetlands.* New York: Van Nostrand Reinhold.

Monterey County Water Resources Agency. 1995. *North Monterey County Hydrogeologic Study.* Salinas, CA: Monterey County Water Resources Agency.

Moriarty, D. J. W., P. I. Boon, J. A. Hansen, W. G. Hunt, I. R. Poiner, P. C. Pollard, G. W. Skyring, and D. C. White. 1984. Microbial biomass and productivity in seagrass beds. *Geomicrobiology Journal* 4: 21-51.

Muscutt, A. D., G. L. Harris, S. W. Bailey, and D. B. Davies. 1993. Buffer zones to improve water quality: A review of their potential use in UK agriculture. *Agriculture, Ecosystems, and Environment* 45: 59-77.

Nichols, F. H., J. E. Cloern, S. N. Luoma, and D. H. Peterson. 1986. The modification of an estuary. *Science* 231: 567-573.

Nixon, S. W. 1981. Freshwater inputs and estuarine productivity. In R. D. Cross and D. L. Williams, eds., *Freshwater Inflow to Estuaries*, 1:31-57. Washington, D. C.: U.S. Fish and Wildlife Service, Office of Biological Service.

———. 1995. Coastal marine eutrophication: A definition, social causes, and future concerns. *Ophelia* 41: 199-219.

Nixon, S. W., and M. E. Q. Pilson. 1984. Nitrogen in estuarine and coastal marine systems. In E. J. Carpenter and D. G. Capone, eds., *Nitrogen in the Marine Environment*, pp. 565-648. New York: Academic Press.

Nybakken, J., G. Cailliet, and W. Broenkow. 1977. *Ecologic and Hydrographic Studies of Elkhorn Slough, Moss Landing Harbor, and Nearshore Coastal Waters, July 1974-June 1976.* Moss Landing, CA: Moss Landing Marine Laboratories Technical Publication.

Odum, E. P. 1969. The strategy of ecosystem development. *Science* 164: 262-270.

Odum, H. T. 1956. Primary production in flowing waters. *Limnology and Oceanography* 1:102-117.

Omernik, J. M., A. R. Abernathy, and L. M. Male. 1981. Stream nutrient levels and proximity of agricultural and forest land to streams: Some relationships. *Journal of Soil and Water Conservation* 36: 227-231.

Paerl, H. W. 1995. Coastal eutrophication in relation to atmospheric nitrogen deposition: Current perspectives. *Ophelia* 41: 237-259.

———. 1997. Nutrient-driven eutrophication and harmful algal bloom dynamics: The roles of atmospheric deposition and groundwater as "new" N (and other nutrient sources). *Limnology and Oceanography* 42: 1154-1165.

Paerl, H. W., M. A. Mallin, C. A. Donahue, M. Go, and B. L. Peierls. 1995. *Nitrogen Loading Sources and Eutrophication of the Neuse River Eestuary, North Carolina: Direct and Indirect Roles of Atmospheric Deposition.* Water Resources Research Institute Report, No. 291. Raleigh: WRRI, North Carolina State University.

Peierls, B., N. Caraco, M. Pace, and J. Cole. 1991. Human influence on river nitrogen. *Nature* 350: 386-387.

Peterjohn, W. T., and D. L. Correll. 1984. Nutrient dynamics in an agricultural watershed: Observations on the role of a riparian forest. *Ecology* 65: 1466-1475.

Pollard, P. C., and D. J. W. Moriarty. 1984. Validity of the tritiated thymidine method for estimating bacterial growth rates: Measurement of isotope dilution during DNA synthesis. *Applied and Environmental Microbiology* 48: 1076-1083.

Reay, W. G., D. L. Gallagher, and G. M. Simmons Jr. 1995. Sediment-water column oxygen and nutrient fluxes in nearshore environments of the lower Delmarva Peninsula, USA. *Marine Ecology Progress Series* 118: 215-227.

Rein, F. A. 1999. Vegetative buffer strips in a Mediterranean climate: Potential for protecting soil and water resources. Ph.D. diss., University of California, Santa Cruz.

Ryther, J. H., and W. M. Dunstan. 1971. Nitrogen, phosphorus and eutrophication in the coastal marine environment. *Science* 171: 1008-1013.

Seitzinger, S. P. 1988. Denitrification in freshwater and coastal marine ecosystems: Ecological and geochemical significance. *Limnology and Oceanography* 33: 702-724.

———. 1990. Denitrification in aquatic sediments. In N. P. Revsbech and J. Sørensen, eds., *Denitrification in Soil and Sediment*, pp. 301-322. FEMS Symposium, No. 56. New York: Plenum Press.

Schemel, L., and S. W. Hager. 1986. Sources of nitrogen and phosphorus to northern San Francisco Bay. *Estuaries* 15: 40-52.

Schindler, D. W. 1981. Studies of eutrophication in lakes and their relevance to the estuarine environment. In B. J. Neilson and L. E. Cronin, eds., *Estuaries and Nutrients*, pp. 71-82. Totowa, N.J.: Humana Press.

Shea, R. E., and W. W. Broenkow. 1982. The role of internal tides in the nutrient enrichment of Monterey Bay, California. *Estuarine Coastal, Shelf Science* 15: 57-66.

Sloth, N. P., L. P. Nielsen, and T. H. Blackburn. 1992. Nitrification in sediment cores measured with acetylene inhibition. *Limnology and Oceanography* 37: 1108-1112.

Smith, R. E. 1974. The hydrography of Elkhorn Slough, a shallow California coastal embayment. M.A. thesis, Moss Landing Marine Laboratories. San Jose State University.

Smith, S. V. 1984. Phosphorus versus nitrogen limitation in the marine environment. *Limnology and Oceanography* 29: 1149-1160.

Smith, S. V., and J. T. Hollibaugh. 1997. Annual cycle and interannual variability of ecosystem metabolism in a temperate climate embayment. *Ecological Monographs* 67: 509-533.

Smith, S. V., W. J. Wiebe, J. T. Hollibaugh, S. J. Dollar, S. W. Hager, B. E. Cole, G. W. Tribble, and P. A. Wheeler. 1987. Stoichiometry of C, N, P, and Si fluxes in a temperate-climate embayment. *Journal of Marine Research* 45: 427-460.

Sørensen, J. 1987. Nitrate reduction in marine sediment: Pathways and interactions with iron and sulfur cycling. *Geomicrobiology Journal* 5: 401-421.

Strathouse, S. M., G. Sposito, P. J. Sullivan, and L. J. Lund. 1980. Geologic nitrogen: A potential geochemical hazard in the San Joaquin Valley, California. *Journal of Environmental Quality* 9: 54-60.

Sundbäck, K., V. Enoksson, W. Graneli, and K. Pettersson. 1991. Influence of sublittoral microphytobenthos on the oxygen and nutrient flux between sediment and water: A laboratory continuous-flow study. *Marine Ecology Progress Series* 74: 263-279.

U.S. Department of Agriculture. 1994. Timing nitrogen fertilizer applications to improve water quality. Washington, DC: US Department of Agriculture.

Valiela, I., G. Collins, J. Kremery, K. Lajtha, M. Geist, B. Seely, J. Brawley, and C. H. Sham. 1997. Nitrogen loading from coastal watersheds to receiving estuaries: New method and application. *Ecological Applications* 7: 358-380.

Valiela, I., J. Costa, K. Foreman, J. M. Teal, B. Howes, and D. Aubrey. 1990. Transport of groundwater-borne nutrients from watersheds and their effects on coastal waters. *Biogeochemistry* 10: 177-197.

van Duyl, F. C., W. van Raaphorst, and A. J. Kop. 1993. Benthic bacterial production and nutrient sediment-water exchange in sandy North Sea sediments. *Marine Ecology Progress Series* 100: 85-95.

Viets, F. G. J., and R. H. Hageman. 1971. *Factors Affecting the Accumulation of Nitrate in Soil, Water, and Plants*. Washington, D.C.: Agricultural Research Service.

Vollenweider, R. A. 1976. Advances in defining critical loading levels of phosphorus in lake eutrophication. *Mem. Ist. Ital. Idrobiol.* 33: 53-83.

Vought, L. B. M., G. Pinay, A. Fuglsang, and C. Ruffinoni. 1995. Structure and function of buffer strips from a water quality perspective in agricultural landscapes. *Landscape and Urban Planning* 31: 323-331

Werner, W. R., M. Silberstein, M. Los Huertos, S. Gliessman, L. Newman, R. Tjeerdema, J. Caffrey, S. Shaw, A. De Vogelaere, and J. Mason. 1997. Agriculture and wetlands: Examining the interface at the Azevedo Ranch and the Elkhorn Slough National Estuarine Research Reserve in coastal central California. Final report submitted to National Oceanic and Atmospheric Administration, Office of Ocean and Coastal Resource Management. Grant #NA370R0102.

13

Chapter Thirteen

Land Use & Contaminants

Bryn Phillips, Mark Stephenson, Michele Jacobi, Gary Ichikawa, Mark Silberstein, Martha Brown

Despite its status as a protected wetland, Elkhorn Slough bears the impact of past and present agricultural practices, boating and shipyard activities, other commercial operations such as junkyard facilities, and residential development.

The Central Coast Regional Water Quality Board currently lists the slough as a 303(d) water quality limited water body.* In 1990, Elkhorn Slough was classified by the California State Water Resources Control Board as a water body where beneficial uses of receiving waters have been impaired, and it is included on the United States Environmental Protection Agency's 304(l) list of impaired waters.†

Concerns over water and sediment quality have generated a number of studies on slough contaminants: their sources and how they make their way into the slough, their movement through the slough's food chain, and their effects on the slough's invertebrates, fishes, birds, and mammals.

This chapter summarizes the research on pollutants entering Elkhorn Slough and their impacts on slough biota. We describe the contaminants that affect the slough, focusing on the sources, transport mechanisms, locations, and biological effects of currently and historically used agricultural chemicals, microbial pathogens, commercial products, and other nonpoint source contaminants. (The runoff and fate of nutrients such as nitrate and phosphate are discussed in chapter 12, "Biogeochemical Cycling.") We conclude with a discussion of management issues, research needs, and recommendations for improving water quality.

Elkhorn Slough Contaminants: An Overview

Water and sediment quality in Elkhorn Slough is influenced by inputs from land surrounding the slough. Agriculture accounts for 4,378 hectares (10,818 acres), or 24%, of land use in the slough watershed, with strawberries making up the majority (35%) of the cultivated acreage (U.S. Department of Agriculture [USDA] 1994). Other crops in the watershed include artichokes, flowers, orchard fruits, and berries. Strawberry production can involve application of as many as thirty-six chemical fertilizers and pesticides (Association of Monterey Bay Area Governments 1988). These agrichemicals can contaminate the slough via freshwater runoff and through sediment erosion, which is a major mechanism of contaminant

* List of Water Quality Limited Segments where objectives or goals of the Clean Water Act are not attainable with the Best Available Treatment/Best Control Technology. A 303(d) listed water body must have a Total Maximum Daily Load analysis conducted to assess contaminant sources and loadings.

† The 304(l) list is a "long-list" of waters not meeting water quality goals of the Clean Water Act after implementation of Best Available Treatment/Best Control Technology, due to either point sources or nonpoint source discharges.

Commercial operations at Elkhorn Slough include harbor activities and Duke Energy's power plant, located at the slough's mouth. Photo credit: Mark Silberstein.

transport into coastal surface waters and slough sediments (U.S. Environmental Protection Agency [EPA] 1992). Current erosion rates on agricultural lands in the Elkhorn Slough watershed are estimated to average 33 tons per acre per year on strawberry land, but can be as high as 145 tons per acre during wet winters (USDA 1994). In contrast, natural erosion rates average 1 ton per acre per year (Soil Conservation Service 1984). *Nonpoint source (NPS) pollution* generated by farm use of chemical fertilizers and pesticides has been identified as a primary cause of water and sediment quality degradation in Elkhorn Slough (Werner et al. 1997). Waste from the watershed's dairy and livestock operations, coupled with surface runoff and erosion, is also a potential source of microbial contaminants in receiving waters (US EPA 1992).

Residential areas in the Elkhorn Slough watershed contribute additional contaminants. Tembladero Slough drains runoff from approximately two-thirds of the city of Salinas and numerous agricultural fields via the old Salinas River channel. This runoff contains several pesticides that originate from agricultural or residential applications (Hunt et al. 1999). Other nonpoint source inputs include Moro Cojo Slough and Carneros Creek, located at the head of the slough. All of these waterways also contribute microbial contaminants originating from residential septic systems and residential livestock (Young 1996). Flooding in the winter of 1995 brought water and sediment from the Pajaro River system into the slough. Events such as this deposit unknown amounts of sediment and nonpoint source contaminants from residential areas.

Moss Landing Harbor and other commercial activities are another source of slough pollutants. The harbor is home to commercial fishing vessels, research vessels, and privately owned recreational boats. Routine vessel maintenance includes use of antifouling paints containing organometallic compounds that, when released into the environment, can contribute to sediment and water contamination (Law et al. 1998). Occupied vessels with unregulated sewage containment may be a source of microbial contamination in typical marinas (US EPA 1992). Further study is needed to determine if other commercial operations located near the slough, including junkyards and car crushers, are impacting slough sediment and water quality (Young 1996).

The most obvious commercial structure at Elkhorn Slough is the power plant located along Highway 1 at the slough's mouth. Operated by Duke Energy, this natural gas-powered generating station formerly discharged part of its cooling water into the slough 0.5 kilometer east of Highway 1. It has been suggested that cooling water drawn into the plant from harbor intakes and discharged into the slough could have transported harbor-related contaminants to slough waters, but no studies have been conducted to investigate this hypothesis (ABA Consultants 1986). This discharge has not been used since 1996, but remains functional in order to discharge storm or circulation water. Cooling water from the plant is now discharged directly to Monterey Bay, and it is unlikely that the slough outfall has any current impacts on water quality.

Bioaccumulation studies that measure the amount of chemicals being absorbed by animal tissues have detected high levels of DDT and its metabolites and other pesticides in both resident and transplanted bivalves in Elkhorn Slough and adjacent waterways. Toxicity tests have determined that slough water and sediment collected from several sites are toxic to a variety of invertebrates and vertebrates. Both the bioaccumulation studies and the toxicity tests demonstrate that in some instances, contaminants in Elkhorn Slough have short-term impacts on individual organisms, with implications for long-term effects on community structure and organisms at higher trophic levels (i.e., higher on the food chain).

Contaminant Sources, Transport, and Location

A number of studies in Elkhorn Slough have examined the sources and locations of contaminants and monitored contaminant concentrations in sediment and animal tissue. Two programs figure prominently. The California State Mussel Watch Program (CSMWP) measures tissue concentrations of chemicals in naturally occurring and transplanted bivalves. The CSMWP has monitored as many as ten stations in the slough since the early 1980s, focusing on contaminants such as pesticides, metals, and other organic compounds that can bioaccumulate in bivalves. Currently, the CSMWP monitors only one station in the slough. The Bay Protection and Toxic Cleanup Program (BPTCP), which operated from 1992 to 1998, has collected additional data at several stations in the slough, including some CSMWP stations. The goals of the BPTCP were to identify toxic hot spots by measuring concentrations of contaminants in sediment, determining sediment toxicity, and examining the benthic (bottom-dwelling) community structure. This program targeted areas such as the old Salinas River channel and parts of Moss Landing Harbor thought to be likely *sinks* for toxic chemicals. Other studies have examined pesticide concentrations in soil, sediment, water, and wildlife throughout the slough, and the California Department of Health Services has been monitoring microbial contaminants for over twenty-three years.

Because of the range of land use practices around the slough, a variety of contaminants could have a potential impact on slough waters and sediments. These contaminants fall into three main categories: pesticides, microbial contaminants, and contaminants associated with harbor operations. The following sections describe the contaminants present in slough waters and sediments, their link to land uses around Elkhorn Slough and throughout the watershed, and possible transport routes for their deposition in the slough.

Pesticides

Pesticide Background Extensive agricultural production adjacent to Elkhorn Slough and in the surrounding watershed has prompted a number of studies on pesticide fate and effects. Pesticides may originate from both agricultural and residential applications. Although many organochlorine pesticides have been banned, these persistent compounds are still detected in soils surrounding the slough (Blankinship and Evans 1993; Werner et al. 1997). Soil erosion transports these hydrophobic chemicals into slough water and sediment (Downing et al. 1998). Contaminants that have entered slough waters have been taken up by the resident and transplanted biota (Phillips 1988; Rasmussen 1994, 1996).

DDT is representative of many of the older persistent chlorinated pesticides banned in the 1970s. Its use in Monterey County fell dramatically following a national ban in 1972: in 1970, 33,371 pounds of DDT were applied to 19,387 acres in Monterey County; in 1973, 14 pounds were applied to 31.1 acres (based on data from the Department of Pesticide Regulation Information Systems Branch). Statewide, DDT use fell from 1,164,699 pounds in 1970 (mainly for agriculture) to 160 pounds in 1974, most of which was used for residential pest control (Mischke et al. 1985).

Agricultural production accounts for the majority of pesticide use in the Elkhorn Slough watershed. Here workers prepare fields for planting strawberries. Photo credit: Mark Silberstein.

Table 13.2. Hazard Rating Coefficient (HRC) for common pesticides used in the Elkhorn Slough.

Pesticide	HRC	Usage in 1994 (kg)	HRC x Usage
2,4-D	0.1794	*	*
Abamectin	**0.5604**	105	59
Acephate	0.0167	1306	22
Benomyl	0.1097	903	99
Captafol	0.1431	*	*
Captan	0.1931	4989	963
Carbaryl	0.2572	257	66
Chlorobenzilate	**0.7286**	*	*
Chloropicrin	0.0139	**18880**	262
Chlorpyrifos	**1.1685**	1491	**1742**
Cypermethrin	**2.7551**	26	72
DCPA	0.0673	1396	94
Diazinon	0.6900	979	676
Dicofol	**1.2751**	2039	**2600**
Dodoine	0.2983	*	*
Fosetyl aluminum	0.0023	1086	2
Haloxyfop	0.0629	*	*
Iprodine	0.0819	1542	126
Malathion	0.1295	**17839**	**2310**
Maneb	0.0160	1419	23
Metham-sodium	0.3158	**22736**	**7180**
Methidathion	0.2144	1057	227
Methomyl	0.2530	1392	352
Methyl bromide	0.0142	5835	83
Napropamide	0.0466	195	9
Pyrethrins	0.2803	8	2
Sulfometuron-methyl	0.0105	*	*
Sulfur	0.000006	**16846**	0
Thiram	0.2191	**15957**	**3496**

Note: High HRC numbers indicate increased potential for contaminant impact on slough fauna. Boldface indicates the five highest numbers in each column. *data not available

Concentrations of total DDT from selected study sites are depicted in figure 13.2. The highest concentrations occur through the old Salinas River channel (382 ppb) and in the south end of Moss Landing Harbor (303 ppb). Rice et al. (1993) found total DDT concentrations in south harbor sediments as high as 963 ppb. These concentrations of total DDT coupled with high concentrations in the Tembladero Slough (650 ppb in 1997) are further evidence that erosion and freshwater runoff in the Elkhorn Slough watershed continue to deposit DDT in slough sediments.

Sediment analyses conducted as part of the routine dredging of Moss Landing Harbor have detected total DDT concentrations ranging from 0 to 444 ppb. Two studies by Toxscan, Inc./Kinnetic Laboratories (1993 and 1996), one by Advanced Biological Testing (1996), and one by Harding Lawson Associates (1997) examined sediments to determine the disposal status of dredged material. Dredged materials that do not meet certain quality standards must be disposed of at upland sites. These studies have documented a large range of DDT concentrations and demonstrated large spatial and temporal variability in sediment DDT concentrations in Moss Landing Harbor. Although sediment samples from many harbor locations contain elevated concentrations of DDT, these levels are lower than the concentration that produces ecotoxicological effect (Swartz et al. 1994).

Other persistent organochlorine compounds are located mostly within Moss Landing Harbor and the Tembladero Slough watershed. Sediment samples collected near Sandholdt Bridge, located at the south end of Moss Landing Harbor, had concentrations of total chlordane as high as 9 ppb (Downing et al. 1998), twice the published sediment quality guideline that predicts probable effect of this chemical on sediment toxicity test organisms (Long and Morgan 1990; Long et al., 1995). The upper Tembladero Slough watershed had a total chlordane concentration of over 20 ppb (Downing et al. 1998). Because Tembladero Slough empties into the old Salinas River channel, it may be the source of pesticide-laden sediments found in Elkhorn Slough. Several areas in Elkhorn Slough have high concentrations of dieldrin, with Sandholdt Bridge exceeding sediment quality guidelines by 50% (Long and Morgan 1990; Long et al. 1995). Six of the seven BPTCP stations in the Tembladero watershed had dieldrin concentrations within the top 10% of stations sampled during the BPTCP (Downing et al. 1998). Again, these sediments might be sources for dieldrin found in Elkhorn Slough. Concentrations of endosulfan in sediments from BPTCP stations were occasionally elevated. Sediment from Sandholdt Bridge and the Tembladero watershed stations had concentrations of endosulfan ranging from 0.15 to 8.44 ppb. There are currently no sediment quality guidelines for endosulfan.

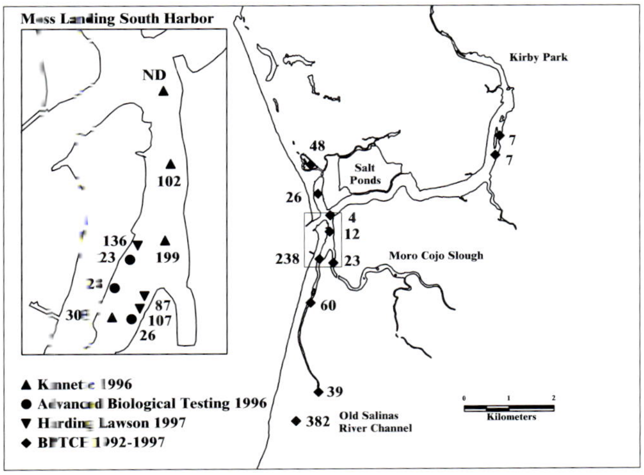

Figure 13.2. DDT concentrations from three independent studies of Moss Landing Harbor and the Bay Protection and Toxic Cleanup Program (BPTCP).

Once eroded sediment and its associated contaminants enter the slough, it is either transported out of the slough mouth into the ocean or deposited in the slough. Deposited sediments can become sinks for further contamination or sources for contaminants in the water.

Monitoring Newer Pesticides The effects of the persistent organochlorine pesticides have been investigated in recent toxicological studies (see "Biological Effects," below), but more research is needed to better understand the sources, transport routes, deposition, and effects of newer, less persistent but more powerful pesticides that may impact Elkhorn Slough. Because these pesticides are not as persistent in the environment, they are more likely to be transported in toxic pulses through surface runoff and less likely to be detected. Kuivila and Foe (1995) demonstrated that there were distinct pulses of pesticides, including diazinon and chlorpyrifos, in the San Joaquin and Sacramento Rivers after rain events. Chlorpyrifos and diazinon are two metabolically activated pesticides that are commonly used for agricultural and residential applications.§

To evaluate the potential impacts of newer pesticide compounds, and to target pesticides for further study, the Hazard Rating Coefficient (HRC) procedure was implemented for Elkhorn Slough (Elkhorn Slough National Estuarine Reserve, unpublished data). The HRC was originally developed to provide a simple means of ranking pesticides based on their potential to impact the estuarine environment (National Oceanographic and Atmospheric Administration [NOAA] 1992). It was designed to evaluate estuarine drainage areas by assessing parameters influencing the aquatic fate and effects of pesticides. As presented by NOAA, the HRC considers four parameters: toxicity to fish, toxicity to crustaceans, bioconcentration factor for fish, and soil half-life. An estimate of the Hazard Normalized Application of the pesticide in the estuarine drainage can be calculated by multiplying the final HRC by the pesticide use in the drainage area. As suggested by NOAA, this system was not designed to provide a quantitative measure of risk from agricultural pesticides, but to rank those pesticides that might affect the ecosystem of an estuarine drainage area.

The Elkhorn Slough HRC was developed to identify which locally applied pesticides might adversely affect the watershed. Fish and crustacean toxicity, bioconcentration factors, and soil half-life were gathered from the literature, converted to subcoefficients, and combined using an evenly weighted, multiplicative aggregation function to calculate the HRC for each pesticide (NOAA 1992). Table 13.2 ranks the coefficients for agricultural pesticides used in the Elkhorn Slough watershed. When the HRC is multiplied by the usage of the pesticide in the watershed, it becomes apparent which chemicals might have an impact on slough biology and should be monitored.

Examination of the HRC values indicates that cypermethrin, dicofol, chlorpyrifos, chlorobenzilate, and abamectin have the highest hazard rating (table 13.2). This ranking changes when the HRC is multiplied by usage in the watershed: metham-sodium, thiram, dicofol, malathion, and chlorpyrifos then become the pesticides of concern. It should be noted that usage data for chlorobenzilate, one of the higher-ranked pesticides, was not available when this list was compiled. Studies conducted by the Marine Pollution Studies Laboratory demonstrate that chlorpyrifos and diazinon in water samples from the Tembladero Slough and Salinas River tributaries cause toxicity to laboratory test organisms (Hunt et al. 1999). These chemicals are commonly used in agricultural as well as residential applications and could potentially impact Elkhorn Slough biota.

§These organophosphate pesticides block the production of the enzyme cholinesterase, which ensures that the chemical signal that causes a nerve impulse is halted at the appropriate time.

Because the HRC is a predictive model of potential contamination impacts, further testing and continuous updating are needed to verify its utility. This validation could take the form of dose-response confirmation of pesticides using toxicity tests and spiked bioaccumulation studies with fish. NOAA (1992) tested the HRC approach by developing a set of coefficients for seven organochlorine pesticides including aldrin, chlordane, DDT, dieldrin, endrin, heptachlor, and lindane. When combined with thirty-five inventoried pesticides, the organochlorine compounds accounted for six of the top seven hazard rating coefficients.

Microbial Contaminants

Coliform bacteria have been common pollutants within Elkhorn Slough since the early 1960s (ABA Consultants 1986). These bacteria are a key indicator of water quality, as their density can be related quantitatively to potential health risks and exposure to pathogens within the water column (California State Water Resources Control Board 1990). Coliforms are microorganisms that commonly live in the intestines of humans and other warm-blooded animals. The presence of coliforms within the water column indicates that fecal material has been introduced, and with it the potential for contamination with pathogenic diseases spread via feces, such as typhoid fever, viral and bacterial gastroenteritis, and hepatitis. Because pathogenic organisms are less dense and usually more difficult to detect than coliform, it is often easier to test for fecal coliforms and then assume the pathogenic bacteria and viruses are present whenever coliform counts are elevated.

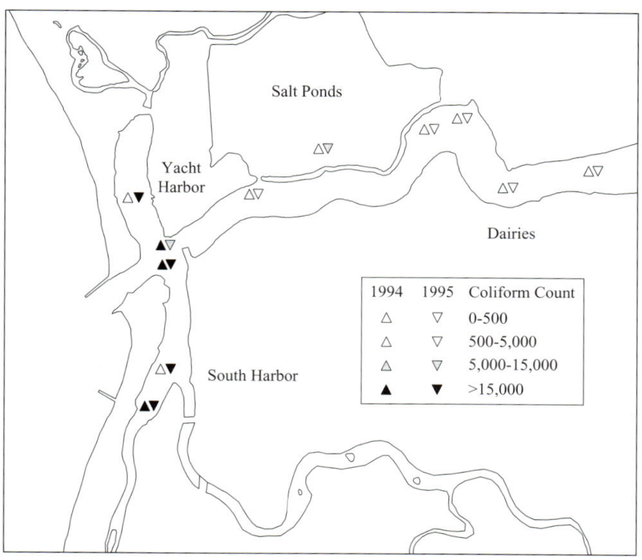

Figure 13.3. Total coliform counts sampled in January 1994 and January 1995.

Coliform pollution can be measured in a variety of ways, and acceptable regulatory levels depend on the specified use of the water body. Generally the Most Probable Number (MPN) method is used to assess total coliform and/or fecal coliform concentrations per 100 mL of water. Water bodies with concentrations of total coliform bacteria greater than 70 MPN or concentrations of 14 MPN fecal coliform are unsuitable for commercial shellfish harvesting, whereas water bodies with concentrations that do not exceed 1,000 MPN of total coliform or 200 MPN of fecal coliform are suitable for recreational use. Coliform bacteria concentrations fluctuate drastically depending on rainfall, nonpoint source inputs, and sample location; therefore, thirty- or sixty-day averages and repeated sampling are often necessary to correctly assess potential pollution problems.

Microbial Sources, Transport, and Locations Bacteria in the slough have been attributed to live-aboard vessels in the harbor, wildlife such as harbor seals, and especially freshwater runoff that transports potential coliform contamination from dairy farming and poorly maintained septic systems (ABA Consultants 1986; Young 1996). Sampling stations located near freshwater inputs have recorded the highest coliform counts. For example, in 1985 and 1986 a station at Hudson's Landing, far from dairies and boats but near freshwater inputs from Carneros Creek, had the highest coliform levels of any station during one sampling period (ABA Consultants 1986). Further evidence that the sources of bacteria were not from vessels in the harbor came from data collected in 1985 that showed the majority of bacteria was not human fecal coliform (ABA Consultants 1986). Data from 1994 and 1995 indicate that there are only moderate levels of bacteria near harbor seal haul-out sites and dairies (Young 1996), and bacteria levels were highest in water samples from the Moss Landing Harbor area (fig. 13.3) and other areas receiving freshwater inputs (Young 1996). Investigations of bacteria levels before and after rain events found that levels were highest after rainfall (Young 1996). Figure 13.4, which tracks coliform counts with rainfall totals at Sandholdt Bridge, demonstrates that total coliform increased during rains. These data suggest that freshwater inputs are the main sources of bacteria to the slough.

Contaminants Associated with the Harbor

Harbor and marina operations such as boat building and maintenance, pier maintenance, and dredging can have an impact on wetland areas (US EPA 1992). Contaminants

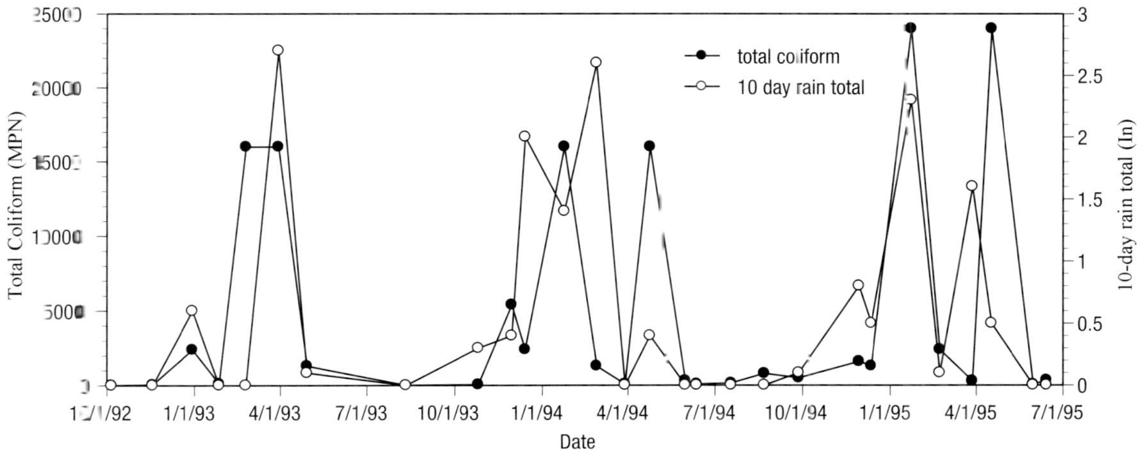

Figure 13.4. Total coliform levels (MPN) at Sandholdt Bridge from October 1992 to July 1995 plotted with 10-day rain total.

originating from antifouling paints used in hull maintenance can contain organometallic compounds such as copper oxides or organotins. Antifouling paints protect the hull by slowly releasing the metallic compound into the environment (Kennish 1992). Organometallic compounds can also be introduced into the environment when paint is removed from boat hulls during routine maintenance.

The effects of copper ion leachates have been well documented in the literature. Copper can affect photosynthesis, growth, and amino acid synthesis in phytoplankton (Thomas and Robinson 1986) and can cause physical abnormalities in local slough fish such as topsmelt (*Atherinops affinis*; Anderson et al. 1991). The effects of organotin compounds include shell thickening in adults and death in larval oysters. TBT (tributyltin, another metallic compound released from antifouling paints) acts as an endocrine disruptor and can lead to the abnormal development of reproductive organs in gastropods (Bryan et al. 1987; Waite et al. 1991). TBT can accumulate in the tissues of marine mammals that eat contaminated prey (Law et al. 1998), suppressing the immune systems (Snoeij et al. 1989).

Although boat maintenance operations are present in Moss Landing Harbor, there are relatively low concentrations of copper and other metals associated with harbor sediments (Downing et al. 1998). However, tributyltin is currently present in the sediment in isolated harbor areas. The previously discussed sediment studies detected concentrations of butyltins in Moss Landing Harbor ranging from approximately 2 to 70 ppb dry weight. The highest concentrations of TBT were found near Gravelle's Boatyard (Toxscan/Kinnetic 1993, 1996; Advanced Biological Testing 1996; Harding Lawson 1997). In the statewide BPTCP database, the highest TBT concentration was 6.21 ppb, which suggests that isolated areas of Moss Landing Harbor contain relatively high concentrations of TBT. Sediment concentrations of TBT at BPTCP stations in other parts of the slough were all low.

Pier pilings containing creosote can release polycyclic aromatic hydrocarbons (PAHs) into the water and, along with fueling operations and surface runoff from roads, can be a source of hydrocarbons (US EPA 1992). Transplanted mussels in Moss Landing Harbor that were placed close to pier pilings did not take up significant amounts of PAHs, but mussels growing on the pilings did (Krzysztof Jop and John Butala, unpublished manuscript). Concentrations of individual PAH compounds released from the pilings were below the chronic effect thresholds for most marine organisms. Bestari et al. (1998) found that creosote-treated pilings in experimental microcosms caused an increase in PAH concentrations in water, but those levels quickly returned to background concentrations due to degradation. Sediment in the microcosms did not demonstrate any increase in PAH concentrations. Low PAH concentrations at BPTCP sediment stations in the slough indicate that PAH contamination from creosote-treated pier pilings and other inputs is probably minimal.

Routine dredging of the harbor channels can resuspend contaminants that are present in harbor sediments (US EPA 1992). The sources and types of these contaminants are varied and include chemicals associated with the harbor itself as well as deposited chemicals from surface runoff and agricultural erosion. For instance, if resuspension of contaminated sediment were to occur during dredging, potentially toxic sediment

Table 13.3. Percent of transplanted California mussels (*Mytilus californianus*) in California State Mussel Watch Program study areas exceeding 95% of all samples measured statewide.

	California Mussel Watch Program Study Area						
	Elkhorn Slough	North California	Humboldt Bay	San Francisco Bay	Los Angeles Harbor	Newport Bay	San Diego Bay
Chemical							
Silver	0	0	0	16	0	0	[illegible]
Copper	0	0	0	0	10	10	[illegible]
Mercury	0	0	0	13	0	18	5
Nickel	0	7	10	5	0	4	0
Lead	0	0	0	1	8	4	5
Zinc	0	0	0	0	8	17	10
Total PAH	0	0	0	6	0	0	2
Total DDT	18	6	0	7	6	13	0
Dieldrin	37	11	0	15	0	1	0
Total Chlordane	0	0	0	3	0	21	0
Tributyltin	3	0	0	2	5	12	6

could be carried into the slough during the tidal cycle. The final section of this chapter discusses dredging as one of Elkhorn Slough's management issues.

Bioaccumulation of Contaminants

Contaminants that are transported into the slough with particulate matter can be monitored with local and transplanted bivalves. These filter-feeding organisms take up and store (bioaccumulate) these contaminants in their tissues. Bioaccumulation studies, such as those employed by the California State Mussel Watch Program, attempt to evaluate the accumulation of contaminants in prey that may subsequently be consumed by organisms at higher trophic levels. Bioaccumulation studies and food web modeling examine the potential effects of contaminants on community and ecosystem structure. Also, by demonstrating levels of exposure at a variety of sites throughout the state, relative comparisons can be made among locations.

Concentrations of contaminants in Elkhorn Slough can be compared to other areas in California using the CSMWP database. Mussel Watch data indicate that transplanted mussels in the slough contain relatively low concentrations of trace metals, PAHs, and total chlordane compared to the rest of the state, particularly more industrial areas (table 13.3). However, Elkhorn Slough mussels contain higher concentrations of total DDT and dieldrin than other sampling locations. These bioaccumulation results track well with contaminants that are in the sediments throughout the state (table 13.4). Compared to other areas of the state, slough sediments contain elevated levels of dieldrin, tributyltin and, to a lesser extent, DDT. DDT readily accumulates in tissues and breaks down slowly, with a half-life of greater than ten years (Woodwell, Craig, and Johnson 1971), and DDT and its metabolites are known to cause detrimental effects on endocrine functions and reproductive viability in birds (Risebrough and Jarman 1985) and fish (Burdick et al. 1964). Organochlorine contaminants can accumulate in benthic invertebrates (Boese, Lee, and Echols 1997) and be transferred to higher trophic levels via contaminated prey.

Concentrations of selected contaminants in several local and transplanted species of bivalves and fish were screened using a Human Health Risk Model based on estimated potential consumption rates (table 13.5; US EPA 1995). This model takes into account several variables, depending on the target group being protected. Using estimated body weight and consumption, screening values are produced that can be compared to tissue loads in bivalves and fish. Boldface values in this table indicate that screening values were exceeded. Concentrations of total DDT were high in bivalves collected from the slough, as well as in bivalves that were transplanted into the south harbor and the yacht harbor. Transplanted

Table 13.4. Percent of sediment samples in California State Bay Protection and Toxic Cleanup Program study areas exceeding 95% of all samples measured statewide

	California Mussel Watch Program Study Area						
Chemical	**Elkhorn Slough**	**North California**	**Humboldt Bay**	**San Francisco Bay**	**Los Angeles Harbor**	**Newport Bay**	**San Diego Bay**
Silver	0	0	0	14	11	0	54
Copper	0	0	0	18	14	7	39
Mercury	0	0	0	33	26	15	26
Nickel	0	0	40	60	0	0	0
Lead	0	0	0	41	26	0	15
Zinc	0	0	0	37	11	0	37
Total PAH	0	0	0	13	5	0	80
Total DDT	4	6	0	4	52	0	0
Dieldrin	30	11	0	20	10	0	10
Total Chlordane	0	0	0	21	26	0	21
Tributyltin	12	0	0	4	8	8	64

bivalves also contained high concentrations of dieldrin. White surfperch collected at Sandholdt Bridge as part of the BPTCP contained detectable concentrations of total chlordane and total DDT as well as several other pesticides and PCBs, but DDT concentrations have been declining in CSMWP samples collected at this station (Stephenson, Martin, and Tjeerdema 1995). Transplanted and naturally occurring mussels in Moss Landing Harbor also contained elevated concentrations of TBT. In some cases these concentrations exceed human health model screening values.

Microbial contaminants can also accumulate in bivalve tissues. The presence of coliform bacteria within Elkhorn Slough has been not only a human health concern but also an economic concern. Commercial shellfish harvesting in Elkhorn Slough began in 1923; by 1931, 45% of California's total shellfish harvest came from this area (ABA Consultants 1986). As development in the areas surrounding Elkhorn Slough increased, so did coliform levels. By 1967 the State Department of Health concluded that no area within the estuary was safe for harvesting shellfish for human consumption. Shellfish collected during 1985 from Parson's Slough to the Highway 1 bridge contained high levels of bacteria and were declared unfit for human consumption by the Monterey County Health Department (ABA Consultants 1986). Sixty percent of Pacific oysters collected from 1988 to 1995 exceeded the wholesale market standard for fecal coliform bacteria of 230 MPN per gram of meat. However, because of the potential for high productivity within Elkhorn Slough, interest in commercial mariculture within the estuary continued. Despite denials of shellfish market permits, the last oyster nursery remained on site until 1990 (Young 1996).

In response to more recent applications for commercial harvesting of oysters and mussels, an Elkhorn Slough Sanitary Survey Report was completed in 1996. Once again, the area was classified as "Prohibited" for the harvest of shellfish for human consumption (Young 1996). The report listed several potential fecal contamination sources, such as Moss Landing Harbor boat facilities, harbor seal haul-out sites, the large numbers of birds within the slough, drainage from several tributaries with high coliform counts, and proximity to cattle operations. Any applicant desiring to culture shellfish from the slough would have to complete a purification effectiveness study for a relaying or depuration (purifying) operation (Bradley 1996). This operation would allow shellfish to be sold if they could be purified in clean water. The Regional Water Quality Control Board has listed shellfish harvesting in the slough as a beneficial use and as such will try to encourage, through implementation of Best Management Practices, a reduction in the number of bacteria in the slough (Karen Worcester, Central Coast Regional Water Quality Control Board, pers. comm.).

Comparing concentrations of contaminants in bivalves with those in sediments and water from similar locations in the slough reveals patterns of spatial contamination and further suggests possible sources of bioaccumulated contaminants. Harbor sites with elevated concentrations of DDT in sediments also have the highest tissue concentrations of DDT in transplanted mussels (figs. 13.1 and 13.2). Suspected sources of DDT and other persistent pesticides include the old Salinas River channel as well as the smaller sloughs that drain local agriculture. High concentrations of TBT were found in harbor mussels that were closely associated with boat storage and maintenance. The locations of the highest levels of microbial contaminants found in bivalve tissues in 1994 and 1995 were in the harbor areas near the freshwater inputs and among the private and commercial vessels (fig. 13.3).

Biological Effects

Once contaminants have been transported into the slough and taken up by organisms, there is potential for biological effects. Harbor dredge studies and the BPTCP have monitored the effects of pollution in the slough through the use of standardized toxicity tests. These tests expose various life stages of algae, invertebrates, and fish to samples of water and sediment in order to determine the effects of contaminants on survival, growth, and reproduction. Although such toxicity tests determine the effects of contaminants on individual organisms, contaminants may also affect the ecosystem in ways that are not directly measured by these tests (Kimball and Levine 1985). In addition to having a wide range of acute or severe short-term impacts, such as mortality, slough contaminants can cause chronic effects that occur over longer periods. These may be expressed as sublethal impacts on reproduction, growth, behavior, or other physiological functions. Thus slough contaminants can affect the short-term survival of individual organisms, and may also have chronic impacts at the community and ecosystem level.

Toxicity Test Results

Most of the BPTCP toxicity test samples were collected in 1992, with additional samples collected at Sandholdt Bridge and the Tembladero Slough in 1996 and 1997. Toxicity at these stations was tested by measuring the survival of laboratory test organisms in sediment and interstitial water (water extracted from sediment samples). Samples were considered toxic if the test organism response in the sample was significantly different from that of a control. The test organism response also had to be less than a statistical threshold based on the overall variability within the toxicity data set for the specific test organism (Phillips, Hunt, and Anderson 2001). There has been consistent toxicity to amphipods at Sandholdt Bridge, Bennett Slough, and other areas in Elkhorn Slough (Downing et al. 1998). There has also been interstitial water toxicity to larval organisms at Sandholdt Bridge and in the harbor. Additional toxicity to freshwater amphipods and daphnids was noted in the Tembladero Slough.

Toxicity identification evaluations have been used to investigate the cause of toxicity at Sandholdt Bridge. Organic contaminants were removed from sediment interstitial water using a solid-phase extraction column and returned to test samples by eluting the column. The addition of an organic binding agent to Sandholdt Bridge sediment also removed toxicity, indicating that the source was one or more organic contaminants. Similar studies in the Tembladero Slough indicate that the cause of toxicity to water column organisms is the presence of the pesticides chlorpyrifos and diazinon. Several studies in Moss Landing Harbor have also measured sediment toxicity and chemistry in duplicate samples. All of these studies have detected varying levels of toxicity in the harbor. With the exception of DDT, most contaminants were at low concentrations or below detection limits. An independent study on the seasonal and annual distribution of organic contaminants in the slough also reported high concentrations of DDT in sediment from south harbor stations (Rice et al. 1993).

Statistical analyses can be used to examine the relationships between toxicity and contaminant concentrations in order to suggest causes for the toxicity. These analyses, conducted by the BPTCP, demonstrated significant relationships between amphipod toxicity and several DDT metabolites. There were also significant correlations with dieldrin, total chlordane, and total DDT. Other Moss Landing Harbor studies have found elevated concentrations of DDT in harbor sediments, with amphipod toxicity coinciding at various study sites.

Compared to other areas in the state, Elkhorn Slough sediments contain elevated concentrations of pesticides and tributyltin (table 13.4), but statistical relationships between

Table 13.5. Total DDT, DDE, Dieldrin and Tributyltin in fish, clam and mussel tissue sampled from the Elkhorn Slough area. Data are from CSMWP and BPTCP studies and are in ppm wet weight except for *Mytilus*. Bold numbers exceed screening values for Human Health Risk Model. TCM = Transplanted California Mussels, RBM = Resident Bay Mussel, TFC = Transplanted Freshwater Clams. Chemical concentrations are listed in ppb.

Site	Tissue Type	Total DDT	DDE	Dieldrin	Tributyltin
Elkhorn Slough	Smelt	nd	0.039	nd	nd
Elkhorn Slough	Perch	nd	0.13	nd	nd
Elkhorn Slough	Shiner Surfperch	**0.63**	0.47	**0.022**	nd
Elkhorn Slough	N. Anchovy	<0.04	0.022	<0.005	nd
Elkhorn Slough	Topsmelt	<0.052	0.032	<0.005	nd
Elkhorn Slough	TCM	**0.133**	nd	**0.018**	nd
Elkhorn /Hwy 1 Bridge	TCM	0.05	nd	**0.006**	nd
Elkhorn /Hwy 1 Bridge	*Mytilus* (mussel)	**0.232**	nd	nd	nd
Elkhorn /Pacific Mariculture	Oyster	**0.073**	nd	**0.003**	nd
Elkhorn /Pacific Mariculture	RBM	**0.084**	nd	0.0005	nd
Elkhorn /PG&E	TCM	**0.103**	nd	nd	nd
Elkhorn /Tidal Pond	TCM	0.06	nd	**0.02**	nd
Moss Landing	*Mytilus* (mussel)	**0.54**	nd	**0.039**	**0.44**
Moss Landing	*Saxadomas* (clam)	**0.238**	nd	nd	nd
Moss Landing	*Zirfaea* (clam)	**0.236**	nd	nd	nd
Moss Landing	*Tresus* (clam)	**0.197**	nd	nd	nd
Moss Landing Harbor	Staghorn Sculpin	0.05	nd	nd	nd
Moss Landing South Harbor	TCM	nd	nd	nd	0.38
Moss Landing Yacht Harbor	RBM	**0.435**	nd	**0.033**	nd
Moss Landing Yacht Harbor	TCM	**0.089**	nd	**0.038**	**0.75**
Parson's Slough	*Mytilus* (mussel)	**0.857**	nd	nd	nd
San Andreas Road	TFC	**1.446**	nd	**0.324**	nd
Sandholdt Bridge	RBM	**0.424**	nd	**0.014**	nd
Sandholdt Bridge	TCM	**0.480**	nd	**0.05**	nd

sediment chemistry and toxicity can only suggest causes. Currently, an environmental risk assessment is being conducted in the Moss Landing Harbor by the Harbor District to determine the exact cause of the toxicity observed there. This assessment is part of an ongoing project to determine how to dispose of dredged sediment. Concentrations of DDT and its metabolites are elevated, but other agricultural chemicals, such as chlordane, also occur in the sediment. Results of toxicity identification evaluations will aid in answering questions related to dredge disposal.

A toxicity study by the Marine Pollution Studies Laboratory demonstrated the biological effects of some local agricultural inputs. This study examined water column toxicity in the Salinas River watershed from 1998 through 1999, and identified consistent toxicity in two agricultural drainages that flow into the Salinas River and the old Salinas River channel. In both drainages there were significant relationships between the occurrence of toxicity and elevated concentrations of the organophosphate pesticides chlorpyrifos and diazinon (Hunt et al. 1999). Identification evaluations demonstrated that these chemicals were causing the observed toxicity: the pesticides and associated toxicity could be removed and added back to the sample using solid-phase extraction columns. Both of these chemicals have been found in concentrations exceeding the effect threshold for laboratory test organisms and above the California Department of Fish and Game chronic water quality criteria (Menconi and Cox 1994).

Table 13.6. Eggshell thickness of Caspian Terns (*Sterna caspia*).

	Parkin (1998)	Ohlendorf et al. 1985	Pre 1947
Eggshell thickness (mm)	0.286	0.334	0.346
SD	0.007	0.018	0.015
n	6	25	5

Food Chain Transfer

Shellfish in Moss Landing Harbor may contain DDT and other pesticides in concentrations above levels safe for human consumption (Joyce Bradley, California Department of Health Services, pers. comm.). In 1985 the Monterey County Health Department issued a "health advisory against eating shellfish in the slough due to pesticide contamination."

Contamination in sediments and invertebrates may also pose a threat to marine mammals through food chain transfer. The California sea otter (*Enhydra lutris*) is a local bivalve predator, and benthic clams—the otter's main prey—contain elevated concentrations of DDT (Kvitek et al. 1988). Recent analysis of five archived tissue samples recovered from dead otters in or near Elkhorn Slough revealed elevated concentrations of butyltin residues. Other sea otters recovered in the Monterey Bay have contained detectable concentrations of DDT metabolites, dieldrin, chlordanes, and PCBs.

Studies to develop a risk analysis model, test sea otter prey from Elkhorn Slough for specific contaminants, and analyze diet to see whether the otters' daily intake exceeds safe levels are now underway (Canright 1999). In addition, in 1999 researchers from UC Santa Cruz and UC Davis, the California Department of Fish and Game, and the Monterey Bay Aquarium collected blood samples from otters occupying three sites: Pleasure Point in Santa Cruz, Cannery Row in Monterey, and Elkhorn Slough. The samples were tested for concentrations of heavy metals, pesticides, and other contaminants, as well as viruses. They were also compared to blood samples collected from various sites in Alaska and from animals from Avila Beach, where an oil spill occurred in the early 1990s (Canright 1999).

Harbor seals may also be at risk if consuming contaminated prey. Moser (1996) detected concentrations of trace elements and organochlorine compounds in hair, blood, and blubber of harbor seals from Elkhorn Slough, and found concentrations of DDE and PCB in blubber samples. Blood concentrations of PCB and DDE were comparable to those of harbor seals from San Francisco Bay (Kopec and Harvey 1995). Blubber concentrations of total DDT and dieldrin were higher in Elkhorn Slough harbor seals than in seals sampled from Oregon and Washington (Moser 1996).

One of the most striking examples of a contaminant affecting slough residents via food chain transfer is that of DDT's impact on the local population of Caspian terns (*Sterna caspia*). After Pajaro and Salinas River floodwaters entered Elkhorn Slough in 1995, the Caspian tern population suffered major reproductive failure. Parkin (1998) had monitored tern breeding since 1993. As the 1995 breeding season began, 26% of laid eggs did not hatch, and 25% of hatchlings died within several days. An additional 10% were lost during the pipping stage. Analysis of eggs indicated elevated concentrations of DDE, PCBs, and toxaphene, which terns can bioaccumulate via fish. DDT is known to cause shell thinning (Risebrough and Jarman 1985), so eggshell thickness was measured. Mean thickness was significantly less than the thickness of tern eggs in other colonies measured before 1947 (before the use of DDT) and during 1981 (Ohlendorf et al. 1985; table 13.6). DDE concentrations in eggs ranged from 3.2 to 26 ppm up to three orders of magnitude higher than fish and invertebrate tissue concentrations summarized in table 13.5.

Management Issues and Research Recommendations

Although several studies have made snapshot analyses of the condition of the slough, there have been no long-term, integrated monitoring programs to answer complex questions regarding water and sediment pollution and its impacts. A comprehensive research and monitoring program to identify and evaluate contaminant sources, transformation, transport, exposure, and effects on slough biota is needed to address a variety of management issues. Here we suggest specific research and management areas that would help us address pollutant issues and give us more complete information on the effects of pollutants on slough organisms.

Pesticide Effects

Results of recent sediment and bioaccumulation studies suggest that banned pesticides such as DDT and its metabolites, dieldrin, and chlordane persist in Elkhorn Slough. Murdoch et al. (1997) state that organochlorine contaminants in sediment will most likely cause chronic effects to benthic organisms rather than acute effects. Although small-scale studies have answered specific questions about contamination in certain areas of the slough, an appropriate and comprehensive monitoring program is needed to determine whether current concentrations of persistent pesticides found in slough water and sediments are causing chronic toxicity to organisms at lower trophic levels, such as benthic invertebrates, and contaminating organisms at higher trophic levels, such as birds and mammals.

The California sea otter could provide a model for such ecotoxological research. Studies by the CSMWP have shown that contaminants in slough sediments are taken up by filter- and deposit-feeding clams and mussels, and are incorporated into shellfish tissue through bioaccumulation. These contaminated benthic clams and mussels are the main prey of sea otters foraging in Elkhorn Slough and thus pose a potential threat to the population. Research on resident otters would help determine whether the ingestion of contaminated bivalves is causing endocrine disruption, immunosuppression, reproductive failure, and other long-term, chronic effects. In addition, analysis of sea otter tissues from Moss Landing/Elkhorn Slough populations could provide information on relative contaminant levels. A focused study could compare tissue concentrations of slough predators with those outside the slough. Some of this work is now under way.

Newer pesticides, which are often difficult to detect due to their transient nature, also pose a threat to the slough ecosystem. A program to monitor these pesticides and other chemicals in slough waters should be established and toxicity studies conducted to determine whether they are causing short- and long-term effects on slough biota. The use of new pesticides in the slough could be further evaluated by using the Hazard Rating Coefficient, which could direct monitoring and research efforts toward pesticides that represent a greater potential risk to slough fauna. Given the difficulty of obtaining funds for monitoring programs and the greater challenge in detecting these newer pesticides, the HRC represents a way to prioritize and focus monitoring efforts.

Toxicity tests conducted on sediments from Elkhorn Slough have demonstrated toxic effects to laboratory organisms, but have not determined causality. Further investigations through the use of chemical analyses and toxicity identification evaluations would help determine the causes of toxicity and suggest potential impacts on the resident biota.

Erosion Control and Harbor Dredging

Erosion control throughout the Elkhorn Slough watershed is one of the region's most pressing management issues. Soil erosion transports agricultural contaminants into slough sediments and water, and because of high erosion rates—up to 145 tons per acre per year, with 57% of the total estimated soil that erodes into the slough coming from land used to grow strawberries (Soil Conservation Service 1984)—there is increasing recognition that erosion control must be implemented to protect the slough. The continued use of Best Management Practices in upland areas could help reduce erosion, but because eroded soils also enter from major freshwater inputs, contaminant inputs from other areas must be considered. These areas include agricultural land around Moro Cojo, Parsons, and Tembladero Sloughs (Blankinship and Evans 1993).

Although erosion control measures seem like an obvious solution, these measures are beneficial only if they are appropriate to the environment in which they are used (Mountjoy 1993). Practices that concentrate and transport runoff or disperse and absorb runoff might not be enough to reduce erosion and agrichemical movement when used alone. A combination of practices that maximize absorption but allow for transport and removal of excessive precipitation is ideal. The cost of these systems might be prohibitive to some area growers, but outreach programs that provide financial assistance could encourage their implementation. Once erosion control practices are in place, a program to monitor their effectiveness should be established. Practices that reduce the use of agrichemical inputs, such as integrated pest management and organic farming techniques, should also be encouraged.

The impact of erosion is particularly visible at Moss Landing Harbor. Because of sediment input during rain events, the harbor needs to be periodically dredged to maintain adequate depth. Discovery of contaminated sediment in the harbor has created the need to transport dredge spoils to appropriate disposal facilities—an expensive and logistically challenging management problem. An environmental risk assessment that includes toxicity identification evaluation studies could help pinpoint the sources and transport routes of pollutants contributing to sediment toxicity in the harbor. If a consistent monitoring program were in place, sediment contamination problems could be characterized early to avoid potential environmental problems. By identifying contaminant sources and pathways, flood and erosion control practices could be directed at target areas to help prevent contamination of harbor sediments.

Bacterial Contamination

The Elkhorn Slough once contained productive aquaculture operations, but high levels of bacterial contamination helped cause their demise. Levels of total coliform are often measured above the limit for recreational use (1,000 MPN) and are consistently above the limit for shellfish culture (70 MPN). There is a large body of evidence demonstrating that the bacteria are entering the slough through freshwater inputs, particularly after rain events, but there is no conclusive evidence of specific sources.

The first step in managing bacterial contamination in the slough is to determine its sources. Tracking bacterial contamination upstream will help determine if the sources are natural or anthropogenic. Once the sources are discovered, a number of practices can be implemented to reduce the input of bacterial contamination into waterways leading to the slough. For smaller livestock areas such as small rural ranches, these practices might include dikes and diversions, sediment control basins, liquid containment structures, or composting facilities. If the bacteria are originating from human sources, steps can be taken to improve faulty or antiquated septic systems through public education and outreach plans.

Harbor Contaminants

Because the Moss Landing Harbor has the potential to introduce different types of contaminants, a variety of practices must be implemented to reduce the input of harbor related pollutants. The most significant inputs from harbor operations are metals from antifouling paints and possibly bacteria and nutrients from boat sewage. Reducing these inputs begins with public education on the environmental impacts of these pollutants. The public must be encouraged to use wastewater disposal facilities, and access to these facilities must be convenient. Containment facilities for boatyard waste must be constructed to eliminate the input of debris from hull maintenance. Although hydrocarbons do not appear to be a problem in Moss Landing Harbor, the construction of waste oil handling sites and the distribution of oil-absorbent pads to boaters will promote the proper management of petroleum products.

Comprehensive Monitoring Program

Addressing the above management issues will require an integrated monitoring effort. Klecka et al. (1999) define monitoring as the long-term and standardized measurement, observation, evaluation, and reporting of physical, chemical, or biological parameters in order to define status, trends, and mass-flows. They go on to say that multimedia monitoring is essential for assessing equilibrium conditions, mass balances, fluxes, persistence, and long-range transport of organic chemicals in the environment. The Water Quality Protection Program for the Monterey Bay National Marine Sanctuary has called for a coordinated monitoring effort to determine the impacts of pollution on sediment and water quality. This monitoring effort could utilize facets of programs such as the Bay Protection and Toxic Cleanup Program (chemical analyses, toxicity testing, and benthic community structure) and the California State Mussel Watch Program (bioaccumulation) at established stations in the slough. Data from these stations would provide information on the sources, transport, and biological effects of contaminants. Modeling tools such as the HRC could be used to help focus sampling and chemical analysis efforts. Additional sampling for microbial contaminants could be included to determine potential sources and impacts. This information, along with data from past studies that have examined different monitoring components, could be integrated into a comprehensive database to add additional spatial and temporal components to the program. The establishment of a monitoring program and database will allow us to better answer questions regarding the fate and effects of contaminants in Elkhorn Slough.

References

ABA Consultants. 1986. *Elkhorn Slough Wetlands Management Plan.* Prepared for the California State Coastal Conservancy and the Monterey County Planning Department. Capitola, CA: ABA Consultants.

Advanced Biological Testing. 1996. Results of chemical, physical, and bioassay testing of sediments from Moss Landing Harbor. Report prepared for Moss Landing Harbor District, Moss Landing, CA.

Anderson, B. S., D. P. Middaugh, J. W. Hunt, and S. L. Turpen. 1991. Copper toxicity to sperm, embryos, and larvae of topsmelt *Atherinops affinis*, with notes on induced spawning. *Marine Environmental Research* 31: 17–35.

Association of Monterey Bay Area Governments. 1988. Work plan for water quality management planning and outreach program for the Elkhorn Slough.

Bestari, K. T. J., R. D. Robinson, K. R. Solomon, T. S. Steele, K. E. Day, and P. K. Sibley. 1998. Distribution and composition of polycyclic aromatic hydrocarbons within experimental microcosms treated with creosote-impregnated Douglas fir pilings. *Environmental Toxicology and Chemistry* 17: 2369–2377.

Blankinship, M. S., and A. D. Evans. 1993. Elkhorn Slough 205(j) uplands water quality management plan: Final report. Prepared by Kleinfelder, Inc., for the Association of Monterey Bay Area Governments.

Boese, B. L., H. Lee II, and S. Echols. 1997. Evaluation of a first-order model for the prediction of bioaccumulation of PCBs and DDT from sediment into the marine deposit-feeding clam *Macoma nasuta*. *Environmental Toxicology and Chemistry* 7: 1545–1553.

Bradley, J. F. (California Department of Health Services, Environmental Monitoring Branch, Sacramento). 1996. Letter dated June 12, 1996, to Rodger Briggs, California Regional Water Quality Control Board, San Luis Obispo.

Bryan, G. W., P. E. Gibbs, G. R. Burt, and L. G. Hummerstone. 1987. The effects of tributyltin (TBT) accumulation on adult dog whelks, *Nucella lapillus*: Long-term field and laboratory experiments. *Journal of the Marine Biology Association, United Kingdom* 67: 525–544.

Burdick, G. E., E. J. Harris, H. J. Dean, and C. J. Henny. 1964. The accumulation of DDT in lake trout and the effect on reproduction. *Transactions of the American Fisheries Society* 93: 127.

California State Water Resources Control Board. 1990. Water quality assessment report for 1990. Sacramento.

Canright, A. 1999. Elkhorn Slough sea otters. *California Coast & Ocean*, spring 1999, 32–37.

Downing, J., R. Fairey, C. Roberts, E. Landrau, R. Clark, J. Hunt, B. Anderson, B. Phillips, C. J. Wilson, F. LaCaro, G. Kapahi, K. Worcester, M. Stephenson, and M. Puckett. 1998. Chemical and biological measures of sediment quality in the central coast region. Final report for the Bay Protection and Toxic Cleanup Program. California State Water Resources Control Board.

Harding Lawson Associates. 1997. Sediment sampling and analysis plan, southern Q-dock and MBARI dock, Moss Landing Harbor District, Moss Landing, California. Report prepared for Moss Landing Harbor District, Moss Landing.

Hunt, J. W., B. S. Anderson, B. M. Phillips, V. de Vlaming, and K. Worcester. 1999. Ambient toxicity in two Monterey Bay watersheds: Implicated chemicals and tributary sources. Abstract, Society of Environmental Toxicology and Chemistry 20th Annual Meeting, Philadelphia.

Kennish, M. J. 1992. *Ecology of Estuaries: Anthropogenic Effects*. Boca Raton, FL: CRC Press Marine Science Series.

Kimball, K. D., and S. A. Levine. 1985. Limitations of laboratory bioassays: The need for ecosystem-level testing. *BioScience* 35: 165–171.

Klecka, G. M., D. Mackay, R. S. Boethling, D. Calamari, C. Cowan-Ellsberry, S. Eisenreich, J. Franklin, C. P. L. Grady Jr., D. G. Graham, B. Hansen, P. H. Howard, K. C. Jones, K. Kannan, R. J. Larson, R. W. Macdonald, T. McKone, D. Muir, T. Parkerton, L. Thibodeaux, D. van de Meent, T. Walllington, and C. Zetzsch. 1999. Evaluation of persistence and long-range transport of organic chemicals in the environment: In *Summary of the SETAC Pellson Workshop on Persistence and Long Range Transport of Organic Chemicals in the Environment, 13–19 July 1998, Fairmont Hot Springs, British Columbia, Canada*. Pensacola, FL: Society of Environmental Toxicology and Chemistry.

Kopec, D., and J. Harvey. 1995. Toxic pollutants, health indices, and population dynamics of harbor seals in San Francisco Bay, 1989–1992. Unpublished report, Moss Landing Marine Laboratories.

Kuivila, K. M., and Foe, C.G. 1995. Concentrations, transport and biological effects of dormant spray pesticides in the San Francisco Estuary, California. *Environmental Toxicology and Chemistry* 14: 1141–1150.

Kvitek, R. G., A. K. Fukayama, B. S. Anderson, and B. K. Grimm. 1988. Sea otter foraging on deep-burrowing bivalves in a California coastal lagoon. *Marine Biology* 98: 157–167.

La Point, T. W. 1995. Signs and measurements of ecotoxicity in the aquatic environment. In D. J. Hoffman, B. A. Rattner, G. A. Burton Jr., and J. Cairns Jr., eds., *Handbook of Ecotoxicology*, 13–24. Boca Raton, FL: Lewis Publishers.

Law, R .J., S. J. Blake, B. R. Jones, and E. Rogan. 1998. Organotin compounds in liver tissue of harbour porpoises (*Phocoena phocoena*) and grey seals (*Halichoerus grypus*) from the coastal waters of England and Wales. *Marine Pollution Bulletin* 36: 241–247.

Long, E. R., and L. G. Morgan. 1990. *The Potential for Biological Effects of Sediment-Sorbed Contaminants Tested in the National Status and Trends Program*. NOAA Technical Memorandum NOS OMA 62. Seattle: National Oceanic and Atmospheric Administration.

Long, E. R., D. L. MacDonald, S. L. Smith, and F. D. Calder. 1995. Incidence of adverse biological effects within ranges of chemical concentration in marine and estuarine sediments. *Environmental Management* 19 (1): 81–97.

Martin, M., J. W. Hunt, B. S. Anderson, L. Espinosa, and F. H. Palmer. 1986. Marine Bioassay Project second report: Acute toxicity tests and protocols for mysids, red abalone, and giant kelp. Prepared for the State Water Resources Control Board, Sacramento.

Menconi, M., and C. Cox. 1994. *Hazard Assessment Report of the Insecticide Diazinon to Aquatic Organisms in the Sacramento-San Joaquin River System*. Sacramento: California Department of Fish and Game, Environmental Sciences Division Administrative Report 94-2.

Mischke, T., K. Brunetti, V. Acosta, D. Weaver, and M. Brown. 1985. *Agricultural Sources of DDT Residues in California's Environment*. A report prepared in response to House Resolution No. 53 (1984). Sacramento, CA: California Department of Food and Agriculture, Environmental Hazards Assessment Program.

Moser, D. G. 1996. Trace element and organochlorine compounds in harbor seals (*Phoca vitulina richardsi*) along the Pacific Coast. Master's thesis, Moss Landing Marine Laboratories. San Jose State University.

Mountjoy, D. C. 1993. Farming practices survey and outreach recommendations for the Elkhorn Slough Water Quality Management Plan: Final report. Prepared for the Association of Monterey Bay Area Governments.

Munn, M. D., and S. J. Gruber. 1997. The relationship between land use and organochlorine compounds in streambed sediment and fish in the central Columbia Plateau, Washington, and Idaho, USA. *Environmental Toxicology and Chemistry* 16: 1877–1887.

Murdoch, M. H., P. M. Chapman, D. M. Johns, and M. D. Paine. 1997. Chronic effects of organochlorine exposure in sediment to the marine polychaete *Neanthes arenaceodentata*. *Environmental Toxicology and Chemistry* 16: 1494-1503.

National Oceanic and Atmospheric Administration. 1992. *Agricultural Pesticide Use in Coastal Areas: A National Summary*. A. S. Pait, A. E. DeSouza, and D. R. G. Farrow, eds. Rockville, MD: Strategic Environmental Assessment Division, Office of Ocean Resources Conservation and Assessment, National Ocean Service, NOAA.

Ohlendorf, H. M., F. C. Schaffner, T. W. Custer, and C. J. Stafford. 1985. Reproduction and organochlorine contaminants in terns at San Diego Bay. *Colonial Waterbirds* 8 (1): 42–53

Parkin, J. L. 1998. Ecology of breeding Caspian terns (*Sterna caspia*) in Elkhorn Slough, California. Master's thesis, Moss Landing Marine Laboratories. San Jose State University.

Phillips, P. T. 1988. *California State Mussel Watch: Ten Year Data Summary, 1977–1987.* Water Quality Monitoring Report 87-3. Sacramento: California State Water Resources Control Board.

Phillips, B. M., J. W. Hunt, and B. S. Anderson. 2001. Statistical significance of sediment toxicity test results: Threshold values derived by the detectable significance approach. *Environmental Toxicology and Chemistry* 20: 371-373.

Rasmussen, D. 1994. *State Mussel Watch Program Data Report, 1987–1993.* Water Quality Monitoring Report 94-1. Sacramento: California State Water Resources Control Board.

———. 1996. *State Mussel Watch Program Data Report, 1993–1995.* Water Quality Monitoring Report 96-2. Sacramento: California State Water Resources Control Board.

Rein, F. 1999. Vegetative buffer strips in a Mediterranean climate: Potential for protecting soil and water resources. Ph.D. diss., University of California, Santa Cruz.

Rice, D. W., C. P. Seltenrich, R. B. Spies, and M. L. Keller. 1993. Seasonal and annual distribution of organic contaminants in marine sediments from Elkhorn Slough, Moss Landing Harbor, and nearshore Monterey Bay, California. *Environmental Pollution* 82: 79–91.

Risebrough, R., and W. Jarman. 1985. Organochlorine contaminants in California peregrine falcons: Birds found dead in 1985. Unpublished report, Long Marine Laboratory, Santa Cruz.

Risebrough, R., B. de Lappe, E. Letterman, J. Lane, M. Firestone-Gilles, A. Springer, and W. Walker. 1980. *California Mussel Watch, 1977-1978. Vol. 3: Organic Pollutants in Mussels,* Mytilus californianus *and* M. edulis, *along the California Coast.* Water Quality Monitoring Report No. 79-22. Sacramento: State Water Resources Control Board.

Ritter, W. F. 1988. Reducing impacts of nonpoint source pollution from agriculture: A review. *Journal of Environmental Science and Health* A23 (7): 645–667.

Snoeij, N. J., A. H. Pennicks, and W. Seinen. 1989. Thymus atrophy and immunosuppression induced by organotin compounds. *Archives of Toxicology* 13: 171–174.

Soil Conservation Service. 1985. Strawberry Hills erosion control program. Report prepared for Soil Conservation Service, Salinas and Aptos.

Stephenson, M. D., M. Martin, and R. S. Tjeerdema. 1995. Long-term trends in DDT, polychlorinated biphenyls, and chlordane in California mussels. *Archives of Environmental Contamination and Toxicology* 28: 443–450.

Swartz, R. C., F. A Cole, J. O. Lamberson, S. P. Ferraro, D. W. Schults, W. A. Deben, H. Lee II, and R. J. Ozretich. 1994. Sediment toxicity, contamination, and amphipod abundance at a DDT- and Dieldrin-contaminated site in San Francisco Bay. *Environmental Toxicology and Chemistry* 13: 949–962.

Thomas, T. E., and M. G. Robinson. 1986. The physiological effects of the leachates from a self-polishing organotin antifouling paint on marine diatoms. *Marine Environmental Research* 18: 215–222.

Toxscan, Inc./Kinnetic Laboratories, Inc. 1993. Chemical analysis and toxicity evaluation of sediments at Moss Landing Harbor. U.S. Army Engineering, District of San Francisco Corps of Engineeers.

———. 1996. Chemical analysis and toxicity evaluation of sediments at Moss Landing Harbor. U.S. Army Engineering, District of San Francisco Corps of Engineeers.

U.S. Department of Agriculture. 1994. Draft watershed plan and environmental assessment, Elkhorn Slough Watershed Project, Monterey and San Benito Counties, California. On file at the USDA Natural Resources Conservation Service office, Salinas, CA.

U.S. Environmental Protection Agency. 1992. *Guidance Specifying Management Measures for Sources of Nonpoint Pollution in Coastal Waters*. EPA 840-B-92-002. Washington, DC: Office of Water.

———. 1995. *Guidance for Assessing Chemical Contaminant Data for Use in Fish Advisories*. Vol. 1: *Fish Sampling and Analysis*. 2d ed. EPA 823-R-95-007. Washington, DC: Office of Water.

Waite, M. E., M. J. Waldock, J. E. Thain, D. J. Smith, and S. M. Milton. 1991. Reductions in TBT concentrations in UK estuaries following legislation in 1986 and 1987. *Marine Environmental Research* 32: 89–111.

Watkins, D., C. Reiner, and D. Crane. 1984. *Toxic Substances Monitoring Program, 1983 Data Report*. California Department of Fish and Game Laboratory Report 84-2. Sacramento: Department of Fish and Game.

Werner, M. R., M. Silberstein, M. Los Huertos, S. Gliessman, J. Newman, R. Tjeerdema, J. Caffrey, S. Shaw, A. De Vogelaere, and J. Mason. 1997. Agriculture and wetlands: Examining the interface at the Azevedo Ranch and the Elkhorn Slough National Estuarine Research Reserve in coastal central California. Final report submitted to National Oceanic and Atmospheric Administration, Office of Coastal Resource Management. Grant # NA370R0162.

Woodwell, G. M., P. P. Craig, and H. A. Johnson. 1971. DDT in the biosphere: Where does it go? *Science* 174: 1101–1107.

Young, J. F. 1996. *Commercial Shellfish Growing Area Sanitary Survey Report for Elkhorn Slough, Moss Landing, California*. Technical Report No. 97-01. Sacramento: California Department of Health Services Division of Drinking Water and Environmental Management.

14

Chapter Fourteen

Management Issues

Jane Caffrey, Daniel Mountjoy, Mark Silberstein, and Cheta Zabin

Conservation is a state of harmony between men and land. By land is meant all of the things on, over or in the earth. Harmony with land is like harmony with a friend; you cannot cherish his right hand and chop off his left. That is to say, you cannot love game and hate predators; you cannot conserve the waters and waste the ranges; you cannot build the forest and mine the farm. The land is one organism. Its parts, like our own parts, compete with each other and co-operate with each other. The competitions are as much a part of the inner workings as the co-operations. You can regulate them—cautiously—but not abolish them.
— Aldo Leopold, Round River Essays

With a watershed area of approximately 182 square kilometers (70 mi^2), Elkhorn Slough provides an ideal microcosm in which to examine interactions between a biologically rich coastal ecosystem and changing human economic, social, and political needs and uses. Elkhorn Slough and its watershed encompass examples of most types of land uses currently affecting coastal and estuarine areas throughout the country. Industrial development, commercial and sport fishing, an industrial and recreational harbor, railway corridor, highways, residential development, small businesses, extensive agriculture, mariculture, recreational hiking and boating, educational institutions, research projects, and protected and restored areas all surround one of the most productive wildlife habitats in North America.

The types of pressures and changes facing the nation's estuaries are manifest in Elkhorn Slough. Many agencies, institutions, organizations, and individuals have risen to the challenge of attempting to define, manage, and balance the often competing interests in the Elkhorn Slough area. Over the last two decades, these partnerships in coastal zone management have researched and tested policies in preservation, restoration, and conservation.

No single agency is responsible for the management of Elkhorn Slough and its watershed; instead, a variety of federal, state, and local agencies safeguard slough resources, habitats, and communities (appendix 1.1). Responsibilities of these agencies fall into two broad categories: protection of native plant and animal communities and habitats (including soil and water resources), and protection of human health or public access to the resources of the slough and watershed. Some potential for conflict between these two management goals exists. Thus, the main challenge, and therefore opportunity, we face in Elkhorn Slough is figuring out how to maintain this spectacular biological diversity in a "working landscape." Can we balance the economic, social, and political needs of human communities while sustaining and restoring natural communities?

Management Strategies

Due to omnipresent human uses, the ecological health of the Elkhorn Slough ecosystem is, for better or worse, a function of human management and manipulation of the slough itself and the surrounding watershed. Decisions made by individual landowners, public agencies, and citizens who recreate, fish, live, work, or travel in the watershed all affect the long-term viability of the ecosystem. A variety of management strategies and policies have been used over the past thirty years to reduce the ecological impact of these human activities. These strategies include educational and research programs, technical and financial assistance programs for landowners and farmers, legislated environmental standards and regulations, land use planning processes, and purchases of conservation easements or entire properties.

A key feature of successful management strategies is the establishment of collaborative partnerships among the many agencies, nongovernmental organizations, citizen groups, and individual landowners concerned with Elkhorn Slough and the watershed. No single entity could have achieved the degree of awareness and protection now evident in the watershed without building strategic relationships across interest group boundaries. For example, reducing the effects on water quality of nitrate entering the slough will require the combined efforts of researchers, farmers, rural residents, environmental advocates, public and private technical advisors, land use planners, and regulatory agencies. These groups do not traditionally gather to find solutions to common resource problems. However, in the Elkhorn Slough watershed, members of these disparate communities have learned to trust one another and recognize that they are often better off working together to find an acceptable solution, rather than waiting for litigation or regulation. This is what the original management strategy ambitiously aimed for when Elkhorn Slough was designated a National Estuarine Sanctuary in 1979.

Workshops addressing specific management issues have been effective in bringing diverse groups together. The Elkhorn Slough National Estuarine Research Reserve (ESNERR) and Elkhorn Slough Foundation (ESF), with support from the National Oceanic and Atmospheric Administration (NOAA), have conducted a variety of workshops on agricultural erosion, mudflat health, water quality, watershed planning, and boating impacts. Local and regional planners, elected officials, resource professionals, local business people, and farmers have discussed these problems and developed solutions.

Selecting appropriate partners and management strategies to address a resource issue requires an understanding of both the issue's ecological causes and the historic social and economic factors that have driven resource use decisions. Many of the chapters in this book have detailed or referenced management issues and concerns about the long-term health of the slough. In this chapter, we outline these issues, discuss responses made to date, and address ongoing management needs. Scientific research has played a key role in identifying and describing the natural resources and cycles of the slough. Science will likewise be key in helping identify issues and solve ongoing problems. This chapter will explore the ways that partnerships and management strategies have been combined to address specific natural resource issues, first in wetland and aquatic systems, then in the upland watershed.

Wetlands and Aquatic Issues and Management Responses

Habitat Loss

Beginning in the late 1880s, farmers and ranchers turned their attention to Elkhorn Slough's wetlands as potential farming areas. Low-lying lands and portions of the slough proper were "reclaimed" by farmers through an elaborate system of dikes and dams. The diversion of the Salinas River between 1908 and 1910 decreased seasonal flooding in the slough area. By 1940, some 50 percent of the wetlands associated with Elkhorn Slough had been converted to productive farmland (Browning 1972).

The Salinas River diversion meant a loss of habitat for freshwater organisms as well, as the slough gradually moved from a brackish marsh to a more tidal estuary (see chapter 4, "Hydrography," and chapter 7, "History of Land Use," for details). The opening of the slough to create Moss Landing Harbor in 1947 accelerated that change. Associated freshwater ponds also disappeared, although the ESNERR maintains five small (0.20 ha, 1/2 ac) freshwater ponds, some of which are created by pumping well water, and eighteen wildlife drinking holes, or "guzzlers." The ponds are maintained for wildlife that requires year-round freshwater and are used by a variety of species. Red-legged frogs (*Rana aurora draytonii*), a federally threatened species, are present in a majority of these ponds and guzzlers. Some nonnative bullfrogs (*R. catesbeiana*), a predator of red-legged frogs, have recently been found in ponds on the reserve. Because bullfrog tadpoles require two years to mature, while red-legged frogs require only one, one control strategy is to allow ponds to dry up during the dry season, or to pump them dry, thus killing bullfrog tadpoles.

Wetlands are still being lost to sediment transported from eroding upland areas. In a 1985 study, Dickert and Tuttle noted the development of more than thirty "sand and mud fans at the mouths of creeks draining the slopes above the slough." These deposits smother tidal flora and fauna and gradually convert marshlands to uplands. Wetland habitat is also lost to erosion due to tidal scour.

Wetland Restoration

Following the establishment of the ESNERR in 1979, management staff developed plans to reverse wetland loss on reserve property. Much of the pastureland that had been created by diking and draining was restored to salt marsh. In the South Marsh restoration area in Parson's Slough, over 81 hectares (200 ac) were dredged to create channels, and the spoil was used to develop elevated areas suitable for colonization by salt marsh species.

Dikes separating the South Marsh restoration area from the main portion of Parson's Slough failed during heavy winter rains in 1982. This led to more extensive tidal flushing in the restoration area than was originally anticipated, and increased problems associated with tidal scour throughout Elkhorn Slough as a whole (see following section and chapters 4 and 7 for more about tidal scour).

The South Marsh restoration has successfully provided additional critical marsh, mudflat, and tidal creek habitats. Shortly after the South Marsh project was completed, 50 hectares (125 ac) in the northern part of the reserve (North Marsh) were restored. Tide gates were installed and set so that flushing of this site occurs only on higher high tides. For much of the year, North Marsh is kept shallow to provide critical feeding and roosting habitat for migrating sandpipers and other shorebirds. Low flushing rates and standing water have led to periodic problems with mosquitoes. Soil erosion from uplands adjacent to this area and sedimentation in the marsh are a continuing concern, as is nonpoint source runoff of nutrients and pesticides.

The newest focus areas for restoration of tidal wetlands are within the Moro Cojo watershed, where marshes were diked and drained for agriculture. The California State Coastal Conservancy played a key role in developing plans for restoration and in funding land aquisitions. Recent acquisitions and development of conservation easements within this watershed are the first steps in increasing wetlands acreage. Restoration efforts have also focused on freshwater and riparian areas along Carneros Creek (see below).

Land acquisition and restoration efforts in the watershed recently received significant funding as a result of the Moss Landing power plant expansion process. To mitigate for the impacts of the plant's expansion, particularly the loss of larvae in seawater taken into the cooling system, the California Energy Commission in consultation with an expert panel developed and approved a mitigation plan as a part of the permitting process. In 2001, power plant owner Duke Energy gave seven million dollars to the Elkhorn Slough Foundation. Five million dollars will be used to acquire and restore areas of the watershed in an effort to create or enhance wetland production to offset the losses to the power plant; two million dollars will be used as an endowment to fund ongoing stewardship of the properties to be acquired.

Separately, a group of regional environmental groups, concerned about the impacts of the power plant, were able to obtain additional funds from Duke Energy for research and monitoring. These funds are being managed by the Monterey Bay National Marine Sanctuary, and will be used to fund studies on slough habitats and communities that might be influenced by the power plant.

Tidal Scour

When the U.S. Army Corps of Engineers dredged a channel to the bay to create Moss Landing Harbor in 1946, Elkhorn Slough changed from a depositional system in which sediments were gradually accumulating to an erosional system (see chapter 2, "Geology"). The channel at the Highway 1 bridge has deepened and widened until it is now 7.5 meters (25 ft) deep and 91 meters (300 ft) wide. Tidal currents at the bridge doubled between 1971 and 1994

(Malzone 1999; see also chapter 4), and tidal creeks have deepened and widened during this same period. Analysis of aerial images has shown that salt marshes along the main channel have been converted to mudflats following the opening of the harbor (Lowe 1999). One consequence of these hydrologic changes has been dramatic changes in the invertebrates, fish, birds, and mammals using these habitats. Marine species, such as harbor seals, sea lions, sea otters, Pacific herring, and northern anchovies, have become increasingly common throughout the slough (see chapters 9, "Invertebrates"; 10, "Fishes"; and 11, "Birds and Mammals").

The Corps recognized that building the harbor would alter the slough's hydrology. However, their proposal to construct tide gates that would have lessened the impact was never implemented (see chapter 4 for details). Since the early 1980s, when the first tidal scour studies were conducted (Oliver and Schwartz 1988), the need for a solution to the tidal scour problem has been increasing. Construction of a submerged rock sill at the Highway 1 bridge or at the mouth of Parsons Slough has been proposed to "fix" the problem; ideally, this would reduce tidal velocities in the channel, thus reducing sediment transport and erosion.

The ramifications of constructing such a sill are unclear. In 1997, a panel of experts on Elkhorn Slough, hydrodynamics, sediment transport, and modeling was convened at the ESNERR to evaluate previous studies, assess the feasibility of constructing a sill or some other structure, and make recommendations for targeted research to address this issue. Following extensive discussion, several points of agreement and questions to address emerged:

- There was a consensus that it is difficult, if not impossible, to predict what the slough will look like in ten, twenty, or fifty years.
- The exact mechanism causing the marsh banks to slump or calve off is unknown and deserving of further study.
- Sills could lead to rapids or dangerous flow conditions at the barrier, posing a risk to recreational boaters, and could restrict the migration of animals into or out of the slough.
- A two-dimensional computer model of water flow could be developed and used to test different management scenarios. However, developing a sediment model to predict sediment transport throughout the slough is much more difficult and would probably not be constructive in any case.

Since that 1997 workshop, Stanford University researchers have made some progress toward developing a computer model of circulation in the slough, and additional current and sediment transport measurements have been made, advancing our ability to develop predictive circulation models. In addition, recent research at Upper Azevedo Pond has shown that areas with restricted tidal circulation can experience periods of hypoxia (low oxygen levels) lasting as long as ten hours (Beck and Bruland 2000). The implication is that reducing tidal circulation can increase hypoxia, which in turn can affect biological communities and alter biogeochemical cycling.

Given the technical difficulties of trying to build a sill or structure and concerns about the impacts of such a structure, one strategy for addressing the impacts of increasing erosion and tidal scour could be to balance the loss of marshes in Elkhorn Slough with the restoration of degraded marshes. To pursue this strategy, accurate measurements of salt marsh and wetland acreage for all regions of the slough are needed. Recent advances in geographic information systems (GIS) technology and support by the Packard Foundation for habitat mapping within the Elkhorn Slough watershed have helped the ESF and ESNERR make progress toward this goal.

Resource Use

Despite signs posted by the Monterey County Health Department warning of shellfish contamination, people continue to harvest clams for fishing bait and probably for food. Studies have shown that certain species of clams once common in the slough are no longer found, and that the overall size of many bivalves is decreasing. Harvesting by humans is believed to be a major factor, although increased predation by sea otters and other factors such as displacement by nonnative species have probably also contributed to declining bivalve populations (see chapter 9 for details). Populations of ghost shrimp (*Callianassa californiensis*), blue mud shrimp (*Upogebia pugettensis*), and in some areas, fat innkeeper worms (*Urechis caupo*) have also declined due to harvesting for bait. A program to restore tide flats at the mouth of Elkhorn Slough met with limited success. Protection and monitoring of these sensitive habitats is essential to preserve invertebrate species diversity. Recreational fishing has long been a popular activity in the slough, although organized fishing events like the shark derbies have not occurred since 1996. Take by humans is not currently thought to have a significant impact on fish populations. Although harbor seal predation could significantly affect fish populations, recent research suggests that most harbor seals leave the slough to feed offshore (see chapter 11). Changes in habitat due to tidal scour and nonpoint source runoff probably have the most significant impact on vertebrate populations.

Introduced Species

Introduced species, which disrupt species composition and food webs as well as causing the loss of native species, are an increasing problem in the nation's estuaries. This is a particularly challenging management issue because it is extremely difficult to eliminate a nonnative species once it becomes established. Many species were introduced into Elkhorn Slough during the period of oyster culturing between 1930 and 1970. For example, the Japanese mud snail (*Batillaria attramentaria*) completely displaced the native horn snail (*Cerithidea californica*) during this period, and is now one of the most dominant and obvious species on intertidal mudflats. Ballast water and fouled boat hulls have introduced additional nonnative invertebrates (see chapter 9). The Australian isopod (*Sphaeroma quoyanum*), which was apparently transported to Elkhorn Slough relatively recently by ship fouling, burrows into salt marsh banks, weakening their structural integrity and possibly exacerbating the problems of tidal scour and marsh erosion.

While Elkhorn Slough does not appear to have as many invasive aquatic species as nearby San Francisco Bay, the easy exchange between the two areas by commercial and recreational boaters puts the slough at risk for further introductions. Regulations combined with an aggressive public education program are needed to reduce the introduction of new species.

Water Quality

Coliform bacteria occur in high levels throughout the slough, particularly at the harbor mouth and after heavy rains. The origin of this contamination is not clear, but levels of fecal and total coliform bacteria indicate that the source is not primarily human waste. This high bacterial level was the major reason the slough's oyster aquaculture ceased (ABA Consultants 1989): coliform levels rose dramatically in the 1960s, leading to the closure of the last commercial oyster bed in 1967 (see chapter 7 for oyster cultivation history). In 1969, the Monterey County Health Department posted signs warning against eating local clams, and the State Health Department denied a permit to an oyster aquaculture firm in 1974 because of continued high bacteria levels. Because of the persistently high coliform levels that restrict commercial aquaculture and shellfish consumption, the Central Regional Water Quality Control Board has listed Elkhorn Slough as an impaired water body.

Many persistent agricultural pesticides (DDT and breakdown products, endosulfan, dieldrin, and others) have not been used since 1972 but continue to be found in agricultural soils, adhered to fine soil particles. Large storm events and poor soil management erode these soils and their adhered contaminants, transporting them into slough waters where they accumulate (see chapter 13, "Land Use and Contaminants"). In 1985, high pesticide levels in mussels led the Monterey County Health Department to issue a health advisory against eating shellfish from the slough. That advisory remains in effect today.

In addition, nutrient concentrations, particularly nitrate concentrations, have greatly increased between the 1970s and 1990s. Peak concentrations occurring during winter runoff can be twenty times higher than concentrations measured in the 1970s. Eutrophication and the associated problems of excessive algal blooms, hypoxia, and fish kills do not yet appear to be a significant problem in the main channel of the slough, although areas with restricted circulation have been impacted (see chapter 12, "Biogeochemical Cycling").

The key to improving water quality is to reduce the amount of pollutants entering Elkhorn Slough by decreasing sources or transport processes. Concern over nonpoint source pollution, particularly from sediments and pesticides, has prompted the creation of several management plans in the region. One of the first was the Elkhorn Slough Wetland Management Plan (ABA Consultants 1989), which proposed recommendations, enhancement, and restoration projects to address water quality problems. Components of the plan include land use controls, enforcement, technical assistance to landowners, and the promotion of best management practices. The plan was funded by the California Coastal Conservancy and Monterey County. The Moro Cojo Slough Management and Enhancement Plan was developed to enhance fresh and brackish wetlands of Moro Cojo Slough through a multifaceted program, which includes establishing small- to moderate-scale freshwater impoundments. Other elements of the plan include regulation, technical assistance, Best Management Practices, research, monitoring, and data exchange activities as authorized under the North Monterey County Land Use Plan (Coastal Act). Most recently, the Elkhorn Slough Watershed Conservation Plan (Scharffenberger 1999) represents a comprehensive and integrated approach to preserving critical habitats, including marshes, riparian zones, and uplands, and reducing stresses on natural resources within Elkhorn Slough.

A number of projects identified in the Elkhorn Slough Wetland Management and Moro Cojo plans have been implemented, including the acquisition of the Blohm and Azevedo Ranches, restoration on several properties, and a research focus on water quality. These plans have also provided support for several Natural Resource Conservation Service (NRCS) projects (discussed below).

Agrochemical-related water quality problems were influential in establishing the U.S. Department of Agriculture (USDA)–NRCS Elkhorn Slough Watershed Project. NRCS hydrologic calculations revealed that typical strawberry farms in the Springfield Terrace area of the watershed yielded as much as twenty times more runoff than the native grassland vegetation. NRCS develops farm plans to control runoff and associated downstream erosion by detaining rainfall and releasing it slowly. Contoured furrows, seeding of deep-rooted cover crops and critical-area plantings, and construction of water and sediment control basins enhance water percolation and buffer peak flows. Water quality is also improved by reducing runoff and filtering it through dense vegetation or detaining it so fine-textured soil particles with adhered chemicals can settle out. These same concepts can be applied to residential development where impervious roof and driveway surfaces also increase runoff.

However, barriers remain to widespread adoption of these practices. Farmers and residential developers are reluctant to give up valuable acreage for the construction of water and sediment detention basins. Only recently has the county required this investment as a precondition for development permits on steep lands. The use of grasses and other vegetation to filter runoff is perceived by many to be a potential source of insect pests and weed seeds, and thus contradictory to "clean" farming. These perceptions are gradually changing thanks to reliable experimental evidence being collected as part of a farmscaping project of UCSC and NRCS (see Upland Habitat section below). Native perennial grasses can control erosion and slow runoff, and do not appear to attract crop pests. Once established, they can also effectively suppress weeds and thus reduce herbicide costs (Rein 1999).

The use of farmscaping and other conservation practices is being encouraged by the Monterey Bay Farmers Clean Water Initiative, a voluntary pilot project funded by the State Water Quality Control Board. Growers who agree to implement a template of practices approved by the project's technical advisory committee, including monitoring water and fertilizer use to maximize their effects and minimize overuse, can qualify for certification and use a label on their products reading "Fields to Oceans, Coastal Farmers Protecting Monterey Bay"(Pajaro Valley Pilot Project 2000).

In addition to these management and voluntary plans, a variety of state and local agencies have a direct impact on protecting water quality. Under the Local Coastal Program, Monterey County Department of Planning and Building Inspection must approve any projects requiring permits (construction, development, land clearing, etc.), although decisions may be appealed to the California Coastal Commission. The Monterey Bay National Marine Sanctuary has permitting authority over areas below mean high tide.

Water quality monitoring programs are useful for determining whether these management strategies are effective. Since the early 1980s, the California Department of Fish and Game's Mussel Watch program has provided critical information on pesticide bioaccumulation in Elkhorn Slough's mussels. However, a more comprehensive program that includes monitoring of newer pesticides is needed (see chapter 13). In 1988, ESF and ESNERR began a volunteer-based program to collect basic water quality data each month. These data, along with nutrient analyses performed by the Monterey County Water Resources Agency, serve as a baseline data set to measure the effect of management changes on water quality. In addition, since 1995 the ESNERR has participated in NOAA's System Wide Monitoring Program, which tracks short-term variability and long-term changes in key water quality parameters in representative estuarine ecosystems and coastal watersheds throughout the country's coastal zone.

Visitor Impact

Recreational use of the slough has increased in the last few decades, along with opportunities for viewing the slough from the water. Visitors can now rent small motorboats at Moss Landing Harbor and kayaks at several local venues to explore the slough on their own, or take part in organized slough tours by pontoon boats or kayaks.

Because boaters can reach areas of the slough off limits to hikers, they can potentially disrupt wildlife and habitats that are otherwise protected from humans. Marine mammals and nesting shorebirds are the most likely to be disturbed by boaters; in 1999, speeding boats are believed to have struck and killed four sea otters in the slough.

Launching and hauling out kayaks and canoes on unprotected slough banks can add to the slough's erosion problems. In January 2000, the popular Rubis Creek kayak haul-out was closed due to overuse, vandalism, and damage to private property. Now that this site on private property has closed, an alternative site needs to be identified and developed to prevent ever-increasing numbers of boaters from indiscriminately hauling out on slough banks.

Local kayaking outfitters, in conjunction with the ESF and ESNERR, developed a pamphlet for boaters, including a map that shows the locations of approved launching and haul-out areas. Annually, ESNERR and ESF sponsor a workshop with local outfitters to discuss ways to teach the boating public about protecting the slough. Boating is prohibited within the boundaries of the ESNERR, east of the railroad.

Some 50,000 visitors come to the ESNERR each year, a number that is expected to increase. Although large numbers of hikers can potentially affect wildlife, the reserve management plan projects that the number of visitors could double without creating a problem. Trails are informally monitored by staff and volunteers, and signage helps keep visitors out of sensitive areas. However, more management and docent programs are needed to continue educating the public as recreational uses increase. The Department of Fish and Game, which manages the ESNERR, is committed to this program.

Hazardous Materials

The slough is vulnerable to accidental releases of hazardous materials such as oil, industrial chemicals, fertilizers, and pesticides. Although Highway 1 is only two lanes wide where it borders the slough, it is a major transportation corridor in this area, with heavy truck, agricultural machinery, and passenger car traffic that passes over a bridge at the slough mouth. The Union Pacific Railroad tracks, which extend for 9.6 kilometers (6 mi) along the main channel of Elkhorn Slough, carry Amtrak passengers and commercial loads, including hazardous materials (U.S. Coast Guard 2000).

The California Department of Fish and Game's Oil Spill Prevention and Response program and the U.S. Coast Guard have developed a plan to respond to any discharge of hazardous materials in the slough. This plan describes the five areas in the slough most likely to be the site of a hazardous material spill, and what the response should be to protect sensitive habitats adjacent to the spill (U.S. Coast Guard 2000). Because of the strong tidal currents, material could rapidly be transported and dispersed throughout the slough. Thus, the response plan emphasizes the need to respond to any spill quickly (e.g., within six hours) to be most effective. A recent drill simulating a train derailment at the mouth of Parson's Slough on the ESNERR brought together participants from ESNERR, the California Department of Fish and Game, U.S. Coast Guard, California State Parks, Monterey Bay National Marine Sanctuary, Monterey and Santa Cruz Counties, Union Pacific Railroad, and private cleanup companies.

In addition to catastrophic spills, the slough is also subject to small-scale inputs such as tires, batteries, and trash. Volunteers from ESNERR and other groups have removed hundreds of discarded tires and other trash from the slough during the annual Coastal Cleanup, sponsored by the California Coastal Commission.

Upland Management Issues and Management Responses

Habitat Loss

Fragmentation and loss of habitat may be the most significant factors affecting many native animal species and threatening unique coastal plant communities. Wooded and brushy habitats, which had been circumscribed for years by fires set by the Ohlone, were cleared by Mexican and Spanish ranchers to provide pasture for cattle. Later, American settlers cut huge tracts of oaks for firewood and to clear land for farming. Native grasslands were altered by the introduction of European grasses and European weeds associated with agriculture, as well as intensive grazing pressure within fenced pastures. Existing agricultural parcels are primarily managed for single crops with no tolerance for other vegetation that is perceived to harbor pests.

Currently, 4,378 hectares (10,813 ac), or 24% of the Elkhorn Slough watershed, is intensively farmed. In comparison with the surrounding Salinas and Pajaro Valleys, this is a relatively minor land use. However, hillside agricultural activities in the erodable Elkhorn Slough watershed on slopes often exceeding 20% frequently alter downstream habitats through sedimentation of riparian corridors and wetlands. Low density residential development has fragmented an additional 1,820 hectares (4,500 ac) of the watershed.

Despite a growing awareness of the need to preserve wildlife habitat, people continue to alter the landscape around Elkhorn Slough in response to economic pressures. Oaks have been cut down to expand cropland, highly lucrative strawberry and flower farms are replacing open grasslands and encroach to the very edge of streams and wetlands, and premium homes are replacing maritime chaparral on ridgetops.

While residential development is still increasing, major industrial development in the watershed probably peaked in the 1950s and 1960s. In the late 1940s and 1950s, building and expansion of National Refractories and the Moss Landing Power Plant by PG&E (now owned and operated by Duke Energy) occurred (see chapter 7). Further industrial development took place in the 1970s and 1980s at Dolan Industrial Park with the addition of car wrecking yards. The National Refractory plant is being sold and its future is uncertain. Duke Energy is modernizing the Moss Landing Power Plant. They anticipate using less water and reducing air emissions. Old oil storage tanks will be removed, and restoration of native vegetation in these areas is planned.

Monterey County influences the development process through land use zoning and ordinances. Oak tree removal is now prohibited by ordinance, and new farmland development on slopes in excess of 10% requires an erosion plan. While the county has strong regulations to protect natural resources and habitats, over the years enforcement has been weak (Scharffenberger 1999). Rural, "out of sight" land use alteration is extremely hard to regulate. A long-term education process that shows local citizens how they are important to the preservation of the natural system is likely to provide better resource protection than is land use regulation. Land use zoning and ordinances slow the pace of agricultural conversion of certain native habitats but lead to more rapid development of former grassland and savanna. Although living oak trees are not cut down, irrigation practices adjacent to the drought-adapted trees often result in root rot and their gradual death. Residential development continues to be permitted adjacent to wetlands and on maritime chaparral ridges. This development, driven by the spread of Silicon Valley commuters seeking the tranquillity of a rural home, is only likely to increase in the future.

Conservation and Restoration Efforts

The establishment of the National Estuarine Sanctuary (now National Estuarine Research Reserve) in 1979 was an early step toward protecting both upland and wetland habitats in the Elkhorn Slough watershed. The Elkhorn Slough Foundation has been a leader in slowing habitat loss, and to date has raised funds to protect an additional 930 hectares (2,300 ac) of sensitive habitats through purchase of property or acquisition of conservation easements.

The ESF completed a comprehensive habitat assessment and management plan in 1999 for the entire watershed with support from the Nature Conservancy (Scharffenberger 1999). Local wildlife and land use experts identified development trends for subregions of the watershed and specified sensitive habitats on GIS maps. Based on this study, the five critical resources within the Elkhorn Slough watershed of highest priority for protection have been identified as: (1) coastal marsh, (2) riparian forest and freshwater wetlands, (3) maritime chaparral and associated oak woodlands and grassland, (4) highly productive cultivated farmlands, and (5) scenic viewsheds surrounding Elkhorn Slough and Carneros Creek.

In addition to protecting remaining native habitats, there is increasing interest in restoring native vegetation on degraded sites. The Nature Conservancy and Elkhorn Slough Foundation are replanting oaks and restoring habitat on some protected parcels, but much of this land still remains drastically altered from the Ohlone days. Active restoration is occurring on the ESNERR, where 6.1 hectares (15 ac) of oaks have been planted, eucalyptus removed from 5.3 hectares (13 ac), and grasslands restored on 1.2 hectares (3 ac). A greenhouse and native plant nursery on the reserve enhance restoration efforts. Future ESNERR restoration work will be guided by a comprehensive vegetation plan (now under development), which can serve as a model for other areas within the watershed, particularly for coastal prairie and oak understory habitats.

Other restoration projects include efforts by the Carneros Creek Association, residents and farmers along Carneros Creek who have begun to replant willows, alders, cottonwood, box elders, and other riparian species to protect the creek banks from erosion and restore more than 1.5 miles of the riparian corridor. Since 1991, the Watershed Institute at California State University Monterey Bay has restored 13.4 hectares (33 ac) in the upper slough (Porter Ranch and Elkhorn Heights) and 44 hectares (109 ac) in Moro Cojo Slough (Calcagno Marsh, ESF property, Blackie Farms, North County High, Tottino Marsh, Ocean Mist Farm, and North County Recreation). The Watershed Institute works with landowners to develop a restoration plan and uses students (both secondary and college), interns, and volunteers to plant native species and monitor projects.

Techniques to restore former agricultural fields purchased either for conservation or for rural home development are also being developed. Restoration efforts have shown that exposed soils must be stabilized and revegetated rapidly once farming ceases; otherwise, they are highly susceptible to severe erosion and colonization by invasive introduced plants. Trial-and-error experiences on several ranches have shown that cultivated slopes must be regraded to remove micro-topographic features that concentrate winter runoff that can cause erosion.

Farmscaping, another strategy for restoring and increasing biotic diversity, is currently being developed and implemented by the NRCS and RCD (Resource Conservation District of Monterey County) conservation planners, researchers from the Center for Agroecology and Sustainable Food Systems at UC Santa Cruz, the Community Alliance with Family Farmers, and innovative local farmers. Farmscaping introduces native plants within monocropped farm landscapes to achieve multiple environmental and economic benefits.

Native flowering shrubs such as ceanothus (*Ceanothus thrysiflorus*), coffeeberry (*Rhamnus californica*), black sage (*Salvia mellifera*), and elderberry (*Sambucus* spp.), and herbaceous perennials such as yarrow (*Achillea millefolia*), creeping wild rye (*Leymus triticoides*), and other native grasses have been planted on the noncropped borders of several strawberry farms to control erosion and reduce pesticide use. Plant selection is based on locally adapted varieties that have floral nectaries to attract insect pest species out of the cropped area, or that nourish beneficial insects that will then prey on the crop pests. By saving farmers money, this strategy offers an economic incentive for reintroducing native habitats within intensively cropped environments. Further research is needed to determine the optimal spacing of these hedgerows to provide pest protection throughout an agricultural field and to evaluate other wildlife benefits of these habitat patches.

Introduced Species

Nonnative plants introduced accidentally or deliberately by European settlers have come to dominate certain upland areas of the Elkhorn Slough watershed. Among these are many of the plants now associated with California coastal landscapes, including poison hemlock (*Conium maculatum*), wild mustard (*Brassica* spp.), sweet fennel (*Foeniculum vulgare*), Bermuda grass (*Cynodon dactylon*), wild radish (*Raphanus sativus*), foxtail (*Alopecurus* spp.), wild oat (*Avena fatua*), Himalayan blackberry (*Rubus discolor*), cape ivy (*Delairea odorata*), Scotch broom (*Cytisus scoparius*), French broom (*Genista monspessulanus*), and eucalyptus (*Eucalyptus* spp.). Most of these species are adapted to colonize disturbed habitats and therefore thrive around agricultural operations and rural development sites. Although Monterey pine (*Pinus radiata*) is native to the central coast area, stands of these pines have been planted in the slough area where they did not naturally occur.

An important management strategy for preventing further spread of invasive introduced plant species is to establish complete vegetative cover on all disturbed soil surfaces such as road cuts, fallow fields, and construction sites. These restoration activities, if properly maintained, eliminate sites for invasive plants to become established.

Nonnative animals, including the European starling (*Sturnus vulgaris*), house sparrow (*Passer domesticus*), red fox (*Vulpes fulva*), field mouse, and earthworm (*Lumbricus terrestris*), have also gained dominance. Although the effect of some of these species on native populations is not thoroughly understood, in some instances the impacts are quite clear: nonnatives prey on native species or displace natives by taking over breeding habitat and competing for food. For example, red foxes have had an impact on nesting waterfowl and may be linked to the decline of the endangered Snowy Plover and disappearance of the endangered Clapper Rail from Elkhorn Slough. Additionally, roaming domestic and feral dogs and house cats continue to pose a problem to wildlife in the slough area, and feral pigs have done some damage to the area in the past, although not in recent years. On the reserve, feral cats and dogs are trapped and removed. Periodic trapping of red foxes and feral cats occurs at the salt ponds in Moss Landing Wildlife Management Area to reduce predation on Snowy Plover nests. Trapping efforts begin before the Snowy Plover nesting season in February and continue into September. Breeding success by the Snowy Plover has been linked to effectiveness of trapping efforts.

Erosion

Farmed lands in the Elkhorn Slough watershed have some of the highest erosion rates in the nation. The average annual rate of soil loss in the area in 1983 was estimated to be 33 tons per acre (Soil Conservation Service 1984) but rates as high as 145 tons per acre have been measured on some farms during wet winters. Sandy soils, heavy winter rains, steeply sloped hills, crop choice, and farming practices all contribute to the high erosion rates (see chapter 5, "Soils"). The rate of natural soil regeneration from underlying parent material averages 3 to 5 tons per acre per year for the loamy sand soils of the watershed. At these rates it is clear that farming or the growth of any vegetation will be limited in a very short period of time. Some farms have already been abandoned where the topsoil depth has dwindled to less than 0.3 meter (12 in) over the sandstone base.

Strawberries, the area's most lucrative crop, have been planted in increasing numbers in the watershed since the late 1970s (see chapter 4). Land in strawberry production is particularly susceptible to erosion due to the annual planting cycle, which requires that fields be prepared in the fall just prior to the arrival of the rainy season, with planting occurring in November and December.

Erosion and deposition of upstream soils is significantly higher, irrespective of slope or soil type, on sites where natural cover has been removed and soil disturbed (Dickert and Tuttle 1985). The highest percentage of bare ground exposure of nine types of land use examined occurred on strawberry and row-crop fields, higher even than commercial or residential areas (Dickert and Tuttle 1985). These crops had ten times more exposed bare ground than areas used for pasture and rangeland, suggesting that upland erosion has increased strongly as lands around Elkhorn Slough have been converted from cattle lands to crops. Between 1931 and 1981, row-crop acreage increased 163%, from 800 to 2,100 hectares (1,975 to 5,194 ac); of that, strawberry acreage increased from 0 to 955 hectares (2,358 ac; Dickert and Tuttle 1985). From 1981 to 1993, crop acreage in the watershed increased another 29% to 2,711 hectares (6,700 ac), with strawberry acreage expanding the fastest (Soil Conservation Service 1994), growing another 53% since 1981 to 1,457 hectares (3,600 ac).

Although nearly all of the local farmers use some method to prevent erosion, those farming on the steepest slopes with the highest erosion potential also tend to have the least amount of capital or access to existing farm information programs (Mountjoy 1996). Eighty percent of the farmers growing strawberries in the Elkhorn Slough watershed began as immigrant Mexican farm laborers. When they decided to begin farming independently they settled in the hills due to the cheaper cost of land and leases. However, this cost savings is often offset by a loss in crop productivity and high maintenance costs associated with erosion problems. This vicious cycle of erosion and reduced profits leads to an inability on the part of many farmers to invest in improved land management. As a result, the erosion-control methods used by these farmers tend to be low-cost, short-term, and relatively ineffective.

Agency and public response to the growing erosion problems in the early 1980s resulted in a Target Area Study conducted by the Soil Conservation Service (SCS) to develop suitable erosion control practices to address the erosion problem (Soil Conservation Service 1984). These methods were promoted over the next ten years by SCS but had limited effect due to staffing constraints and a lack of cross-cultural outreach. Only ten ranches implemented and maintained the recommended practices.

In 1993, the Resource Conservation District of Monterey County again focused attention on the erosion problem in the Elkhorn Slough Watershed in response to downstream water quality problems. USDA funding for an eight-year watershed project was approved in 1994. By targeting assistance to the farmers' constraints and available resources, the project offered an alternative to the conventional technology-driven diffusion approach that had characterized the previous ten years of SCS assistance.

The Elkhorn Slough Watershed Project, a joint effort of the USDA-NRCS and the Resource Conservation District of Monterey County, has tackled the erosion problem by conducting socioeconomic assessments of farmers' needs, developing culturally appropriate technology and outreach, and providing cost-sharing financial incentives. A bilingual staff of seven has worked alongside more than 100 farmers to implement conservation plans on a total of 959 hectares (2,368 ac), many of which had the worst erosion problems in the watershed. The project team works closely with other agencies, environmental organizations, and the private sector to more effectively offer services. Recommendations have been provided to many farmers and their advisors, brokers, and landlords, and successful ideas have begun to spread within the farming community. The efforts of the Watershed Project team have resulted in an estimated 46,673 tons of soil annually being prevented from eroding downstream. This is equivalent in volume to a football field piled two stories high with soil, and represents a 35% reduction in the total annual human-induced erosion in the watershed.

Flooding

The benefits of controlling upland erosion and runoff are realized downstream in the form of reduced flooding of low-lying farmland and valley residences and reduced transport of sediments into streams, wetlands, and roads. Natural erosion and runoff processes have shaped the hills and valleys of the region over geologic time and account for the fertile soils of the Carneros Valley and numerous smaller tributaries to the slough. These valleys were ditched and drained early in the twentieth century to allow for expanded farming and rural settlement and to rid the area of mosquito-infested "swamps." Periodic maintenance of the ditches kept them free of what sediment did accumulate. However, since the 1970s, hillside agriculture and rural residential development have expanded, leading to increased runoff and erosion of sediment from farms and home sites. This sediment began to accumulate more rapidly in lowland channels of smaller streams and along ditch banks. Environmental regulations have discouraged the habitat disturbance caused by routine clearing of stream channels and wetlands. As a result, many of the former drainage channels in the watershed have completely filled in, which in turn reduces their flood-carrying capacity.

Efforts to reduce flooding along Carneros Creek are an excellent example of how cooperative work by landowners, agencies, and nonprofit organizations can make a difference in land management issues. Above-average rainfall events in 1997, 1998, and 1999 caused Carneros Creek to overflow its sediment-filled banks and spread out across farmland and into backyards. A group of landowners along the creek organized themselves into the Carneros Creek Association, a Coordinated Resource Management and Planning (CRMP) group. Recognized by local, state, and federal agencies, the CRMP has become a voice for related natural resource issues in the watershed.

Initially the group organized to dredge the stream channel and reduce flooding, but through monthly educational forums put on by the Resource Conservation District of Monterey County, CRMP members realized that flooding was interconnected with other natural resource issues. They could not achieve a long-term solution to the flooding problem without addressing erosion and upland management. Their proposed dredging activities would alter riparian habitat and affect water quality in the slough if agricultural runoff was allowed to flow more directly rather than ponding on vegetated fields. In addition, opening the channel straight to the slough might release valuable freshwater that should be ponded on the land to increase recharge and offset the problem of groundwater overdraft and saltwater intrusion.

The interconnection of these issues could have led to confusion, political gridlock, and ongoing studies. However, the community educational process has turned into action in several areas. The Carneros Creek Association has lobbied county government to enforce existing erosion ordinances and evaluate the erosion and runoff consequences of future development. They organized free prison labor and donations of equipment from private contractors not only to clear the sand but also to restore the riparian vegetation along almost a mile of stream channel where flooding was affecting homes. Community members have developed a native plant propagation nursery to provide plants to other landowners to stabilize stream banks and create riparian habitat. Several landowners along the stream in flood-prone areas are considering the option of selling conservation easements to establish freshwater wetlands on their former agricultural lands. These wetlands would serve the multiple functions of filtering runoff before it reaches the slough, increasing groundwater recharge capacity in the watershed, and establishing wetland habitats. And finally, community members have initiated a volunteer water quality monitoring program to track changes and identify problem areas.

Thanks to these joint efforts of community members, agencies, and nonprofit organizations, for the first time in recent history county government is paying attention to the otherwise unheard communities of northern Monterey County. The combination of code enforcement and availability of technical assistance should result in more rapid adoption of conservation management practices by all land users.

Groundwater Recharge

Even while flood problems have increased, a number of studies have shown that since the 1950s the Elkhorn Slough watershed has been in a state of groundwater overdraft: more water is pumped out of groundwater reserves than is percolating back in. Recharge of groundwater supplies in the area comes solely from that percentage of rainfall that reaches the groundwater aquifer, a function of rainfall intensity and duration, soil surface permeability, and geologic strata. Estimates of annual recharge rates for the area vary from 3 to 5.3 cm (1.2–2.1 in). Two comprehensive studies of the area put the recharge volume from the watershed at 8,055 acre-feet in 1977 and 20,508 acre-feet in 1983 (the 1983 study included an additional 5,600 acres of land area, and followed an above-average rainfall year). In those same years, extractions totaled 23,000 acre-feet and 36,300 acre-feet, respectively.

Recharge rates may have declined with the increase in impervious surfaces associated with home development, paving, and plastic mulching of agricultural crops. In addition, soil compaction and the loss of deep-rooted vegetation on farm fields can reduce soil permeability and rainfall retention. Many local residents have also observed that surface springs and shallow wells (less than 3 meters (10 ft) deep) that once produced large volumes of freshwater have been declining since the early 1940s—further evidence of reduced recharge and changes in hydrology. A 1988 study by ABA Consultants suggests that this phenomenon may be linked to a combination of the Salinas River diversion, the "reclamation" of slough lands, and overpumping of groundwater.

In a 1995 report, the Monterey County Water Resources Agency confirmed that the area was "severely overdrafted" and that annual groundwater use was twice that of annual recharge. The agency calculated that if the area were to be fully developed with existing land use plans, water demand would increase to 300 percent of sustainable yield or more.

This chronic overdraft has led to falling groundwater levels and resulted in a saltwater intrusion problem. The horizontal aquifers underlying the watershed open out into the ocean waters of Monterey Bay. As freshwater pressure from inland aquifers is intercepted and reduced by pumping, ocean water pressure pushes inland to compensate. The Water Resources Agency reported that in the Springfield Terrace area seawater intrusion has occurred at least as far inland as Highway 1 and possibly farther.

Elevated levels of chloride ions in groundwater pumped immediately adjacent to Elkhorn Slough appear to be the result of vertical leakage from the slough itself. Monterey County Water Resources Agency found this saltwater to be chemically similar to slough water and to be most highly concentrated at a depth just below sea level (Monterey County Water Resources Agency 1995). Additional support for this theory comes from a sudden and significant increase in saltwater in several wells adjacent to the slough following the destruction of tide levees in 1983. Overall, the leakage from the slough appears to be responsible for 80% of the saltwater intrusion in the area; the remaining 20% comes from the bay and affects aquifers as deep as 45 meters (148 ft) below sea level.

In its report, the Water Resources Agency points out that remedies for saltwater intrusion are limited. Prevention of further leakage would require a near cessation of groundwater pumping, alterations to Elkhorn Slough hydrography to reduce tidal influence, and perhaps a return of the slough to brackish or freshwater conditions, none of which seem, at this time, to be likely.

Nitrogen Contamination

Nitrate contamination of groundwater has also become a problem in the Elkhorn Slough area, although individual wells vary in nitrate concentration from far below worrisome levels to nearly nine times the safe drinking water standard. High nitrate intake interferes with the blood's ability to absorb oxygen molecules and is especially dangerous to infants and the elderly.

Nitrate contamination comes from both agricultural leaching of chemical and organic fertilizers and septic systems in the watershed. In 1996 the problem appeared to be worst in the Springfield and Highland subareas, both of which have high agricultural use, and in shallow wells. In general, elevated nitrate concentrations were not found in wells deeper than 38 meters (125 ft) below static water level (Monterey County Water Resources Agency). Nitrate breaks down very slowly once in the water supply, thus each year's loading of nitrate is added to the already existing concentration. In its 1995 study, the Water Resources Agency predicted an increase at buildout (maximum allowable development) of 200 times the current annual load. Even if loading remained at the current rate, water quality can be expected to degrade significantly. Although abandoning wells in localized problem areas and drilling deep wells are immediate solutions, contamination of the larger regional aquifers will eventually occur, prompting the need for water treatment. Such treatment is expected to be expensive and difficult to implement.

Current efforts by the agricultural community are focusing on raising awareness of the nitrate problem and promoting self-monitoring technologies for farmers. The Regional Water Quality Control Board has funded several local programs by the Resource Conservation District, local water agencies and University of California Cooperative Extension to show farmers how they can save money and reduce nitrate loss by doing simple soil, water, and plant tissue analyses.

Watershed Permit Coordination

Another strategy to improve management of watershed resources is the Elkhorn Slough Watershed Permit Coordination Program. Farmers and landowners had been reluctant to implement resource-enhancing projects due to the burden of obtaining permits from numerous regulatory agencies. This disincentive to conservation and restoration activities was overcome through the efforts of Sustainable Conservation, a nonprofit environmental organization, and the NRCS, which convened six regulatory agencies and obtained watershed-based permit agreements for ten common conservation practices.

Since 1988, twenty-eight farmers and landowners have obtained and implemented conservation plans from the NRCS that comply with the permit conditions of the various agencies without having to pay application fees or obtain individual permits. The sediment removal and bank restoration work in Carneros Creek was conducted under this permit program. In addition, the program has allowed nine landowners to invest in the construction of a total of twelve water and sediment control basins without the additional time and cost of obtaining individual permits.

When regulatory disincentives to natural resource enhancement are removed, landowners have demonstrated that they are willing to invest financial and material resources to improve water quality, wildlife habitat, and soil conditions. The permit coordination program now serves as a statewide model for other watersheds where permitting has become an obstacle to restoration activities.

Conclusions and Recommendations

In general, there has been a mismatch between research and management efforts in the Elkhorn Slough watershed. On the one hand, scientists are conducting interesting estuarine studies that have few direct applications for managers. On the other hand, management efforts and grant funding tend to focus on implementing conservation strategies in terrestrial systems, especially in the uplands of the watershed, with little scientific basis for decisions about where to prioritize efforts or about the effectiveness of methods for protecting estuary resources. Obtaining grant funding for applied research on key questions is a high priority, with the greatest need being to identify the biggest threat to slough communities (tidal scour vs. pesticide runoff vs. nutrient runoff vs. visitor impact vs. invasive species vs. the power plant). Thereafter, managers would know where to focus their efforts. There is also a need for large-scale, replicated experiments to test the effectiveness of different management strategies, which would help to direct management.

Applied research is also needed in other areas to support effective management—for example:

- Tidal scour has been one of the most difficult management issues to address. Would artificial shoals or planting of pickleweed or eelgrass reduce tidal currents? Small-scale testing of different flow reduction methods could be used to guide larger-scale remediation.
- What sort of vegetation in buffer zones will trap nutrients or contaminants coming off farm fields? Although vegetated buffer strips reduce sediment transport down slopes, buffer strips of barley or native bunchgrasses have not been found to effectively remove excess nutrients from agricultural runoff (Los Huertos 1999; Rein 1999).
- How often is water from the old Salinas River advected from Moss Landing Harbor into the slough proper? If this occurs frequently, then the old Salinas River could be a significant source of agricultural runoff high in nutrients and pesticides. This factor should be part of the development of nutrient and pesticide budgets for the slough, pinpointing the sources and removal mechanisms. Thus, removal of the most significant contaminant sources could be prioritized.
- Exchanges between groundwater and Elkhorn Slough need to be examined. Is surficial groundwater a source of high-nitrate water to the slough? Conversely, how quickly does high-nitrate, salty water from the slough penetrate groundwater aquifers?
- How do stochastic climatic variations, such as El Niño, affect watershed processes such as erosion, sediment and nutrient transport, and riparian and wetland habitat succession?
- How do adjustments in tide level in areas with tide gates, such as Porter Marsh or North Marsh, affect plant and animal communities? Can water levels be regulated to sustain the marshes as well as provide roosting and feeding habitats for waterbirds and other species?

This chapter has included numerous examples of diverse groups working together to address particular management problems within the Elkhorn Slough watershed. Although a commitment to such cooperative efforts is essential, developing action plans and assessing their success requires information. Understanding most issues, such as the impacts of invasive species, tidal scour, and water quality, requires both in-depth research and long-term monitoring. Research and monitoring provide critical information for making good management decisions and determining whether management actions have been successful.

Although substantial progress has been made in protecting sensitive habitats and resources within the Elkhorn Slough watershed over the last thirty years, much more work needs to be done. It has become increasingly obvious that working partnerships among individual stakeholders; nonprofit organizations; local, state, and federal agencies; and educational institutions are the most effective way to protect the slough and watershed. The Elkhorn Slough Conservation Plan (Scharffenberger 1999) exemplifies a comprehensive management plan that recognizes complex interrelationships within an ecosystem. Our hope is that the plan will be fully implemented in the coming years and that wise management and thoughtful conservation will protect Elkhorn Slough for future generations.

References

ABA Consultants. 1989. *Elkhorn Slough Wetlands Management Plan: Hearing Draft Report.* Prepared for Monterey County and the California Coastal Conservancy, Capitola, CA: ABA Consultants.

Beck, N. G., and K. W. Bruland. 2000. Diel biogeochemical cycling in a hyperventilating shallow estuarine environment. *Estuaries* 23: 177–187.

Browning, B. M. 1972. *The Natural Resources of Elkhorn Slough: Their Present and Future Use.* California Department of Fish and Game, Coastal Wetland Series No. 4.

Dickert, T. G., and A. E. Tuttle. 1985. Cumulative impact assessment in environmental planning, a coastal wetland watershed example. *Environ-men-tal Impact Assessment Review* 5: 37–64

Leopold, A. 1966. *A Sand County Almanac: With Essays on Conservation from Round River.* San Francisco/New York: Sierra Club/Balantine Book.

Los Huertos, M. 1999. Nitrogen dynamics in vegetative buffer strips receiving nitrogen runoff in Elkhorn Slough Watershed. Ph.D. dissertation, University of California, Santa Cruz.

Lowe, P. 1999. Marsh loss in Elkhorn Slough, CA: Patterns, mechanisms, and impact on shorebirds. M.S. thesis, Moss Landing Marine Laboratories. San Jose State University.

Malzone, C. M. 1999. Tidal scour and its relation to erosion and sediment transport in Elkhorn Slough. M.S. thesis, Moss Landing Marine Laboratories. San Jose State University.

Monterey County Water Resources Agency. 1995. North Monterey County hydrogeologic study.

Mountjoy, D. C. 1996. Ethnic diversity and the patterned adoption of soil conservation in the Strawberry Hills of Monterey, California. *Society and Natural Resources* 9: 339–357.

Oliver, J. S., and D. L. Schwartz. 1988. Photograph and map library for Elkhorn Slough Reserve: Erosion and sedimentation in Elkhorn. Final Report. On file at the Elkhorn Slough National Estuarine Research Reserve.

Pajaro Valley Pilot Project. 2000. Monterey Bay Farmers Clean Water Initiative. http://www.pvpilot.org, 203 (p. 194) (p. 203)

Rein, F. 1999. An economic analysis of vegetative buffer strip implementation case study: Elkhorn Slough, Monterey Bay, California. *Coastal Management* 27: 377–390.

Scharffenberger, T. 1999. Elkhorn Slough Watershed Conservation Plan. Elkhorn Slough Foundation and The Nature Conservancy. Unpublished report, on file at the Elkhorn Slough National Estuarine Research Reserve.

Soil Conservation Service. 1984. *Strawberry Hills Target Area: Watershed Area Study Report, Monterey, California.* Davis, CA: U.S. Department of Agriculture, Soil Conservation Service.

———. 1994. *Elkhorn Slough Watershed Project: Monterey and San Benito Counties, California: Watershed Plan and Environmental Assessment.* Davis, CA: U.S. Department of Agriculture, Soil Conservation Service.

U.S. Coast Guard. 2000. Area Contingency Plan. Vol. 2, sec. 9974.2: Cen-tral Coast sensitive sites. Elkhorn Slough geographic response plan.http://www.uscg.mil/d11/msosf/dprtmnts/plan/2000files/Vol%202/9974.pd

15

CHAPTER FIFTEEN

Summary

Jane Caffrey and Chela Zabin

A common thread throughout this book has been the changes that have occurred at Elkhorn Slough, including changes in the physical environment, in plant and animal communities, and in human uses and attitudes toward natural areas and resources. Estuaries are naturally dynamic environments, where changes occur on a variety of timescales. These range from the hourly rise and fall of the tides; to decadal variations in climate patterns that can influence freshwater and sediment supply, such as the El Niño/La Niña cycle; to uplifting, faulting, and other geological processes that occur over tens to hundreds of thousands of years.

In contrast to these natural fluctuations, many of the most extreme changes in the slough over the last century are a direct result of human activities. Our increasing demands on natural resources have introduced new stresses, triggering impacts such as pollution, sedimentation, and introduction of nonnative species. These impacts are not unique to Elkhorn Slough, but are in fact common to estuaries worldwide.

In this chapter we summarize the major findings of this book, outline some of the often-overlooked values of wetlands, and discuss several of the challenges facing Elkhorn Slough in the future.

Human Use of the Slough

Humans began using Elkhorn Slough and its watershed nearly 8,000 years ago. Chapter 6, "Archaeology and Prehistory," described the wide variety of food items that the native Ohlone gathered from the slough and uplands, ranging from clams and oysters, to acorns and blackberries, to sea lions and elk. Because these relatively small tribelets used a diversity of species, the Ohlone probably had little effect on the natural resources of the slough and watershed. Their most significant impact was on the region's grasslands, where deliberately set fires altered the composition of vegetation communities, reducing the number of brushy or shrubby species. During the Spanish and Mexican periods, use of the slough also centered on grasslands for cattle grazing.

Over time, activities such as whaling, the sea otter fur trade, duck hunting, salt production, and the shellfish industry drew on a wider range of the region's natural riches. In general, the consequences of overharvesting these types of resources can be easily understood, although development of sustainable resource use can be difficult. The shark derbies provide a good case study of overharvesting in the slough. Compared to many other marine and estuarine areas, there is little evidence of overexploitation of the fin-fisheries in Elkhorn Slough, except for the shark derbies that led to the disappearance of the shovelnose guitarfish in 1972 and declines in the numbers of sharks caught throughout the 1980s and 1990s, until the derbies ended in 1996.

In contrast to this example of overharvesting, indirect effects of human activities in the region have been much more difficult to evaluate and anticipate. The creation in 1946 of a direct connection between the slough's main channel and Monterey Bay via Moss Landing Harbor triggered a cascade of unexpected impacts. The effects of this development on the physical environment and biotic communities have been discussed in detail throughout the book. For instance:

- Tidal currents in the main channel have increased and the slough has become an erosional rather than depositional environment, resulting in a significant loss of salt marshes along the slough's main channel.
- As tidal scour has converted marshes to mudflats, macroalgae have replaced pickleweed as the primary producers.
- Some of the major changes in the invertebrate community may have occurred as a result of the tidal opening, including the decline of once-abundant species and the apparent local extinction of phoronid worms. Human harvesting has likely reduced populations of ghost shrimp and several large clam species.
- Between the 1970s and 1990s, species diversity within the fish community decreased throughout most of the slough. In addition, fish assemblages in the tidal creeks changed dramatically, becoming less distinctive and more similar to assemblages in the main channel.
- As lower Elkhorn Slough has taken on an increasingly marine character, it has become more attractive for harbor seals and sea otters. Increasing numbers of sea otters may be responsible for the declines in large gaper clams at the mouth of the slough between the 1970s and 1990s.
- Shorebirds have also been affected by the habitat conversions caused by tidal scour. Although the more extensive mudflats provide a greater area for feeding, this change has come at the expense of roosting areas within the marshes that may be particularly critical for migratory shorebirds.

None of these profound changes in slough habitats, species composition, and distribution were predicted at the time the harbor was created, either by the Army Corps of Engineers or local residents.

Earlier habitat alterations, such as the isolation of wetlands by the Southern Pacific Railroad in 1872, diking and draining of wetlands by farmers in the late 1800s and early 1900s, and diversion of the Salinas River, probably also had significant impacts. Background data are not yet available to determine the scale of these impacts; however, paleoeco-logical studies (pollen grain analysis, biomarker compounds, and isotope analysis) would help decipher what happened during this period.

Beyond these hydrologic changes, human activities have had other effects on the slough and watershed's natural communities. People moving to the region brought with them new plant and animal species. Grasslands were permanently altered by the deliberate introduction of cattle and nonnative plants for forage and gardens, as well as by accidental introductions. Similarly, oyster culturing in the early 1930s brought a suite of nonnative invertebrates to the slough. Unintentional introductions continue today, pointing up a need for better education to prevent the spread of species capable of displacing native plants and animals and disrupting ecosystems.

Other indirect effects of increasing population growth and intensification of agriculture have brought equally profound changes to Elkhorn Slough. These include nutrient (particularly nitrate) and pesticide runoff, increased soil erosion, and a growing demand for freshwater. Peak nitrate concentrations in the slough have increased tenfold since the 1970s. In the same period, Monterey County's population has increased by 45% and row crop acreage has increased by approximately 25%. In addition, production practices have changed, resulting in increased use of fertilizers on crops such as strawberries. Farming practices, particularly strawberry production on steep slopes, have also led to increased soil erosion in the watershed. This is unfortunate for two reasons: productivity is reduced until farms are finally abandoned, and soils washing into the slough not only smother natural communities, but also can carry high levels of pesticides including DDT.

Despite the national ban on DDT applications in 1972 it remains one of the most persistent and problematic pesticides in Elkhorn Slough, showing up in significant concentrations in mussels and sediments, particularly during wet years. Soil eroding from farm fields transports DDT-laden sediments to the slough, where it enters the food chain. In 1995, heavy rains triggered a sobering example of pesticides' impacts. Floodwaters and sediment from the Pajaro and Salinas Rivers spilled into Elkhorn Slough, raising levels of DDT and other pesticides in Caspian Tern prey. As a result, the tern colony suffered more than a 50% reproductive failure due to eggshell thinning and chick deaths. While the effects of organochlorine pesticides are fairly well understood and periodic monitoring for these compounds has been conducted, very little research or monitoring has been done on the newer pesticides in use.

As intensive farming has expanded and human populations have grown, so has the demand for freshwater for irrigation and household use. Artesian springs and wells were once common in the watershed, but have disappeared in the wake of agricultural and housing development. Groundwater pumping is occurring twice as fast as aquifer recharge, leading to saltwater intrusion. Nitrate contamination of groundwater occurs in aquifers as deep as 38 to 42 meters (125–148 ft). This is clearly an unsustainable situation, likely to be exacerbated as the region's population grows.

Clearly, humans have altered Elkhorn Slough and its watershed in many different ways, creating complex management issues that have no quick or easy solutions. However, progress has occurred, particularly as disparate groups have come together to work on reducing soil erosion and flooding and on addressing water quality issues and other problems. These efforts need to be fostered, and collaboration among groups that have often been adversarial needs to be encouraged. Education and outreach to the community within the watershed and the region have been critical to building support for the slough's protection and promoting conservation practices. A variety of groups, including the Elkhorn Slough National Estuarine Research Reserve, Elkhorn Slough Foundation, the Watershed Institute, Natural Resource Conservation Service, Resource Conservation District, and the Monterey Bay National Marine Sanctuary, have developed educational and outreach programs.

Value of Wetlands and Natural Environments

Although much of our discussion has centered on the important role that Elkhorn Slough and the surrounding watershed play in supporting wildlife, humans also depend on these resources. In some cases, the economic link to the slough, its upland areas, and the bay is clear: Moss Landing Power Plant requires cooling water to operate its generators. Commercial and sport fishing and related businesses rely at least in part on the slough as a fish nursery. Upland slopes are farmed for strawberries, herbs, and row crops. Numerous businesses, including local restaurants and boat rental agencies, rely on the scenic attraction of the slough and the harbor. Easy access to a wide variety of organisms benefits scientific and educational organizations, which make a significant contribution to the local economy. In other cases, the slough's benefits are less tangible: they are educational, emotional, and spiritual. Elkhorn Slough provides people with a place to relax and unwind, view nature, and spend time with families and friends.

Some services that natural ecosystems provide are so essential to our survival that we take them for granted. These include the hydrologic cycle, which provides water for drinking and growing crops, and biogeochemical cycles, which provide oxygen through plant photosynthesis and waste removal by bacteria. Grassland, woodland, and chaparral plant communities store water, moderate the local climate, and reduce soil erosion. Similarly, wetlands can enhance groundwater recharge, improve water quality, and reduce flooding through water retention (Mitsch and Gosselink 1986).

Traditionally, goods and services that nature provides have not been included in calculations of the local or national economy. Although many believe that the preservation of natural areas is a moral issue, much of the debate over preservation places the issue in an economic context, often reducing it to such simplicities as "owls vs. jobs." An economic summary of the benefits of natural areas estimated that, worldwide, ecosystems provide at least $33 trillion worth of services each year (Costanza et al. 1997). Of the different biomes compared in that study, estuaries contribute the greatest number of goods and services, worth approximately $22,832 per hectare per year (Costanza et al. 1997). Among the services estuaries provide are disturbance regulation, nutrient cycling, biological control, habitat/refugia, food production, raw materials, recreation, and cultural services. The papers summarized by Costanza et al. used various methods of evaluating natural services, but most were based, either directly or indirectly, on people's "willingness-to-pay" for such services, or on estimates of the increased cost to productivity if the natural service or good were not available. For a number of reasons, these methods of valuation are most likely underestimates, since, for example, they do not take into account all the services and goods provided by nature, and people's "willingness-to-pay" for a service they are used to thinking of as being free is likely lower than its the "true" value. After all, as Costanza writes: "The economies of the Earth would grind to a halt without the services of ecological life-support systems, so in one sense their total value to the economy is infinite."

In fairness, we must mention that West Coast estuaries—at least those in southern California—are generally thought to be less productive than those on the East Coast, in part because they are so much smaller. In a 1978 paper that combined work at Mugu Lagoon in Santa Barbara with published reports from other California estuaries, Onuf et al. concluded that these estuaries do not contribute as significantly as East Coast estuaries in terms of primary productivity, export of organic matter to the ocean, and as nurseries to commercially important fish stocks. Yet because of their rarity, Onuf proposed that California estuaries are valuable as habitat for endangered wildlife and migrating birds, and as educational and aesthetic resources for people. With the loss of more than 90 percent of the state's original wetlands, the remaining areas have become even more important. For these reasons alone, he stated, California estuaries are worthy of protection.

Challenges for the Future

This book has highlighted a continuing need for both basic and applied research in Elkhorn Slough:

- Fundamental questions about plant ecology in the slough remain to be answered, particularly the relationships between nutrient inputs and primary producers.
- Terrestrial species, both plant and animal, have not been systematically studied in the Elkhorn Slough watershed. Major research areas include the way habitat fragmentation affects plant communities and animal populations, and what the interactions are between invasive and native species.
- Critical applied research needs to be conducted on new pesticides: what are their exact sources, transport mechanisms, persistence, and effects on slough biota?
- The effect of changing land use practices on sediment and nutrient transport must be evaluated.
- A better understanding of bird and mammal habitat requirements, population sizes, ecological interactions, and response to human disturbance will be helpful for guiding preservation and restoration efforts.

In addition to specific research topics, systematic monitoring of water quality, erosion, critical habitats, and important populations needs to be implemented. Establishing baseline data followed by long-term monitoring is the only way to judge the success of restoration efforts and other management initiatives.

Increasing development within the watershed makes preservation of cultural resources a more difficult challenge. Only 10 to 15 percent of the greater Elkhorn Slough region has been surveyed for archaeological sites. A systematic survey would pinpoint sites that are deteriorated or threatened by erosion and potentially make it possible to evaluate these sites before they are lost forever. In addition, little work has been done examining the connections between San Francisco Bay and Elkhorn Slough. More detailed geological as well as archaeological research would provide insights into the first human settlement of the region. The early history of Elkhorn Slough (20,000–1 million years B.P.) is yet to be revealed, from the response of the slough to changing sea levels during this period to the plant and animal communities present before human colonization.

References

Costanza, R., R. dArge, R deGroot, S. Farber, et al. 1997. The value of the world's ecosystem services and natural capital. *Nature* 387: 253–260.

Mitsch, W. J., and J. G. Gosselink. 1986. *Wetlands*. New York: Van Nostrand Reinhold.

Onuf, C. P., M. L. Quammen, G. P. Shaffer, C. H. Peterson, J. W. Chapman, J. Cermak, and R. W. Holmes. 1978. An analysis of the values of central and southern California coastal wetlands. In P. E. Greeson, J. R. Clark, and J. E. Clark, eds., *Wetland Functions and Values: The State of Our Understanding*, pp. 186–199. Minneapolis, MN: American Water Resources Association.

Afterword

Elkhorn Slough as the Cosmic Center of the Universe

Little did I realize when I started this effort in 1996 that it would become such a long and involved project. I am grateful to the authors who volunteered their time and kept the faith throughout. Martha Brown, Breck Tyler, Kirsten Carlson, and Anne Canright's dedication, enthusiasm, and talent brought this project to a higher level. It has been exciting to see this book come together, and in the process learn more about the slough. Despite the ever-increasing demands on and insults to Elkhorn Slough, it remains a special place.

People quickly become attached to Elkhorn Slough—it inspires strong feelings and great devotion. What is it that makes the slough so appealing? No doubt each person has his or her own explanation, but several possible characteristics might explain some of the attraction. The slough is a very knowable, accessible place—you can easily get close to the marshes, mudflats, and water. In several hours you can walk from ridgetop to the slough's edge, or with the right tides you can canoe or kayak its length. Everything is right before your eyes. It's also easy to see the connections between water and land—gentle, rolling hills rise up and surround the slough.

Mark Silberstein has often claimed that Elkhorn Slough is the cosmic center of the universe, and I am inclined to agree with him. On a quiet day or after the school children have left the reserve, one can look out over rolling grassland and see hovering white-tailed kites or red-tailed hawks hunting for mice or gophers. Toward evening, the marsh and slough waters reflect the colors of the setting sun, and there is a sense of peace and serenity. It's a place to refresh one's spirit.

Thanks to its many attractions, Elkhorn Slough is an increasingly popular tourist destination. Our desire to get away from it all puts a growing burden on fragile natural resources. Can we protect the unique character of the slough as more people visit and live in the region? For our own sake and for that of our descendents, I hope we can.

Jane Caffrey
Gulf Breeze, Florida
August 2002

Conversion Table

English Equivalents of Metric (SI) Units

Length

1 centimeter	0.39 inch
1 meter	3.28 feet
1 meter	39.37 inches
1 kilometer	0.62 miles

Area

1 square kilometer	0.39 square mile
1 hectare	2.47 acres

Volume

1 cubic meter	35.3 cubic feet

Mass

1 gram	0.035 ounce
1 kilogram	2.2 pounds

Yield and Rate

1 metric tonne per hectare	.446 ton per acre
1 meter per second	2.24 miles per hour

Conversion Factors for Metric (SI) and English Units

To convert Column 1 into Column 2 multiply by	Column 1 Metric Unit	Column 2 English Unit	To convert Column 2 into Column 1 multiply by
Length			
0.394	centimeter, cm	inch, in	2.54
3.28	meter, m	foot, ft	0.304
0.621	kilometer, km	mile, mi	1.609
Area			
0.386	square kilometer, km^2	square mile, mi^2	2.59
2.47	hectare, ha	acre, ac	0.405
Volume			
35.3	cubic meter, m^3	cubic foot, ft^3	2.83×10^{-2}
Mass			
3.52×10^{-2}	gram, g	ounce, oz	28.4
2.20	kilogram, kg	pound, lb	0.454
Yield and Rate			
0.446	metric ton (Megagram) per hectare, $Mg\ ha^{-1}$	ton per acre	2.24
2.24	meter per second, $m\ s^{-1}$	mile per hour	.447

Glossary

Action limit: Defines when action will be taken to remove contaminated shellfish from markets.

Assemblage: Species that co-occur at a specific location. In the case of archaeological sites, assemblage refers to the mix of animal and plant species, tools, and other artifacts found at a single site.

Bioconcentration factor: A factor that estimates the potential for a pesticide to accumulate in aquatic biota.

Cation exchange capacity: A measure of soil's ability to store cations, such as calcium, magnesium, and potassium, at negatively charged sites on clay and humus particles. These positively charged cations are less likely to be leached away in water but remain available to plant roots.

Demersal: Located on or just above the slough floor.

Disjunct: Separated from the main body of distribution.

Epifaunal: Living on the slough floor.

Euryhaline: Able to tolerate a wide range of salinities.

Eutrophication: The increase of nutrients in lakes and wetlands either naturally or artificially by pollution.

Infaunal: Living in the soft mud and sand substrate of the slough bottom.

High-energy tidal inlet: Depositional sedimentary environments are the continental, oceanic, or coastal surroundings in which transported particles such as sand, silt, and mud accumulate. The relative energy associated with depositional environments generally refers to the amount of water motion or wind energy occurring as sediments accumulate. Coastal tidal inlets, river mouths, and beaches are generally considered high-energy depositional environments, while deep ocean basins, lakes, and lagoons are low-energy depositional environments. The depositional energy eventually changes over geologic time as geographic, climatic, and relative sea-level conditions also slowly change.

Higher High Water and Higher Low Water: See Mixed semidiurnal tide.

Humus: Many organic matter particles can only be partially decomposed; these resistant organic compounds are known as humus. Humus tends to accumulate lower in a soil profile and can act to stabilize organic matter–clay mineral structures called colloids. In general, colloids are thought to be important in imparting clodding ability to soils.

Interrogatorio: An official questionnaire sent by the Spanish government to eighteen missions in 1811. The document included thirty-six questions about the native populations that touched on topics of religion, language, and pre-contact culture. The written responses to this questionnaire, submitted to the Spanish government, are an important source of descriptive information for those California Indian groups who were within the area of mission influence.

Kleptoparasitize: A feeding strategy that involves stealing food from other animals.

Lower Low Water and Lower High Water: See Mixed semidiurnal tide

Mean Lower Low Water (MLLW): The average of the two low tides in our area. The approximate twenty-year average of the two lower tides is called zero tide and forms the basis for tide charts.

Mixed semidiurnal tide: Semidiurnal tides have two highs and two lows each day. Elkhorn Slough experiences a mixed semidiurnal tide because one of the two tide cycles is more extreme (higher high and lower low) than the other. The highest tide level, termed Higher High Water (HHW), is followed by the lowest tide level, Lower Low Water (LLW), then by a moderately high tide (Lower High Water, or LHW), and finally a moderately low tide (Higher Low Water, HLW).

Moiety: Division of society into two social categories or groups, usually by a rule of patrilineal or matrilineal descent.

Nonpoint source (NPS) pollution: Pollution that occurs when water runs over land or through the ground, picks up pollutants, and deposits them in surface waters or introduces them into groundwater. Pollution that cannot be traced to a single point or outfall.

Obsidian profile: The cumulative results of obsidian analysis in a particular region or area which shows the sources of obsidian that were most common (obsidian from nine different sources in the North Coast Ranges, eastern Sierra Nevada, and northern Mojave Desert was traded into the Monterey Bay area at different times) during various periods, and the relative frequency of obsidian trade. Changes in the obsidian profile indicate a change or disruption of trade with outside groups.

Organic material: Made up primarily of decomposed plant material, including pickleweed (*Salicornia virginica*) or peat horizons with unrecognizable peat remains, as well as some microscopic protozoa and invertebrates.

Otolith: The ear bone of a fish. These small bones grow by incremental addition of layers to the outer surface. Cross-section of the bone reveals alternate layering that can be read like tree rings to determine the age of the fish, and in some cases, the season of its death.

Phytotoxins: Chemical substances that prevent germination and growth of other plants.

Porewater: The water present in the spaces between sediment particles.

Residence time: The amount of time freshwater spends in the slough before entering the ocean.

Semidiurnal tides: See mixed semidiurnal tides.

Sinks [contaminant sinks]: Sites where contaminants are likely to accumulate. These sinks may later become sources of contamination through the process of diffusion from sediments to surface waters.

Throughfall: The process of precipitation passing through the plant canopy. This process is controlled by factors such as plant leaf and stem density, type of the precipitation, intensity of the precipitation, and duration of the precipitation event. The amount of precipitation passing through varies greatly with vegetation type.

Trophic guilds: Assemblages of species that rely on the same prey types.